A Feminist Glossary of Human Geography

A Feminist Glossary of Human Geography

Edited by

Linda McDowell

and

Joanne P Sharp

A member of the Hodder Headline Group
LONDON • NEW YORK • SYDNEY • AUCKLAND

First published in Great Britain in 1999 by
Arnold, a member of the Hodder Headline Group,
338 Euston Road, London NW1 3BH

http://www.arnoldpublishers.com

Co-published in the United States of America by
Oxford University Press Inc.
198 Madison Avenue, New York, NY 10016

British Library Cataloguing in Publication Data
A catalogue record for this book is available from the British Library

Library of Congress Cataloging-in-Publication Data
A catalog record for this book is available from the Library of Congress

Production Editor: Julie Delf
Production Controller: Iain McWilliams
Cover Design: T. Griffiths

ISBN 0 340 74143 0

1 2 3 4 5 6 7 8 9 10

Typeset by Scribe Design
Printed and bound in Great Britain by MPG Books, Bodmin, Cornwall

What do you think about this book? Or any other Arnold title?
Please send your comments to feedback.arnold@hodder.co.uk.

List of Contributors

MB   Morag Bell
University of Loughborough

AMB   Alison Blunt
University of Southampton

LB   Liz Bondi
University of Edinburgh

AB   Alistair Bonnet
University of Newcastle

SB   Sophie Bowlby
University of Reading

HC   Hazel Christie
Heriot-Watt University

LD   Lorraine Dowler
Pennsylvania State University

ND   Nancy Duncan
University of Cambridge

CD   Claire Dwyer
University College London

JF   Jo Foord
University of North London

FG   Flora Gathorne-Hardy
University of Cambridge

NG   Nicky Gregson
University of Sheffield

PJ   Peter Jackson
University of Sheffield

CJ   Craig Jeffrey
University of Edinburgh

AJ   Andrew Jones
University of Cambridge

RK   Rebecca Klahr
University of Cambridge

LK   Larry Knopp
University of Minnesota

NL   Nina Laurie
University of Newcastle

JL   Jo Little
University of Exeter

LM   Linda McDowell
London School of Economics

CM   Cathy McIlwaine
Queen Mary & Westfield College, University of London

DM   Doreen Mattingly
San Diego State University

PM   Paula Meth
Sheffield Hallam University

KM   Karen M Morin
Bucknell University

HN   Heidi Nast
DePaul University

CN   Caroline New
Bath College of Higher Education

RHP   Rachel Pain
University of Northumbria

HP   Hester Parr
University of Dundee

BP   Bronwen Parry
University of Cambridge

LP   Linda Peake
York University, Canada

DP	David Pinder *University of Southampton*	FS	Fiona Smith *University of Dundee*
RP	Rosemary Pringle *University of Southampton*	WS	Wendy Somerson *University of Washington*
SAR	Sarah Radcliffe *University of Cambridge*	MS	Matthew Sparke *University of Washington*
SR	Suzanne Reimer *University of Hull*	KT	Karen Till *University of Minnesota*
JR	Jenny Robinson *Open University*	BW	Bronwen Walter *Anglia Polytechnic University*
JS	Joanne P. Sharp *University of Glasgow*	JW	Jane Wills *Queen Mary & Westfield College, University of London*

Acknowledgements

Thanks to all our friends, colleagues, family and students – too numerous to name individually – who not only helped us in this particular task but who have also made a difference to how we think about things. The original idea was Laura McKelvie's at Arnold: we are not sure whether thanks or curses are the most appropriate acknowledgement.

Introduction

One of the most exciting aspects of working as a geographer in the last few years has been the speed with which the discipline has changed. Perhaps somewhat to the surprise of many of us, geography found itself right in the middle of many of the debates that have dominated not only the social sciences but also the humanities throughout the 1990s. Whether the focus was the increasing connections between places through economic globalisation or cultural innovation, the movement and migration of huge flows of capital and labour, the paradoxical reassertion of local power, politics and identities, or the wholesale challenges to theory and knowledge mounted by amongst others, feminists, postmodernists and post-colonial theorists, geographers had something to say. Indeed, the new sets of concepts that came to the fore in a wide range of disciplines – situated knowledge, travelling theory, borderland identities – were almost all associated with a recognition of the significance of location and geographical difference. A wide range of new ideas and concepts, more or less familiar to a geographical audience, entered the vocabulary of the discipline and changed the nature of academic debates.

In this glossary, we have tried to capture the nature of the most significant of these changes in the context of feminist debates. As we argued in *Space, Gender, Knowledge* – our companion volume to this one (McDowell and Sharp 1997) – feminist theories and debates have changed and expanded in an exponential way since they first became visible in the academy and in wider political movements in the postwar period. Indeed it is sometimes hard to remember the early struggles in geography departments across the world to gain acceptance for feminist work. Feminist students and teachers, courses, book collections and journals are evidence of the current vibrancy and dynamism of feminist geography. This very dynamism, of course, makes it difficult to capture and pin on the page definitions of the key terms in these debates. But in the company of a group of energetic contributors from three subcontinents, this is what we have tried to do.

The terms that we have included reflect the changing nature of feminism from its emergence in European and US universities in association with the second wave women's movement from the late 1960s. The changing emphases and the multiplicity of feminist issues and debates are reflected in the definitions; key terms in debates about the nature of feminism itself, about femininity, nature, the political economy, state policies and symbolic meanings are included, as well as some of the links backwards to the arguments about women's emancipation in earlier centuries and to the feminist theorists and thinkers writing in the first half of the twentieth century. Here too you will find reflections of the hard debates between white Western, often heterosexual, feminists, and their critics who felt they were marginalised in literatures that excluded their interests as women of colour, as lesbians, as Third World women in the Third World and in the West. All this work reveals the ways in which the debates and political movements never stand still but change through hard and exhilarating scholarship.

We make no claims that our entries are entirely authoritative or complete. They are provided in the spirit of a guide to rapidly expanding literatures, as an aid to clear thinking and as a stimulus for further exploration. Indeed, we know that they are stimulating and helpful as we ourselves learned a great deal in co-ordinating the entries of the contributors. They reflect the knowledge and experience of this wide range of scholars working on issues about gender at the end of the twentieth century. Each entry has the initials of its author at the end, although the project felt like a collective one. Relevant terms which are also included in the glossary are indicated in small capitals, either as they appear in the entry, or listed at the end. We decided not to list readings at the end of each entry but have, instead provided a complete bibliography to the book, but the key references are indicated by author and date within each entry. If nothing else, the length of the bibliography indicates the health of feminist scholarship.

While this glossary is unique in its focus on feminist geography, there are a number of related glossaries that may be useful. The most obvious is that invaluable guide for all geographers: *The Dictionary of Human Geography* edited by Ron Johnston, Derek Gregory and David Smith and published by Blackwell, which should appear in its fourth edition as this glossary is published. There are two companion volumes to this one, also published by Arnold: *A Glossary of Feminist Theory*, edited by Sonya Andermahr, Terry Lovell and Carol Walkowitz (1997) and a *A Glossary of Contemporary Literary Theory* edited by Jeremy Hawthorn (second edition 1997). Other feminist dictionaries and glossaries, both general and particular, include *The Dictionary of Feminist Theory* by Maggie Humm, published by Harvester in 1989, Elizabeth Wright's *Feminism and psychoanalysis: A Critical Dictionary* published by Blackwell in 1992 and *An Encyclopedia of Feminist Theories* edited by Lorraine Code published in 1998 by Routledge. These give an indication of the changing nature of feminist debates. Finally many academic presses publish dictionaries or encyclopedias of key social, political and philosophical thinkers which are a useful complement to glossaries of terms, although many of them are less than complete in inclusion of feminist thinkers.

LMcD
JPS
August 1998

A

Abject A term from psychoanalysis which French feminist theorist Julia Kristeva (1982) has used in her work; it literally means the cast out. It defines a horror of physical engulfment and is used to capture the importance of maintaining a boundary between the SELF and OTHER in which the self feels a disgust or sense of repulsion about its own wastes and residues (excrement, decay, infection) and defends the boundaries of the embodied self against what are seen as impurities. Women's bodies, especially the maternal BODY, are troubling as the boundaries are permeable; in this way women are defined as less than pure subjects in PHALLOCENTRIC systems. The geographer David Sibley (1995a) has drawn on Kristeva's definition of abjection in his analyses of geographies of exclusion showing how the expulsion of 'filth' in both a literal and metaphoric sense is important in constructing spatial BOUNDARIES.   **LM**
See also FRENCH FEMINIST THEORY; PSYCHOANALYSIS.

Abortion The termination of a pregnancy. An abortion can occur spontaneously during pregnancy (miscarriage) or may be due to deliberate outside intervention (induced abortion). Typically the term is used to mean the latter and is the focus of this definition. The political and legal framework of abortion varies spatially. In many European countries, the USA and Japan, abortion is available on request. In other nations, such as Britain, abortion laws are more restrictive. In the Arab states and Latin America abortion is illegal.

Feminists claim that women will not be liberated until they have legal access to medical assistance for an abortion (Warren, 1980). The availability of legal abortion, therefore, is an indication of how far a nation's health-care system is 'woman-friendly'. Some researchers have suggested, however, that feminist debates about abortion are ETHNOCENTRIC (see Osman, 1983). For example, the archetypal slogan of the feminist abortion campaign, 'A Woman's Right to Choose', is inappropriate in China, where women struggle for the right to refuse abortion (Hadley, 1996). Paternal rights and loss of PATRIARCHAL control have also been questioned by scholars, policy-makers and the media (see McNeil, 1991).

It is difficult to determine the extent of abortion because of the legal codes of confidentiality and the prevalence of illegal abortions. Contrary to popular belief, the liberalisation of abortion laws in some nations has not led to more abortions but a shift from illegal to legal practices. In fact, in most Western countries, the number of abortions has declined. This can be explained by demographic (fewer women of childbearing age) and cultural factors (lone mothers are increasingly accepted in society).   **RK**
See also BIRTH CONTROL.

1

Abstraction Distillation of reality to a more simplified form for theorisation, or, more often, for modelling (e.g. economic man as an abstraction of the range of human motivations becomes the basis for economic models and predictions). Although clearly such abstractions are easier to work with, they are stereotypes. In his work, Marx recognised 'levels of abstraction' ranging from the general to the historically specific.

Importantly, abstractions tend to be apparently gender-neutral, which means that male characteristics tend to become universalised in models of social process. As a result of feminist and POSTMODERN critiques, recent work in geography has insisted upon the importance of context and DIFFERENCE in social theory, which challenges the usefulness of abstraction.   JS
See also MARXIST GEOGRAPHY; STRUCTURALISM.

Abuse Physical, sexual, emotional or psychological harm inflicted upon a person. The boundaries between abuse and acceptable behaviour, and with HARASSMENT and VIOLENCE, are culturally and geographically variable, but abuse commonly describes behaviours in domestic settings dealt with as medical or social welfare issues rather than criminal justice issues. This context means that abuse has been constructed as a problem of individual pathology or family dynamics, theories challenged by feminists who have pointed to its highly gendered nature: men are more likely to be abusers, though not exclusively. The domestic abuse of women, child abuse and elder abuse all share a common basis in gendered POWER relations and norms about AUTHORITY and control over others' BODIES (Aitken and Griffin, 1996). The geography of abuse, usually within the family or other homespaces, such as residential institutions for children or older people, means settings in which there may be a culture of tolerance, in which abuse is unlikely to be reported and is difficult to detect. Feminists have been at the forefront of campaigns to raise public attention and bring about legal and regulatory change to prevent abuse and to aid survivors. Child abuse provided a 'moral panic' in the late twentieth century, and parents' fear means stringent restrictions are placed on many children's mobility and independent experiences of space (Valentine, 1997a). Elder abuse receives less public attention and state commitment, reflecting a different cultural precedent of responsibility for the aged (Penhale, 1993). There is a serious mismatch between public discourses of abuse and private danger – labelling abusers as 'monsters' allows them to be distanced from 'normal' people (Cream, 1993b). This underlies the common attachment of FEAR about abuse of women, children and older people to strangers in PUBLIC space.   RHP
See also VIOLENCE.

ACT-UP The Aids Coalition To Unleash Power (ACT-UP), is a gay community-based resistance organisation founded in New York in 1987. Developed out of a need to adopt a more radicalised and political response to government bureaucratic complacency, incompetence in dealing with the AIDS crisis and the refusal to release experimental but potentially life-saving drugs such as AZT, ACT-UP is prepared to employ strategies of 'direct action' – militant demonstrations, outing of public figures etc. in its fight against AIDS-related homophobia. Slogan: Action=Life, Silence=Death.   BP

Aesthetics/aestheticisation of everyday life Aesthetics refers to understandings of concepts of art, beauty and taste. It is assumed that certain objects, people and LANDSCAPES embody a positive aesthetic, and as a result should be valued and, where necessary, protected. Questions of status and distinction (Bourdieu, 1984) revolve around the ability to recognise – or possess – such objects. The 'aestheticisation of everyday life' refers to situations where issues of CLASS conflict and exploitation can be hidden beneath naturalised issues of beauty, NATURE, art and taste (see Duncan and Duncan, 1997, for an example of this). In his book *Distinction* Pierre Bourdieu (1984) historicises questions of 'taste' to demonstrate how systematic social discrimination and exclusion is generated through the aesthetic.

Feminists have denaturalised aesthetic concepts to show the complicity of such aesthetic values with patriarchal power (e.g. the BEAUTY MYTH) and have critically analysed aesthetics which encode the landscape as a female body (see Nash, 1996; Rose, 1993b).   JS
See also GAZE; VISION/VISUAL/VISUALITY.

Ageism Ageism is a form of oppression experienced by people who are regarded by others with aversion on account of their perceived or actual age. It is frequently associated with elderly groups. Ageism operates at many levels, from seemingly innocent assumptions about a person's capabilities through to systematic discriminatory rules inscribed in legislation. As women are often a majority among the oldest groups in many societies, it raises questions about female dependence in old age.   FG

Agency The concept of agency implies volition, free will and MORAL choice on the part of the individual. Agency refers to the POWER of actors to operate independently of the determining constraints of social structure. Agency is most usually juxtaposed with structure in a BINARY opposition: the agency of individual humans to act versus the determination of structures. STRUCTURATION theory attempts to transcend this binary to explain a situation where both agency and structures interact.

Recent feminist and POST-STRUCTURAL critiques have challenged the location of agency only in the figure of the human individual. Haraway's (1991) figure of the CYBORG and Latour's (1993) arguments about non-human agency have suggested understandings of society where agency is displaced from the human to (bio)technology and, in geography, to animals (see Philo, 1995a).   JS

AIDS/HIV AIDS is an acronym for acquired immune deficiency syndrome. AIDS can be clinically defined as a syndrome caused by exposure to the

3

human immunodeficiency virus (HIV) which destroys a subgroup of lympho-cytes resulting in the suppression of the body's immune response. What this clinical definition cannot contain are all of the complex social and cultural implications of the term AIDS, which, as Wilton (1997: xii) suggests, has now come to encompass 'notions of clinical disease, social disease, contamination, exclusion, discrimination, hostility, economic/material inequality, religious doctrine, political expediency, moralism/morality, SEXUALITY, deviance, criminality, risk, blame, dysfunction and death'.

Classically and inaccurately characterised as a sexually transmitted 'GAY' disease (Edwards, 1992) AIDS first came to global attention in the early 1980s as epidemics swept through gay communities in New York and San Francisco. Beset by moral panic, fundamentalist groups in the USA adjudged AIDS to be a plague visited upon HOMOSEXUALS in retribution for their 'unnatural' behaviour, generating hysterical homophobic reactions and calls for the spatial segregation of those living with HIV and AIDS.

The social construction of AIDS as an exclusively 'gay disease' ensured that responses to the disease were deeply politicised from the outset. Perceived as being confined to what was deemed a socially marginalised COMMUNITY, the Reagan administration made little commitment to research-ing the epidemiology of the disease or developing possible treatments. Its neglect of the issue has been since described by the gay activist Larry Kramer as tantamount to 'a deliberate genocide of gay persons'.

AIDS is now understood as a chronic, but not necessarily fatal, condition which is acquired through the exchange of infected bodily fluids. It can be contracted through unsafe sexual relations, blood transfusions, the sharing of contaminated needles, and can be transmitted between mother and child in pregnancy, and post-partum through breast feeding. The disease is now thought to have originated in Africa where transmission rates amongst HETEROSEXUAL populations remain alarmingly high, rivalling those in South East Asia.

Both feminists and LESBIANS have objected to the homosexualisation and implicit gendering of AIDS, arguing that it has contributed to a systematic neglect of the experiences and needs of other affected groups such as women, both heterosexual and homosexual, intravenous drug users and heterosexual men. Wilton (1997: 2) has argued that 'it is undoubtedly the case that the HIV/AIDS related needs of women are marginalised, ignored and denied and that women's subjugation to men is both effected and re-inforced by the ways in which the pandemic has been gendered', whilst Stephens (quoted in Treichler 1994: 141) suggests that AIDS can consequently be understood as 'a paradigm for the condition of women in our society'.

Most of the work conducted within geography on the spread of AIDS has, predictably, centred around classic disease diffusion models. This approach has recently been criticised for the way in which it artificially distances itself from the lived EXPERIENCES of gay men, constructing them as simply vectors of the disease rather than as active participants in the containment of the disease (see Brown, 1995).

The strategies of resistance implemented by political activists within GAY AND LESBIAN COMMUNITIES – safe sex campaigns, information dissemination programmes and access to treatment campaigns – have significantly reduced transmission rates in the WEST, although benign neglect, the absence of political will and the prohibitive cost of new anti-viral therapies continue to depress survival rates in most developing countries. BP

Alterity A term signifiying a radical difference – alterity is used in philosophical, feminist and POST-COLONIAL contexts to signify a radical other. This implies a position entirely outside that of the identity to which it is opposed and usually signifies some kind of binary opposition. However, there have been many criticisms of the possibility of alterity, especially within post-colonialism, where the generalised HYBRIDITY of the post-colonial subject has been counterposed to this 'Manichean' (dualist) BINARY vision of DIFFERENCE (Bhabha, 1994). In feminist thought, and in philosophy as well, the tenability of alterity has been much criticised. The idea that the 'outside' to a given identity may indeed be constitutive of it has been developed through psychoanalytic and philosophical frameworks by, amongst others, Ernesto Laclau (1990). For feminists there have been lengthy debates concerning the feasibility of speaking of woman or the feminine as radically 'other' than MASCULINITY or the masculine. Emphasising difference can often involve an ESSENTIALIST position, as what WOMAN is becomes limited to defining features – e.g. BODIES, SEX, MOTHERHOOD, nurturing (Fuss, 1989). Irigaray argues that what woman is has so far been defined by the masculine order – in this framework she is simply the 'other of the same'. To find a way to bring into being a radically different FEMININE, not determined by the masculine order and without predetermining what this would be, is her ambition (Irigaray, 1985a and b). Suggesting that women occupy a position of alterity can also be evidence of cultural HEGEMONY. In some communities experiencing racial discrimination it has been suggested that the common points of interest and support between men and women are just as important, if not more so, than those uniting women (Davis, 1982). JR
See also FRENCH FEMINIST THEORY.

Androcentric Meaning literally 'male-centred', an androcentic approach or argument is one which privileges the experiences, actions, values and concerns of men whilst largely ignoring or marginalising those of women. First used by the American feminist Charlotte Perkins Gilman in the early part of the twentieth century to draw attention to MASCULINE bias, the term has since been used to highlight and criticise the process by which a male-centred 'world-view' is constructed and promoted as normative despite the fact that over 50% of the world's population is female.

Although some feminists have argued that androcentrism is employed explicitly as an instrument for enforcing and perpetuating the ideology of PATRIARCHY, others would argue that its greater danger lies in its implicitness.

In this view androcentrism is problematic precisely because it simply *assumes* that the perspectives of men are of greater significance and relevance than those of women or, alternatively, and even more insidiously, that there are no important distinctions between them. Those who espouse androcentric views are thus often said to be 'GENDER-BLIND'.

The aim of much feminist scholarship has been to illustrate the ways in which androcentrism has informed and directed the development of social and public policy and determined the production of academic knowledge. FEMINIST GEOGRAPHERS, for example, have sought to highlight the ways in which theoretical concepts, research agendas, research methodologies and representational strategies both reflect and reproduce the conventional male-centred paradigm, promoting in its place an alternative EPISTEMOLOGY which de-centres masculinist perspectives by making visible those of a range of traditionally disenfranchised 'OTHERS'. BP

Androgyny From the Greek *aner/andros* meaning 'man, male' and *gyne* meaning 'woman', an androgynous person is one who displays both stereotypically 'MASCULINE' and 'FEMININE' traits. Not to be confused with physical hermaphroditeism, an anomalous biological condition in which a person is born with both male and female primary and/or secondary sexual characteristics, an androgynous persona is usually adopted as part of a deliberate and considered strategy for transgressing conventional gender roles.

Now inextricably linked to a wider body of social and cultural theory surrounding the SOCIAL CONSTRUCTION of gender identities, the concept of androgyny is of interest to feminists as it disrupts taken-for-granted assumptions about what it is to be a man or a woman, drawing into question any necessary link between sex and gender roles: maleness and masculinity, femaleness and femininity. The presence of both masculine and feminine attributes has the effect of blurring the distinctions between genders or of cancelling them out altogether. Located as it is, at the midpoint between the binary opposites of masculinity and femininity, androgyny has thus been championed as offering a new ideal 'ungendered' or 'monogendered' personality.

Paradoxically though, whilst androgyny is thought to be transgressive of conventional sex roles, it is itself formulated in terms of the discredited concepts of masculinity and femininity which it in turn seeks to destabilise. BP

Anthropology Anthropology is the study of what it is to be human. Because the main concern of the field is to examine physical and cultural variation among humans through time and space, anthropology has significantly contributed to discussions of GENDER and SEX as analytical categories. American anthropologists adopt a holistic perspective to integrate the 'four fields' of physical anthropology, archaeology, cultural anthropology and linguistics. In Europe, physical anthropology developed independently of ethnology (the study of peoples), the latter of which was tainted after World War II by its association with Nazi eugenics; social anthropology is now viewed as a distinct subdiscipline. This section outlines a brief history of cultural anthropology (see FEMINIST ANTHROPOLOGY for discussions of recent research topics).

In the nineteenth century, anthropology, like GEOGRAPHY, was closely tied to the history of COLONIALISM and NATIONALISM (Stocking, 1987). Evolutionists compared less sophisticated, or 'primitive,' 'RACES' (thought to be closest to ancient human groups) to culturally and intellectually advanced (white) 'races'. Their approaches varied: the Anthropological Society of London promoted a polygenetic view of human variation (genetic differences exist between 'races'), whereas Lewis Henry Morgan and Edward Burnett Tylor, influenced by ENLIGHTENMENT thought, promoted a monogenetic view of humanity. Tylor (1958) viewed culture as a single, cumulative attribute of humankind that followed a set evolutionary course of progress; as groups advanced, they grew more RATIONAL and SCIENTIFIC. Victorian scholars lumped BIOLOGY and CULTURE together in their analysis of human populations, treating sex as a biological category that determined social roles.

Franz Boas and his students, most notably Ruth Benedict (and her student Margaret Mead), criticised evolutionist approaches and argued that cultural differences explained human variation. Cultures were not viewed by these scholars as biologically inherited but rather as socially learned. Distinguishing between biological and cultural processes, these scholars recorded differences between cultures by reconstructing the histories of specific cultural complexes in geographical areas. The cultural construction of reality, including gender, was explained in relation to an *a priori* superorganic (see Kroeber, 1917). Although Elsie Clews Parsons (1916) also challenged Victorian evolutionism by foregrounding cultural difference to establish the notion of a universally oppressive PATRIARCHY, Mead (1928, 1935) was the first to use ETHNOGRAPHY to distinguish between biological sex and sociological gender roles (the latter she called 'sex temperament') (Visweswaran, 1997). At this time, Bronislaw Malinowski promoted a functionalist approach to culture by examining organised institutions that were related to the biological and other needs of humans. His fieldwork in the early part of the twentieth century resulted in a dominant mode of ethnographic research called participant observation (Evans-Pritchard, 1937; Malinowski, 1922). At this time, social anthropologists in Europe, such as Alford Radcliffe-Brown, were influenced by the functionalism of French sociologist Emile Durkheim.

During the 1960s and 1970s, STRUCTURAL and HUMANIST approaches to culture became more dominant. Claude Levi-Strauss argued that common intellectual processes determined by the structure of the human mind lay beneath all cultural products (Levi-Strauss, 1963, 1977). In contrast, Victor Turner and Clifford Geertz advocated humanist approaches by examining the performative aspects of culture (Turner, 1957) or viewing culture as 'webs of signification' spun by individuals and groups (Geertz, 1973). Thus, scholars examined the structural symbolic position of women in societies through

7

ethnography (Goodale, 1971) but continued to view women as a universal category until the early 1980s. Sex OPPRESSION was likened to racial oppression, and Gayle Rubin's idea of a sex/gender system – 'a set of arrangements by which the biological raw material of human sex and procreation is shaped by humans, social intervention and satisfied in a conventional manner' (1975: 165) – described how different societies gendered biology.

Research in cultural anthropology since the late 1970s has been influenced by critical social theory, including FEMINISM, POSTMODERNISM and POST-COLONIALISM. Current work examines: Western scientific views of the world; how the history of colonialism has influenced anthropology; the problems of participant observation and ethnography as methods for representing cultures (Clifford, 1988; Clifford and Marcus, 1986; Collier and Yanagisako, 1989; Haraway, 1984; Moore, 1988); the problems of gender ESSENTIALISM (Butler, 1990; Rosaldo, 1980); and the social construction of 'WHITENESS' and the meaning of multiracial IDENTITY (Ajani, 1994; Frankenberg, 1993a and b).   KT

Assembly lines Assembly lines of world market factories, involving labour-intensive tasks on the factory floor, have been dominated by women, particularly in the garments and electronics industries. In this 'global feminisation of labour', female assembly workers are preferred on grounds of various gender stereotypes, including their 'NIMBLE FINGERS' (Elson and Pearson, 1981), their capacity for concentration and their docility (Chant and McIlwaine, 1995).   CM

Authentic/authenticity Authentic and the quality of authenticity are RELATIONAL concepts. When used to denote an ideal type, the authentic is often defined as the 'real' or the original, as more genuine than its counterpart; the inauthentic, a fake or copy. Authenticity is always normative; to be authentic is to be more 'truthful' in character than imitations.

In philosophical thought, Heidegger described the authentic individual as one who understands the existential structure of his or her life. The authentic individual, aroused by *angst*, takes responsibility for his or her life and chooses his or her IDENTITY. In contrast, to be inauthentic is to absorb a situational identity uncritically.

More recently, humanists and social scientists have employed the term to refer to times and ways of life before the advent of CAPITALISM and MODERNITY. To take one example, the original quality of a hand-made object or work of art has been contrasted by the Frankfurt School to mass-produced objects and images. Critical of the 'superficial' nature of popular culture, these theorists argue that the progressive subject must turn away from (inauthentic) products of mass culture and turn toward more 'authentic' objects to avoid the 'colonisation of consciousness' by the culture industries. In tourism literature, scholars describe the search by modern individuals to find 'authentic' or 'real' places that are untouched by the Western world and hidden from the view of outsiders. Postmodernists like Jean Baudrillard and Umberto Eco use the term 'authentic' to refer to the 'real' as opposed to the simulacrum. In the electronic age of POSTMODERNITY, the copy (simulacrum) becomes the original or even better than the original. Postmodern 'hyperreality' describes a condition in which there are only simulacra because the origins (the authentic) are lost, not recoverable or never existed.

Critics of these uses of the term 'authentic' often take a constructivist approach to culture and argue that there are no pre-given or inherent meanings of objects, events or texts. Rather, signification occurs by individuals and groups through the acts of reading, consuming, producing and experiencing texts. Although any cultural production can be defined as a copy that makes some reference to prior events and TEXTS, each production is simultaneously an original insofar that it is adopted to new circumstances. Critics also argue that the very term 'authentic' reifies a romanticised category of 'pristine' and 'pure' places, times, peoples, and artifacts. In this usage, non-Western representational sites (the authentic) are seen as chronotopes (time-space formations) where individuals are in touch with the 'real' world and their 'real' selves. Some Marxist scholars may use the authentic to exoticise, feminise, and essentialise a non-Western sphere that has been penetrated, consumed and conquered by the MASCULINE forces of WESTERN capital.   KT

See also CRITICAL THEORY.

Author, death of/authorship In his 'death-of-the-author' thesis, Roland Barthes (1977) argues that the concept of the TEXT is prior to that of the author. Texts result from an 'interplay of signs', and as such the meanings of any text can proliferate without any privilege to the author's intended meaning. For Barthes, the author (and the notion of 'intent') is seen as a mere construct and is insignificant – hence the death of the author. By shifting away the focus of analysis from authorship to intertextuality, Barthes's POST-STRUCTURAL approach is used to examine the production of meaning through the interaction between texts (which includes cultural productions and landscapes (Duncan and Duncan 1988)). This thesis has been influential in literary criticism, but is more skeptically regarded by some philosophers.   KT

Authority A term that is closely related to that of POWER, in the sense that it is the institutionalised or embodied legitimate use of power over others. Max Weber defined three types of power: traditional, legal-rational and charismatic. It is the second of these three types that is assumed to dominate in modern societies associated with the rise of rational individualism after the ENLIGHTENMENT, and it is this form of authority that is typically the focus of social science analyses of the exercise of power. However, as Weber himself argued, men's authority as head of a household is a specific form of traditional power. Feminist scholars have pointed out the bias towards analyses of the PUBLIC arena and have documented the multiple ways in which men have authority over women in the PRIVATE sphere both in a personal sense and as embodied in institutional structures such as the legal system and the WELFARE STATE. More recently there has been a greater recognition that women may exercise authority over others, especially children in the FAMILY or domestic sphere but also over other women and sometimes men. In historical analyses of HOUSEHOLD production, for example, women have authority in the social relations of household production (Hufton, 1995). Similarly, in more recent work on the body there is some attention to female authority. Although the primary emphasis in this work is on the control exercised over women in the construction of 'docile bodies' through, for example, idealised representations and disciplinary regimes such as diet and exercise (Bordo, 1993), cross-cultural analyses have also documented the variety of ways in which women themselves have legitimate authority over other women's bodies, often in an active sense. **LM**

See also BODY; LAW.

Autobiography A type of personal NARRATIVE in written, oral, photographic, filmic or other media, autobiography is an introspective and interpretive definition of one's life. It is a particularly important outlet for authors barred from more traditional or public venues of writing, thus feminist researchers have used it to recover the lives and EXPERIENCES of women and the larger historical and cultural meaning of their lives (Heilbrun, 1988; Stanton, 1987). Whereas literary theorists distinguish autobiographical writing from the closely related genres of memoir, journal, diary, personal essay and TRAVEL narrative, geographers generally do not.

Readings of PLACE through autobiography have been studied by humanistic and feminist geographers. Masculinist and heterosexualised meanings of the home place in particular became the foundation of the life story in humanistic geography of the 1970s. Feminist geographers use autobiographical material to challenge this meaning of HOME, as well as interpret meanings of LANDSCAPE, NATURE, IDENTITY, COLONIAL and settler societies and travel, among other social spaces (Blunt and Rose, 1994; Jones *et al.,* 1997; Kay, 1991).

American and European feminists have also drawn on autobiography to construct theories of gender difference, arguing in the 1980s that women's writing contained characteristic 'gynocentric' features, such as focus on human relations and domestic, emotional and personal information (Rich, 1986a). Later critics abandoned such totalising notions of women but searched for other sources of identity and forms of difference, such as bell hooks' (1989) postulation of black women's 'double-consciousness'. POST-STRUCTURALISTS, after Foucault, insist that rather than authorial intention, autobiography displays fragmented SUBJECTIVITIES that are the product of the author negotiating complex material and discursive webs of POWER and meaning (Mills 1991). **KM**

B

Backlash See FEMINISM, POST-FEMINISM

Beauty myth A phrase coined by Naomi Wolf in her 1991 bestseller, *The beauty myth*. The book documents mythologies of female beauty in the West, embedded in laws and normative cultural practices. She argues that these were created in post-industrial times to keep feminist advances at bay and women in low-paying service-sector jobs. Wolf makes no mention of the racist and class bases of the mythologies, her empirical focus being middle-class HETEROSEXUAL white women.   HN
See also AESTHETICS/AESTHETISATION OF EVERYDAY LIFE; BODY.

Binary/binary oppositions The social construction of SEX and GENDER is based on the belief in a binary or categorical distinction between men and women. Despite the fact that feminist scholars, as well as biologists and geneticists, have convincingly demonstrated the flawed nature of binary assumptions and an essentialist or biologically based division of humanity into only two categories, the belief in binary gender divisions has remained a key element of contemporary social practices. This distinction is also an unequal one in which the social attributes associated with men – masculine traits – are assumed to be superior to those associated with women or femininity. Thus, women, and feminine characteristics, are defined as inferior, irrational, emotional, dependent and PRIVATE, closer to NATURE than to culture in comparison with men and masculine attributes that are portrayed as superior, RATIONAL, SCIENTIFIC, independent, PUBLIC and cultured. Women, it is commonly argued, are at the mercy of their bodies and their emotions, whereas men represent the transcendence of these baser features: mind to women's BODY.

As many feminist scholars have demonstrated, the belief in CATEGORICAL difference, which is binary and also hierarchical, is deeply embedded not only in individual behaviour and in the construction of gender IDENTITY but also in the structures and practices of Western thought and in institutional practices. In the development of the social sciences, for example, the public attributes of the market or the state are the basis for study by economists and political scientists, respectively, whereas the 'private' decisions taken within the home, say, are the province of sociologists or psychologists. Pateman and Grosz's (1987) edited collection provides a very clear introduction to the binary structure of Western social science. This binary division is also deeply implicated in the social production of SPACE, in assumptions about the 'natural' and built environments and in the sets of regulations which influence who should occupy which spaces and who should be excluded. Thus, in common with the other social sciences, binary categorisations structure geographical scholarship (for further explanation and examples, see Jones *et al.*, 1997; McDowell, 1998; Massey, 1994a; Rose, 1993a; WGSG, 1984, 1997). It is clear, then, that feminist geographers have an ambitious task: not only mapping gender inequalities but tearing down and re-erecting the structures of the discipline, the very ways in which space and place are theorised, studied and explained.   LM

Biology/biological determinism Biology is, literally speaking, the science or study of life; but the term is also used for the structures and processes studied by the life sciences. These include biochemistry, physiology, ethology, genetics and developmental biology, which study living things under very different aspects, within the broad paradigm of neo-Darwinian evolutionary theory. Biological explanations may be phylogenetic, in which case they explain the evolutionary origin of some feature or behaviour characteristic of a species. They may be functional, focusing on the feature's adaptive value in particular environments. They may be causal, describing the mechanisms, maybe at a cellular or even a molecular level, involved in particular physiological processes or as necessary conditions for particular behaviour. Last, explanations may be developmental, tracing the foetal and infantile stages through which a trait becomes specialised. All of these types of biological explanation may be applied to social life. They become deterministic when they deny to social structures and practices any causal influence of their own.

Biological determinism in the human sciences is the view that social life is determined, or entirely caused, by its biological substratum. It is often linked to a form of reductionism in which wholes are treated as no more than the sum of their parts. Most people would agree that what is possible for humans is determined by our embodiment. What human BODIES can and cannot do maps out the range of possibilities for social life. Eating and excretion must be allowed for, and humans cannot readily socialise by hovering at 20 feet above the ground. Language is a biological capacity of the human species, and babies are innately social – programmed to respond to the human face and voice. Biological determinism goes further than this, denying social causes or, at best, giving them the status of environmental triggers which reveal but do not really cause an already present condition or tendency.

For example, inner-city riots have been attributed to overcrowding or to social selection of a population genetically predisposed to violence. Children's inattention in class is likely to be understood as a 'syndrome' known as 'attention deficit disorder', treatable by the drug Ritalin, rather than in sociopsychological terms. SEX roles, despite their cultural variability, have frequently been seen as biologically determined. In the early nineteenth century women were believed incapable of benefitting from higher education because of their smaller brains, or it was believed that if they did succeed, their reproductive organs would atrophy (Sayers, 1982). More recently male VIOLENCE has been attributed to the effect of hormones on the developing male brain, while women's failure to leave violent relationships is explained in terms of an. addiction to a hormonal surge accompanying the violence (Birke, 1986). The larger proportion of women diagnosed as mentally ill is not attributed to the stressful effect of women's OPPRESSION, nor to sexism in psychiatric diagnosis, but to hormonal and neurological differences between the sexes (Russell, 1995).

Sociobiology became popular in the 1970s as a sustained attempt to explain social life in biological terms, which in its strong forms is biologically determinist. Evolutionary biologists now agree that genes are the units of natural selection, although it is species that eventually change as a result. Since the genes which produce the best adapted organisms in given environmental conditions are most likely to be reproduced, genes can be thought of as competing, although in actuality they have no purposes. For 'strong' sociobiologists such as Dawkins, organisms are merely the garments in which genes are clothed, and cultures are largely determined by the selection of the most successful reproductive strategies. Much of the argument has been over 'altruism', and whether it can ever be adaptive. But sociobiology also makes big claims about GENDER relations, which have brought it into conflict with FEMINISM.

For mammals with a long gestation period, especially humans, the genes of female organisms are more likely to be transmitted if a relatively small number of young are fed and well protected during early development. Females can only have a limited number of children, so the best reproductive strategy is to get men to become useful providers – a strategy believed to be genetically based and to include remaining sexually attractive and so on. Males, on the other hand, are likely to maximise the numbers of their descendants by spreading their seed around, gambling on at least some children growing up. The female strategy calls for commitment, the male strategy for sexual freedom – for males. Thus both the 'double standard' and women's primary responsibility for CHILDCARE are seen by 'strong' sociobiologists as biologically determined aspects of CULTURE.

This version of biological determinism has been criticised within biology (e.g. Gould, 1981), as well as by feminists and other social scientists. Only a few of these criticisms can be mentioned. The degree of cultural variability around human sexual behaviour leads anthropologists to accuse 'vulgar' sociobiologists of ethnocentrism (Leacock, 1980) and even of interpreting primate behaviour in terms of their own ETHNOCENTRIC sexual prejudices, which are then used as 'evidence' for analogous behaviour in humans (Haraway, 1991). The 'double standard' is not UNIVERSAL, but sociologists point out that behaviour that is universal is not therefore necessarily inherited. In any case humans, like some other animals, actually select and modify their environments, so 'adaptive' ways of behaving must be relative to culture as well as to 'NATURE'.

Biological determinism has been used to justify racist, homophobic, misogynist and other oppressive policies. However, the evidence against it is not based on its past destructive applications but on its unworkable attempt to deny causal powers to social and psychological structures. Obviously, social life has biological effects: the class system brings about differential rates of mortality and morbidity. Human beings could be described as biologically social, since 'dependence upon social life is built into our very anatomical and physiological constitution, and into our developmental rhythms' (Benton, 1991: 23). Though our sociality was doubtless established partly through natural selection, no causal explanations that ignore its power can be considered valid.   CN

Birth control The practice of managing human fertility and reproduction, for example through contraception. Malthus (1817) generated an interest in birth control in Western Europe in his writings on the problems of overpopulation. The term was devised by Margaret Sanger (1916, cited in Moore and Moore, 1986) in her paper 'The woman rebel'. Feminist theory still promotes Sanger's principle that the right to reproductive freedom is crucial to women's EMANCIPATION.

With the resurgence of feminist organising in Britain and the USA in the 1960s, struggles around a woman's right to control her own fertility were a key part of the movement. The right to contraception, including abortion and chemical (the pill) and mechanical (the diaphragm) methods, was demanded. The politics of the pill, which had been developed in the USA in the early 1960s and tested on poor and often migrant women, was the subject of the doctoral dissertation undertaken at the University of London by the geographer Julia Cream (1993a; see also Cream, 1995b). Contraceptive methods which are in the hands of women rather than men (the pill and diaphragm rather than the sheath) give women more control over their fertility. However, chemical methods have side-effects which may have been both underestimated and underpublicised. A male contraceptive pill has been developed and tested in the West but seems unlikely to be popular.

Some feminists have criticised the women's movement for its emphasis on contraception, arguing that for many women, especially those who are members of ethnic or religious minorities, or women in populous nations who may be subjected to forcible birth control, the right to give birth is as important as the right not to conceive.   RK
See also DEMOGRAPHIC POLICY; SECOND WAVE FEMINISM/SECOND WAVE WOMEN'S MOVEMENT.

Birth rate The number of live births per 1000 of the population over a given period, usually a year. The ratio is normally adopted as an indicator of fertility and ranges from low values typical of developed countries such as 10 per 1000, to high values typical of developing countries of 50 per 1000. The ratio is affected by socio-cultural and economic characteristics, particularly women's rights, contraception and healthcare.   PM

Bisexuality Defined as a condition in which an individual experiences desire for members of the same SEX and of the opposite sex. Bisexuals have been classically constructed as individuals who are situated 'halfway' between the binary opposites of HETEROSEXUALITY and HOMOSEXUALITY. However, bisexuality has not been pathologised in the way that homosexuality has, which may account for the fact that it is conventionally constructed in terms of a series of acts rather than as a mature sexual IDENTITY.

Whilst perhaps now seen as occupying an ideal location on the sexual continuum, bisexuality has been historically constructed as a subversive activity. Heterosexuals and homosexuals have been mistrustful of bisexuals, characterising them, as Hemmings (1995: 46) suggests, as 'the double agent[s] of sexual politics selling out to the highest bidder'. The political commitment of bisexuals has been particularly scrutinised by some GAYS and LESBIANS who suspect them of being 'temporary' homosexuals, able to decamp to the safe confines of heterosexuality whenever maintaining a visible gay identity becomes overly problematic. Bisexual women and men have also been constructed as subversive by some heterosexuals who hold them responsible for bringing HIV into their community.

The advent of QUEER THEORY signalled a radical questioning of the oppressive heterosexual–homosexual binarism, which as Smyth (1992: 20) suggests 'constantly privileged the hetero perspective as normative, the homo perspective as bad and annihilated the spectrum of sexualities that existed in between'. Within heterosexual and homosexual communities an increasing number of people now feel able to make a claim to bisexuality. Bisexuality is thus in the process of becoming rehabilitated, even celebrated, as a valid sexual and political identity. BP

See also COMPULSORY HETEROSEXUALITY; LESBIAN CONTINUUM.

Black The terminology of 'colour' has a complicated history which is bound up with racism. The terms used to refer to 'people of colour' have varied from offensive terms which are no longer tolerated, such as 'nigger' (nowadays, reappropriated by people with African heritage, this term is permissible – an example is the rappers NWA), through 'coloured' people, a common term in Britain in the 1950s and 1960s, which has been replaced by more acceptable language such as 'black' or by more accurate terms such as African Caribbean or African American, which indicate origins rather than skin colour. The term black is contested too, as it sometimes is used to refer not only to people of African origins but also people from the Indian subcontinent, in part to indicate their common experiences of racist attitudes and behaviours. In the 1970s and 1980s 'black is beautiful' was a slogan used by the black pride movement to subvert the connotations of inferiority in the term when used by the 'majority' WHITE population. However, as the term became commodified by the fashion, sport and music industries it has tended to become less frequently used. As Ann DuCille comments in her book on the commodification of blackness, 'I am not convinced that the commercial beauty culture has ever presented black people as beautiful per se; instead CAPITALISM has appropriated what it sees as certain signifiers of *blackness* and made them marketable' (1997: 27, original emphasis). Thus what is being sold is a version of ALTERITY. Further, as the African American writer bell hooks (1991) has warned, there are dangers in merely subverting or inverting terms, in celebrating characteristics that have been imposed by the structures of what she terms white heteropatriarchal society. These dangers have parallels with the similar inversion involved in radical feminists' celebration of some of the characteristics of FEMININITY. LM

Black feminism Feminist struggles against the racism and sexism experienced by black women. In a painful critique of the earliest SECOND WAVE FEMINIST demands, black women pointed out both the class-based and race-blind assumptions of both feminist theory and practice, recalling the question of runaway slave Sojourner Truth 'Ain't I a Woman?' (hooks, 1982). The work of black feminists has been crucial in the develpment of what is often now known as the 'politics of intersectionality' (Brah, 1996) showing that RACE and GENDER are both important axes in understanding the position of WOMEN OF COLOUR and that, further, race and gender are mutually constructed parts of IDENTITY rather than separable or additive dimensions of inequality. There has been a vigorous expansion of black feminist work in many areas, from social criticism (Brah, 1996; hooks, 1991, 1992a and b, 1994; Lorde, 1984) to literary composition (Morrison, 1981, 1992; Walker, 1984a and b).

In Britain the term black feminism tended, at least initially from the early 1970s onwards, to be used in an inclusive sense, as a generic term for all non-white feminisms. However, with the growing interest in the late 1980s and 1990s in IDENTITY POLITICS more generally, there has been a debate about the similarities and differences between what North American feminists refer to as 'women of colour'. In the USA, for example, the work of Chicana feminists, including Gloria Anzaldúa (1987), has drawn attention to the MARGINAL position of Latina women, which has both parallels with and differences from African American women's struggles. In Britain, the influence of POST-COLONIAL theorists has been important in the development of work on Asian women, both in Britain and in the Indian subcontinent (Brah, 1996). Brah (1992) has, for example, resisted the appropriation of all difference under the label 'black'. However, the early work of black feminists has been vital in the recognition by feminists of the significance of DIFFERENCE in all its aspects – including age, class, ethnicity, physical ability and sexual identity. LM

Black masculinity A racist characterisation of assumed rampant HETERO-SEXUAL traits among black men, which in its extreme form in the American south led to convictions for rape, lynchings and hangings. These traits were subverted and celebrated in blaxploitation films of the 1970s such as *Shaft*. Feminist writers, including bell hooks (1994), have been criticised by some for drawing attention to the problematic characteristics of black masculinity, including VIOLENCE against women, sexist representations, especially in areas such as rap music and in the new black cinema, and irresponsible attitudes towards children on the grounds that feminism is a distraction from the greater injustices of racism.   LM

Body Historically, women's bodies have been perceived as weak, unstable, leaking, incomplete, threatening to engulf men and acting as a source of sexual temptation to them. As a result, women have been regarded as unruly and unable to govern themselves. The body has thus necessarily been central to feminist politics but its conceptualisation has posed major difficulties. Following de Beauvoir, many feminists tended to see women as prisoners of their bodies and to view the body *per se* as problematic and as something to be transcended. Some, like Shulamith Firestone (1971), looked hopefully to medical science as a means of reducing the reproductive differences between men and women. Others drew on social construction theory to establish a firm distinction between sex as a biological given, and gender which was a social creation through which the differences between the sexes were exaggerated and hierarchised. If sexual inequalities were based on socialisation processes rather than on biology, it was thought that they were more open to change. The ideal for many feminists was an androgynous or genderless society in which biological differences would have little social meaning.

While the sex–gender distinction gained considerable currency in the social sciences and in everyday language, not all feminists accepted it. They argued that the key problem was not masculinity and femininity as these were socially constituted but the deep hatred and fear of the female body. Adrienne Rich's book, *Of Woman Born* (1979), was one of the first to view the female body as a source not of weakness but of strength. Rather than denying the relevance of bodily differences, difference feminists celebrated women's reproductive, sexual and nurturing capacities and sought sexual autonomy rather than sexual equality on the sameness model. These accounts were criticised for their essentialism, which is exactly what social construction theory had sought to escape. Difference feminists replied that a genderless society would effectively subsume women into a single masculine norm.

As Moira Gatens has observed, both sides of this sexual equality versus sexual difference debate within feminism made very similar assumptions about the body as a biological entity which either has or does not have certain ahistorical characteristics and capacities (Gatens, 1996: 129). They also tended to reproduce a set of assumptions about the duality of mind and body. Through most of the twentieth century the individual has most frequently been theorised in terms of consciousness, thus reinforcing the positive valuation of mind and (by omission) the negative valuation of body. The body has been understood as a passive object, to be controlled by the mind, appropriately the domain of biology or medical science. Drawing on Nietzsche, as well as phenomenological traditions, feminists began to present the body as actively involved in the production of subjectivity and to treat the body itself as the product of lived experience. In a well-known essay, 'Throwing like a girl', Iris Marian Young (1989a) argued that the modalities of feminine bodily comportment, motility and spatiality are based not on anatomy or physiology or any mysterious female essence but on the particular situation of women in a male-dominated society. As Gatens points out, embodied subjectivity entails not simply the material or anatomical body but an imaginary body, or a body which is organised around introjected images. This psychic image of the body is a product of the culturally specific meanings, values and symbols of the subject's social milieu. The phenomena of phantom limbs and hysterical paralysis illustrate the way in which it is the imaginary body, not the anatomical body *per se*, which is experienced (Gatens, 1996).

Feminists have been concerned both with the material body and with the body as represented in medical, philosophical and cultural texts. Feminist philosophers have demonstrated that the illusion of perspectiveless knowledge in Western systems of thought is actually created on the basis of a fantasy that knowledge-producers are disembodied, a fantasy underpinned by the metaphor of woman as body. By making the feminine represent body or nature, masculinity is freed of its own body and is able to produce and operate within culture and knowledge. Michele LeDoeuff has argued that women's bodies have been characterised as negativity, as otherness, as lack, against which the fullness, the centrality, the completeness of philosophy is contrasted (Mackenzie, 1986: 144). Much feminist debate has centred on the use of the female body as a metaphor and the possibility of introducing the body of woman into language. This project was largely inspired by Luce Irigaray and her notion of ÉCRITURE FEMININE. Following on from Adrienne Rich's paper, 'Notes towards a politics of location', feminists have also been challenged to think about where they lived and hence the material conditions from which they spoke. Rich argued we must begin not with a continent or a country or a house but with the geography closest in – the body (1987: 212).

Feminists have drawn extensively on Michel Foucault's work to argue against the notion of the natural body as the basis on which identities and social inequalities are built. Rather than taking consciousness as the starting point, Foucault takes the body as a surface of inscription and understands subjectivity as produced within discourses which are inscribed directly upon bodies. The conclusion we must draw from his work is that very different bodies will be produced under different discursive regimes. For example, the disciplinary regimes of the military, the school, the hospital and the prison produce particular kinds of corporeal subjectivity. While Foucault himself paid very little attention to the female body, his concern with the effects of power on the body has been an important starting point for feminist analysis of particular regimes of power. Much feminist work has been concerned with the processes whereby 'normal' bodies are constructed, including dietary and exercise regimes, dress and bodily adornment. Following Foucault, Judith Butler argues that at stake in the reformulation of the materiality of the body is that the body must necessarily be connected to the geographically and historically specific norms within which we each locate, evaluate and understand our bodies (Bell and Valentine, 1995: 26). In turn we must see these norms as not just impinging on bodies but as that which produces, as Butler (1993) puts it, the very matter of bodies, that which constitutes their materiality.

Rather than demonstrating how men and women become masculine and feminine subjects, feminists began to ask how bodies become marked as male and female. In this way, the body became amenable to sociocultural accounts which emphasise its plasticity and variability. If the body is granted a history, the traditional associations between the female body and the domestic sphere and the male body and the public sphere can be acknowledged without resorting to biological essentialism. The emphasis has shifted from consciousness to the experiences, discourses and practices that constitute embodied subjectivities. Male and female bodies are not prediscursive and immutable natural entities upon which gender traits are mapped. The meaning of the body, its very experience of itself, its capabilities, traits and characteristics are products of the discourses in which it is articulated.

While we are always contained and limited by the historical conditions within which we live, the body may, nevertheless, be seen as an entity in the process of becoming. In the West, at least, the body is becoming increasingly a phenomenon of options and choices. There is both increased potential to control our own bodies and to have them controlled by others. As science facilitates greater intervention, it also destabilises our knowledge of what bodies are and runs ahead of our ability to make moral judgments about how far science should be allowed to reconstruct the body. Transplant surgery and virtual reality both threaten to collapse the boundaries between bodies and between technology and the body. Much recent writing on the body is concerned with analysing the nature, limits and modes of existence of the cyborg body (Haraway, 1991, 1997) which combines animal and machine.

The emphasis on the discursive construction of bodies also raises interesting questions about how bodies are connected to each other. The figure that Gilles Deleuze uses to map bodies, individuals and potentially all social interaction is the rhizome: a model of interconnections that do not posit a core or universal root. This conceptualisation of bodies as constituted and connected across networks has also challenged the dominant understandings of sexuality and desire. The deficit model, in which desire is produced by some kind of lack or gap between the desiring subject and the desired object, is inadequate for a wired-in, networked and intertextual body. Instead Deleuze, and feminists drawing on his work, talk of desire as a productive mechanism, a continual process of stimulation, connection and production. This cartographic conception of bodies is being extended beyond desire to consider the variety of connections between bodies. If Western feminisms have tended to produce the body as sexuality, this may be at the expense of other bodies which eat, work, write and reach out to make political connections. Biddie Martin has observed that emphasis on sexuality has privileged white Western bodies. Against this, the writings of feminists of colour, and those actively engaged in anti-racist struggles, have been important in focusing on the disconnections between different types of body. Martin has framed the challenge of using our bodies as a passage, not over, not by, not around, but through the troubled intersections of gender, sexuality, class and race (1988: 78).

Many feminists have expressed dissatisfaction with post-structuralist theory for its very abstract notions of the body. At the same time, it may be argued that it has opened the way for more detailed understandings of the types of bodies produced in specific historical and spatial contexts. It treats individual human bodies as always parts of larger assemblages and thus provides a conceptual frame in which to take account of the variety of ways in which individual bodies and their capacities are affected in the larger assemblages of family, work and sociopolitical life. And it allows us to consider bodily specificity without getting bogged down in essentialism. And it turns upside down the old metaphor of the 'body politic' to reveal the sexual specificity of the political arena.

In recent feminist work by geographers issues about the body and embodiment have been addressed in a number of different ways. In work on LANDSCAPES, for example, Catherine Nash (1996a and b) has examined the ways in which images of femininity and sometimes masculinity are mapped on to the landscape; in urban geography the similarities and mutual constitution of cities and bodies have begun to be explored (see Pile, 1996); in economic geographies the ways in which the workplace is constructed to marginalise the body and to construct female bodies 'out of place' have been explored in different types of work (see for example Halford *et al.*, 1997; McDowell, 1997b); whereas other geographers have examined the material experiences of pregnancy (Longhurst, 1995, 1996), sickness (Dyck, 1995a and b, 1996; Moss and Dyck, 1996) and disability (Parr and Philo, 1995). RP

Border/borderlands Spatial metaphors have become a fashionable means of thinking about POSITIONALITY and knowledge. The concept, or metaphor, of border/borderlands may be especially provocative. Two key advantages of this term, when applied metaphorically, are that it shifts attention from the individual to socially constructed places where DIFFERENCE and conflict are lived out and that it seems to provide a means of holding change (mobility, deconstruction) and limited statis (dwelling, placement, normative ideals) in creative tension (Pratt, G., 1992: 244).

The work of Gloria Anzaldúa has been particularly influential in defining the 'borderlands' as an actual physical location as well as a discursive space (Anzaldúa, 1987, 1997). Growing up on the international border between Mexico and the USA, Anzaldúa came to express her perception of what it is to be a 'border woman'. She simultaneously develops the borderland as a metaphor for the culturally HYBRID and as a theatre for radical political action. Above all, her poetry expresses a borderland aesthetic involving a topic and style of communication characterised by division, hybridity and mixedness. Anzaldúa's prominence relates to a much wider literary movement centred on the US-Mexico border and exploring the multiple meanings of the line, the border and the boundary as lived space (Soja, 1996).

The fact that Anzaldúa's work is included in a recent reader of feminist geography (McDowell and Sharp, 1997) signals a wider move within geography to champion the borderland intellectual. This involves a specific focus on those who consciously place themselves at the margins or borderlands of dominant discourses. The danger here is that the subjectivities and silences which inevitably characterise an individual's positionality are overlooked in the enthusiasm for the 'view from the edge'. This is evident, for example, in David Harvey's lack of attention to Raymond Williams's neglect of race and racism in his tribute to this man's border identity (Harvey, 1996; McDowell, 1998).

A different strand of work has examined how specific PLACES come to be constructed as borderlands, in the sense of real or imagined locations of ambiguity or threat. Merriman, for example, has shown how, in nineteenth-century Paris, the *faubourgs* and emerging suburbs (*banlieues*) came to be constructed as zones of foolishness and emotion, 'floating worlds' where CITY and country met in an 'awkward embrace' (Merriman, 1991). Contrasted to the emerging urban order of central Paris, this borderland was said to contain the wandering, marginal people (*les marginaux*). The PROSTITUTE and gypsy became the most salient symbols of these excluded, propertyless, mobile individuals (Seigel, 1986, Merriman, 1991). Although writing or painting with very different political and social commitments in mind, Victor Hugo, Emile Zola and Vincent Van Gogh, like Anzaldúa, were concerned with creating a borderland aesthetic in their depictions of the *banlieue* as a pitiful, amphibious, borderland. The border and borderlands remains, then, a powerful imaginative and theoretical tool for thinking about how difference and conflict is constructed, lived and represented in various historical contexts.   CJ
See also FLÂNUER/FLÂNEUSE; MARGINS.

Boundary See BORDER

Breadwinner A gender role ascribed to men who go out to work and provide for their family. Assumed to be opposite but complementary to the role of women who stay at home and perform DOMESTIC LABOUR and CHILDCARE. Together male breadwinners and female caregivers form the traditional nuclear FAMILY.   HC

Bricolage A term used by Claude Levi-Strauss in *The Savage Mind* (1966) to refer the intellectual practice of making do with 'whatever is at hand' to create mythical thought. As a STRUCTURALIST, Levi-Strauss sees structure as more important than the content which can be built up of the 'remains and debris' of past events. The term has been used to describe a POSTMODERN eclectic, even anarchic, aesthetic in which a creative juxtaposition of hetero-geneous fragments makes ironic, playful or subversive reference to dominant meanings. Subversive juxtaposition is used to undermine modernist values of purity, coherence, unity and integrity. **ND**

Built environment A broad term applied to development of the natural environment and to the resulting residential, commercial, transportation and (some) leisure spaces created. In common usage the term built environment implies the transformation of green or natural landscapes although green spaces, especially recreational spaces, may also be considered part of the built environment where they occur in towns and cities. Similarly, canals, rivers and even beaches may be incorporated in the 'built' environment.

Feminist studies have drawn attention to the masculine nature of the built environment, arguing that the design of various aspects of the built environment as well as the broad evolution of towns and cities has prioritised men's needs over those of women (see Boys, 1984, 1998; Little, 1994; Little *et al.*, 1988; McDowell, 1983; Spain, 1992). Such work has commonly portrayed the built environment (of 'developed' countries) as divided into 'PUBLIC' (economic and commercial) spaces and 'PRIVATE' (domestic and residential) space – spaces of production and spaces of CONSUMPTION. Within this broad framework, the built environment has been seen to reflect the gender division of space; private spaces have been seen as female, and public spaces as male. The distribution of differ-ent functions and uses within the built environment and the organisation of transport has helped to reinforce the gender division of space, making movement between different spaces difficult and restricting women's ability to participate in the public world of paid employment. At the local level, studies of women's use of the built environment have examined the design and layout of residential spaces (Matrix, 1984; Roberts, 1991). The evolution of suburban housing estates with their emphasis on privatised spaces of the family home and their lack of services has, again, reflected traditional assumptions of the gendered use of space and of male and female roles in society.

Women's use of the built environment has also been considered in the context of safety. Many areas within the built environment are considered hostile and frightening to women – especially at night. Aspects of the design of, in particular, urban areas are thought to reinforce the fears felt by women. The separation of different land uses, the domination of major roads, inade-quate lighting and the poor design of car parks (especially multi-storey), for example, are frequently seen as insensitive to the dangers faced by women from the threat of male violence.

In some towns and cities more attention is now being paid to the gendered use of space and to the problems faced by women resulting from the design and development of the built environment (Greed, 1994). Attempts are being made in some areas, through the introduction of mixed-use development and the promotion of safety-conscious design, to break down some of the barri-ers to women's use of the built environment. While planning measures

concerning the distribution of services and the location of employment and residential spaces are important, it is also clear that gendering of the built environment is as much about the social construction of gender roles and gender relations as it is a matter of physical design.   JL
See also CITY; FEAR OF CRIME; SAFER CITIES; VIOLENCE.

Bureaucracy For Max Weber bureaucracy is the characteristic feature of modern society. As he theorised it, bureaucracy 'has a "rational" character: rules, means, ends, and matter-of-factness dominate its bearing' (Gerth and Mills, [1948] 1991: 244). According to Weber's 'ideal type', bureaucracies are based on impersonality, functional specialisation, a hierarchy of authority and the impartial application of rules. There are well-defined duties for each specialised position, and recruitment takes place on criteria of demonstrated knowledge and competence. AUTHORITY, established by rules, stands in contrast to the 'regulation of relationships through individual privileges and bestowals of favour' that characterised TRADITIONAL structures. Above all, there is a separation of the PUBLIC world of RATIONALITY and efficiency from the PRIVATE sphere of emotional and personal life, a division which is mapped on to masculine and feminine attitudes. Weber recognised that technical rationality made possible the dynamism and productivity of modern life but he had no illusions about creating an ethically rationalised world in which human values would rule. Feminists have criticised bureaucrats for their reliance on masculinist attributes but have also recognised that the clearly defined hierarchical structure and rules of bureaucracies allow women some redress in comparison with organisations that reward individualised attributes and activities.   RP
See also CORPORATE CULTURE.

C

Capital/capitalism Capitalism is a particular form of societal organisation rooted in a particular form of economic production. At its core is the capitalist labour process whereby workers are separated from the means of production and their labour power is reduced to a commodity. Diagramatically, this can be represented as:

$$M - (MP + LP) - P - M^1$$

(where M is money, MP is means of production, LP is labour power and P is product).

As Marx put it, labour power is the lifeblood of the capitalist production process: 'By the purchase of labour-power, the capitalist incorporates labour, as a living ferment, with the lifeless constituent of the product' (1954, 180). Capitalism is characterised by competition between companies for market share, ensuring that the accumulation of capital takes place.

Capitalism is often thought to develop in stages, characterised by a crisis and a reaction to it to counter the falling rate of profit. Early mercantile capitalism in the West was succeeded by monopoly capitalism in which there is an extreme concentration of ownership in the hands of corporations. Some now argue that a form of 'late capitalism' is developing in advanced industrial societies, associated with a new mode of accumulation and social regulation, that is sometimes called POST-FORDISM or neo-Fordism. The geographer David Harvey (1989), influenced by the work of cultural critic Fredric Jameson (1984), has argued that the features of a postmodern society are primarily the cultural logic of this form of late capitalism rather than a significantly new form of social organisation. Marx himself argued that capitalism would eventually, and inevitably, be superseded by SOCIALISM, as capitalism was destroyed by class conflict and its own internal contradictions. The developments in Eastern Europe and the moves towards a form of 'market socialism' in China seem to have led to a consensus about the inevitability of the dominance of the market and of capitalist social relations.

The relationship between capitalism and women's oppression has been hotly debated. In *The Origin of the Family, Private Property and the State* Friedrich Engels argued in 1884 that the origins of the family and women's subordinate role within it was an inevitable consequence of the rise and requirements of a bourgeois society. This argument became a tenet of MARXIST and SOCIALIST FEMINISM which demonstrated the necessity of DOMESTIC LABOUR for the daily and generational reproduction of the labour force in a capitalist economy. Feminists working within a political economy perspective explored not only domestic labour but also women's occupational segregation in the labour market, whilst disagreeing about the extent to which capitalism and PATRIARCHY constituted a single or dual system. Socialist feminists have also addressed the extent to which the current feminisation of capitalist labour markets will allow women greater independence from men, thus challenging patriarchal relations (McDowell, 1991).   JW
See also CLASS; REGULATION SCHOOL; OPPRESSION.

Care Usually used in the sense of looking after others because of personal connections, love, reciprocity or mutual obligations rather than for monetary motives, although the rise of the caring professions in sectors such as social work and the health services has blurred this distinction. The attributes of caring are often associated with femininity and femaleness and are part of the reason for the stereotypical association of certain tasks and occupations with women. The feminist philosopher Carol Gilligan has argued that there is a specifically female orientation to others which she calls the caring ethic, compared with a masculine orientation to others which is distinguished by rational and instrumental relationships.   LM

See also CHILDCARE; ETHICS; MORALS/MORAL REASONING; PROFESSIONALISATION.

Cartography The making of maps. Cartographers aim to render the world accurately on the two-dimensional surface of a map. Creators of geographical information systems (GIS) now claim their mapping should be considered scientific as it is managed through computer and satellite systems which render errors smaller and smaller.

Recently, however, geographers have begun to criticise the ability of the cartographer to represent the world completely (MIMESIS) and even to question this as a goal. The scientific or objective cartographer attempts what Haraway (1988) has called 'the god-trick of seeing everything from nowhere' in that 'he' is disembodied from the landscape being represented. This masculinist GAZE is objectifiying and distancing. Maps do not represent indeterminacy: clear lines demarcate one thing from another, whether this be rock type or ethnicity. Furthermore, maps do not simply represent a place. Once accepted and circulated – once they are recognised as legitimate knowledge – maps have influence over how the SPACE is perceived and what action takes place within it.

Brian Harley (1992) has argued that cartography is a particular form of knowledge, and that mapping represents an exercise in POWER. Mapping has a history linked intimately with both mercantile CAPITALISM and COLONIAL-ISM. Places were mapped and named in colonial exercises, hence awarding the mapper (or his state) ownership and knowledge. Some critical theorists have criticised GIS for the same reasons, adding that their intense interest in questions of error and order leads them to objectify people, to see them existing in 'Cartesian space and technical, chronological TIME, rather than lived space, or PLACE, and human or narrative time' (Curry, 1995: 79). It should be noted, however, that some computer cartography is now experimenting with fuzzy logic and indeterminacy in attempts to map uncertainty.   JS

See also SPATIAL SCIENCE; SCIENCE/THE SCIENCE QUESTION.

Castration anxiety/castration complex The castration complex is a key element within FREUDIAN THEORY. It is intertwined with the OEDIPAL complex, through which sexual difference, gender inequality and HETERO-SEXUALITY are inscribed. Prior to the onset of the oedipal phase of development, Freud views the little girl as indistinguishable from the little boy. But, within this account, recognition of sexual difference, symbolised by the presence or absence of the PENIS, has very different implications for the sexes. The little boy sees that he has a penis; his anxieties relate to the fear that he might lose it, specifically that his father might castrate him because of his desire for his mother. But the little girl sees that she has no penis and assumes that she has already been castrated (also see PENIS ENVY). Her castration complex entails coming to terms with her loss.

Within LACANIAN THEORY, the castration complex is reformulated and understood in relation to language rather than BIOLOGY. Within this frame, both the little boy and the little girl are castrated in the sense that they are subjected to rather than in possession of language, which is represented by the PHALLUS. However, while both sexes experience a fundamental LACK signified by castration, this operates asymmetrically. Through the conflation of penis and phallus (biology and CULTURE) the MASCULINE position remains one of superiority. In both the Freudian and the Lacanian versions differences between men and women are understood in primarily VISUAL terms.   LB

Casualisation Casualisation refers to particular forms of employment practice. Strictly speaking it implies that workers are employed casually, without contracts or regular conditions of employment. Historically, women have been viewed as a reserve army of LABOUR (after Marx) that can be used as casual labour and brought into the labour market at times of great need. However, in recent times, casualisation has also been used to denote a range of labour contracts that are associated with the deregulation of the labour market. As Allen and Henry (1997: 180) suggest, 'Conventional forms of employment have given way to a proliferation of contractual forms involving a wider variety of working times, benefits and entitlements that are more reminiscent of the way things were before the post-war economic era.' In this context, casualisation refers to the way in which full-time, permanent (often unionised) jobs have been replaced by new forms of employment contract. The greater use of part-time, temporary and zero-hour contracts ensures greater flexibility for employers but usually incurs greater risks and insecurity for employees. Moreover, many of these new jobs are going to women, the majority of whom work part-time (Beechey and Perkins, 1987; McDowell, 1991b). Employment protection (against unfair dismissal) and employment benefits (such as pensions, maternity agreements and sick pay) are often unavailable to women who work part-time hours or on temporary contracts.

Will Hutton (1996) has suggested that these changes in the labour market are reshaping British society, widening the gap between rich and poor, secure and insecure. In *The State We're In*, Hutton presents Britain as a 30:30:40 society; the first 30% are the disadvantaged, without permanent work, engaging in occasional part-time, temporary and casual work; the second 30% are the marginalised and the insecure, including part-time workers, many self-employed people, temporary staff and the working poor (those who earn less than 50% of average earnings) – and a huge number of women workers are in this category; and the final 40% are described as the privileged, those in full-time employment, in more secure work with better conditions. Hutton argues that the 'new' labour market, characterised by part-time and temporary contracts, poor conditions and insecure WORK is encroaching on the privileged 40%. In this sense then, casualisation is a more general, societal

phenomena, because changes in the labour market have an impact upon wider society.   JW
See also FEMINISATION OF THE LABOUR FORCE.

Categories/categorical The construction and application of categories has traditionally been a central part of how geographers have sought to form theories, communicate ideas and define the self and others. In this sense, they are integral to the process by which we seek to reduce the seemingly limitless complexity of the world 'out there' to a manageable complexity of words, ideas and theories.

   Some feminists regard categories by their very nature as finite and hence antithetical to the diversity that feminism seeks to encourage (Penrose, 1992). The debate within feminist geography, however, has centred around which (rather than whether) categories will provide the basis for a progressive, radical politics and how categories integral to geography as an art and science should be recategorised (Pile, 1994; Massey, 1994a). Drawing on certain strands of POSTMODERNISM, a number of academics have addressed the specific difficulties associated with rigid, oversimplistic, DUALISTIC categories in exploring the relationship between sex and power in the construction of knowledge (McDowell, 1990; Penrose, 1992; Pile, 1994). An enthusiasm for thinking about DISCURSIVE practices, partly derived from French POST-STRUCTURALISM, has fuelled a more general search for spatial categories which move beyond dominant, dualistic constructions (Bhabha, 1990; Pile, 1994; Soja, 1996). There has been a parallel effort to reinvent categories conventionally regarded as geographical. Doreen Massey, for example, argues for 'PLACE', as a category, to be reimagined as an 'articulated moment in networks of social relations and understandings' – a relational space which is partly determined by GLOBAL structures and processes (Massey, 1994b: 154). The construction, DECONSTRUCTION and reconstruction of categories remains central, then, to geographical and feminist enterprise.   CJ

Cathexis See GENDER

Chauvinism Defined by the *Oxford English Dictionary* as male displays of excessive loyalty to other men and prejudice against women, chauvinism was named after the Frenchman M. Chauvin. While less commonly used today, the feminist movement during the 1970s adapted the term to 'male chauvinist pig' (MCP).   SAR
See also SEXISM.

Chicano/chicana Chicanos are people of Mexican descent living in the USA. Chicana is the feminine form of the word, used specifically for women, while chicano refers both to men and to the group. Unlike the more mainstream terms 'Mexican-American' and 'Hispanic,' chicano is a political and ideological term that was chosen self-consciously to connote a shared experience of internal colonisation and cultural MARGINALISATION. Generally, the term chicana has feminist connotations, often describing women who struggle to overcome the dual barriers of RACISM and SEXISM with little recognition or POWER.   DM

Child abuse See ABUSE; VIOLENCE

Childcare Responsibility for the day-to-day and longer-term needs of children, or minors, whose status is defined by statutes determining, among other things, the age of criminal responsibility, the age of legal sexual activity and the school-leaving age. Childcare is primarily a responsibility of individual women, undertaken in the HOME either by a female relative, usually the child's mother, for 'love' not money, or by a paid childcare worker in her own home or the child's home – the former worker is usually termed a child-minder in Britain, the latter a nanny – or in a specialist childcare facility, such as a day nursery, crèche or kindergarten. Both state and private collective provision for children under five years old in Britain is poor, although in 1998 the Labour Government announced a scheme for financial support for lower-paid families to purchase forms of regulated childcare provision, which may expand the places available. The geography of childcare provision has been investigated by feminist geographers (England, 1997; Gregson and Lowe, 1995; Pratt, 1997; Tivers, 1984).   LM
See also CARE; ETHICS; LABOUR: DOMESTIC, SWEATED, WAGED, GREEN, SURPLUS.

Child labour The employment of children, under the minimum statutory school-leaving age (wherever sanctioned), in gainful occupations. The term implies an element of economic compulsion and EXPLOITATION, involving a time and energy obligation denying full participation in leisure and educa-tion, which can damage children's health and development (Fyfe, 1989; Lavalette, 1994). Child labour is increasingly sought by capitalists in their search for a cheaper, more FLEXIBLE and subordinate work force (UNICEF, 1997).   RK
See also CAPITAL/CAPITALISM; WORK/WORK FORCE/EMPLOYMENT.

Chora A term derived from philosophy which has been used by feminist theorists in a number of different ways. Usually ascribed to Plato, the term refers to an 'inscriptional space' – a hypothetical moment in the production of a representation, whereby 'Forms' are translated into particular known objects through the medium of a 'receptacle', the *chora*. The *chora* is imagined as always receiving, accepting, welcoming, but never contributing anything to the nature of the object or concept in question. Plato spoke of this receptacle as FEMININE – like a mother, or nurse. However, it is in the nature of the *chora* that it is unable to be represented by anything particu-lar: hence Plato's attempt to read it as feminine can be seen as an attempt to authorise a particular reading of it (Butler, 1993: 44). In her critique of masculinist philosophy Irigaray (1985b) makes use of this association of the *chora* with the maternal and the exclusion of the *chora* from the MASCULINE process of representation in philosophical traditions – despite its constitutive role in that process – as well as its role as 'mimic': 'Properly speaking, one can't say that she mimics anything for that would suppose a certain intention, a project, a minimum of consciousness. She [is] pure mimicry' (1985: 307). She adopts this (feminine) position, as MIMIC – playing the same back to the philosopher, in order to displace and challenge his authority and in order to represent the unknowability of the feminine (for an example of this tactic in geography, see Rose, 1996a).

The *chora* has also figured in the writings of Julia Kristeva (1984), who associates this place where knowledge and naming are not yet established with the semiotic, a pre-oedipal phase in the development of subjects in which the SYMBOLIC ORDER and LANGUAGE are not yet established. Heterogeneous, in the sense that the semiotic (*chora*) is already shaped by the symbolic order (law), this phase forms the basis for language, symbolic ordering and geometry – but is governed by a different ordering, in which the mother's body and her mediation of sociality is central. This phase is often associated with the emergence of processes of abjection. Kristeva's willingness to use the *chora* in this ontological way has been strongly criticised as accepting the association between femininity, maternity and the outside of reason, in contrast to Irigaray. However, Kristeva's emphasis on the heterogeneity of the semiotic and the *chora*, and on the mother as the introducer of the symbolic into the semiotic, suggests a more active reading of both the *chora* and the feminine, and quite a different feminist politics of knowledge from that implied by Irigaray's mimicry. JR

Citizenship The term 'citizenship' plays a central role in political life. Generally, citizenship refers to the contractual relationship between an individual and a territorially based state-like body which defines a person's eligibility to certain RIGHTS that are enforceable through collective institutional arrangements. Thus, the notion of citizenship is also associated with calls for DEMOCRACY, EQUALITY and social JUSTICE. Yet, rather than being carved in stone, eligibility and the nature of these entitlements varies over TIME and SPACE. For example, Marshall (1963) has traced the historical extension of citizenship rights from the sphere of CIVIL SOCIETY, through to political and then economic rights. There is also a very large literature examining the relationship between a citizen's rights and duties or responsibilities towards the wider collective (Miller, 1995).Within geography, there is an exciting charge around the notion of citizenship. The fact that citizenship is rooted in some kind of place-based membership has given rise to numerous fields of enquiry, such as NATION-STATE building and the repercussions of NATIONALISM and the fragmentation of nation states. More recently, the political and geographical dimensions of citizenship have been explored at different SCALES, from the notion of global citizenry presented within various environmental debates to the question of citizenship and local governance. This has yielded the idea of multiple citizenship. For example, a woman from Catalan might consider herself Catalonian, Spanish, a member of the European Union and a global citizen. A common theme within these debates has been to interrogate the 'tangled material and immaterial spaces of citizenship' and the ways in which a sensitivity to space is central to understanding the *de jure* concept of citizenship as well as people's *de facto* ability to exercise their rights as citizens (Painter and Philo, 1995; 108). This need to discern between the normative promise of citizenship and the specific, geographical experiences of different people has been emphasised by feminists (Smith, 1995). Their aim has been to develop citizenship as a critique which reveals people's multiple experiences of citizenship and the processes which marginalise and exclude different individuals and social groups (MacKian, 1995).

A crucial argument within these debates is the need to extend the concept of citizenship into areas of social and cultural life which have traditionally been seen as irrelevant, private or incidental, such as issues of GENDER, RACE and ETHNICITY, SEXUALITY, AGE and a person's differential physical and mental abilities. The fusion between citizenship as critique and an appreciation of the ways in which space and place shape people's experiences promises a powerful tool for analysis of the ways, for example, women experience discrimination in terms of waged and unwaged work which then constrains their ability to participate in political life. The importance of this kind of critical analysis is heightened by the fact that political parties employ ideological concepts such as the 'active citizen', 'citizen's charters' or 'citizen empowerment' whilst simultaneously introducing changes which impact negatively on vulnerable groups, limiting their chances of self-determination as citizens.   FG

See also POLITICS; STATE; DEMOCRACY; PERSONAL IS POLITICAL; WELFARE STATE; SOCIAL MOVEMENTS; ENLIGHTENMENT/ENLIGHTENMENT THEORY.

City Refers to an order of urban settlement, usually defined according to population size or administrative status. Rather than treating it as a static category, however, most geographers and urban theorists conceptualise the 'city' as a dynamic realm in relation to its social context and the processes through which its spaces are constituted. They seek to understand cities as part of the uneven social relations that structure societies, and have highlighted issues of POWER and politics in analyses of issues ranging from the production of material and symbolic LANDSCAPES, contestations over urban identities and meanings, to the formation of the city as an 'imagined environment' where attention centres on 'the DISCOURSES, symbols, metaphors and fantasies through which we ascribe meaning to the modern experience of urban living' (Donald, 1992: 427). Early feminist studies highlighted inequalities of access to resources in cities and the spatial constraints faced by women in daily activities. They focused on gender divisions and urban form and on patriarchal assumptions in PLANNING and design that produce a 'man-made environment' (Matrix, 1984; Roberts, 1991). Alternative feminist visions and projects for house and city design were explored by Hayden, among others, who addressed 'What would a non-sexist city be like?' (Hayden, 1980, 1981, 1984; see also Greed, 1994; Sandercock and Forsyth, 1992).

Many feminist geographers have considered gender inequalities in the city through empirical studies of the gendered use of urban space, often working with, but also criticising the simplistic use of, the idea of a 'divided city' based on the historic separation of home life from waged employment. More recently this has led to extensive research on the links between labour in the HOME and waged WORK (for reviews, see McDowell, 1983, 1993a and b; MacKenzie, 1989). The geographies of women's FEAR in cities have also been studied (Valentine, 1989), and historical research on the gendered spaces of cities has been accompanied by comparative analysis involving different societies (Spain, 1992). In a wide ranging account, Elizabeth Wilson has addressed the 'difficulties' surrounding women's presence in cities generally, focusing on how dominant urban discourses have constructed women as 'an irruption in the city, a symptom of disorder, and a problem' (1991: 9). But she also presents the city in a relatively EMANCIPATORY light, as a space of freedom and TRANSGRESSION for women. Although her arguments have been criticised by some for romanticising urban life, they connect to increasing interest in urban geographical imaginations and the ways in which women 'write' the city (Heron, 1993). Wilson's approach also resonates with expanding feminist and POST-STRUCTURALIST-informed research on the relationships between urban space, gender and SEXUALITY, including studies of GENTRIFICATION and the social construction of femininities (Bondi, 1991) and of the BODY and the city, where the latter is construed as 'the condition and milieu in which CORPOREALITY is socially, sexually, and discursively produced' (Grosz, 1995: 90). Recent work has been developing concerns with REPRESENTATION, power and spatial politics in the city (Deutsche, 1997; Meskimmon, 1997), and revisiting earlier inquiries into urban design and architecture in the wake of contemporary critiques of dualistic EPISTEMOLOGIES (Boys, 1998). In this sense there is no doubt that feminist geographical studies have been a vital means by which 'the city as an object of analysis has been unbound' (Jacobs, 1993: 827).   DP

See also BUILT ENVIRONMENT; FLÂNEUR/FLÂNEUSE; MODERN/MODERNISM/ MODERNITY/HIGH MODERNITY.

Civil society The term was first introduced into modern social science via the analyses of Hegel and Marx. Hegel referred to civil society as an intermediate institution between the FAMILY and the POLITICAL relations of the STATE, that would ultimately be transcended, whereas Marx considered it as the ensemble of socio-economic relations and forces of PRODUCTION. Gramsci (1971) served to popularise the term, referring to it as that area of social life which appears as the realm of the private citizen and individual consent, but which is really a bastion of class HEGEMONY.

Civil society is now commonly understood as occupying the indeterminate and never fixed spaces between the state and the family, that is, between the PUBLIC (as defined in terms of political and criminal rights) and the PRIVATE realm. The development of new social movements, the changing practices of MOTHERHOOD, the privatisation of state functions and the creation of new regulatory bodies all contribute to the shifting and the permeability of the boundaries between the civil society, the state and the family. It is the sphere in which manifestations of CITIZENSHIP, in terms of the rights and obligations of citizens, are practised and transformed through participation in voluntary associations, TRADE UNIONS, professional associations, mass media, SOCIAL MOVEMENTS and so on.

Feminists are divided on the extent to which civil society is primarily a female sphere. While some argue civil society provides the space for feminist action, even positing the notion of WOMEN'S MOVEMENTS as producing a feminist counter-public sphere, and that women are the primary intermediaries between the state and civil society, others claim the means of entry into civil society (i.e. citizenship), has served women unfairly (Pateman, 1988a) and that women's movements do not operate only within its boundaries.   LP

Class/class relations Class refers to systems of social stratification based upon material relationships to PROPERTY and employment. At its simplest, Marx understood capitalist society to be divided between the ruling and the working classes; those who own the means of production employing those who do not. In this sense, class is RELATIONAL in that each class depends on the other, and the relations between the two classes are key axes of POWER relations in capitalist society. The relationship between class structure and class consciousness and AGENCY is the most controversial aspect of class analysis. While Marx distinguished between class structure (class in itself) and class agency (class for itself), the links between class position and consciousness are individually, historically and spatially variable (see Thompson, 1980).

Feminists have taken up the concept of class to ensure that women are included in social analysis. Rather than treating married women as appendages of their husbands, being assigned to the same social class as their partners, sociologists have argued that women should be surveyed in their own right (Crompton, 1993; Crompton and Mann, 1986). Moreover, feminists have extended the notion of class to include the mode of reproduction, as well as production, arguing that the unpaid DOMESTIC LABOUR performed by women in the HOME is another facet of class relations (see Davidoff, 1995).   JW
See also FALSE CONSCIOUNESS; IDEOLOGY.

Clitordectomy The mutilation or removal of parts of the clitoris and/or labia, usually carried out on teenage girls in some African and Middle East societies, affecting sexual pleasure and often making intercourse, and especially child-birth, painful or dangerous; now outlawed in many states.   LM

Closet A silenced, oppressed or marginalised perspective, viewpoint or experience, usually socially constructed as 'private'; typically involves socially reviled sentiments or, particularly, sexual DESIRE; argued by literary critic Eve Sedgwick (1990) to be central to twentieth-century Western thought; can be seen as both a space of OPPRESSION and RESISTANCE.   LK

Closure See NARRATIVE; NARRATIVE CLOSURE

Collective consumption Refers to goods and services provided and managed collectively such as public TRANSPORT, education, health and, in some formulations, housing. The concept was influential in reconceptualisations of the urban question during the 1970s by the French sociologists Manuel Castells and Jean Lojkine (see Pinch, 1985; Saunders, 1986). It has been criticised for gender-blindness in failing to consider services provided by DOMESTIC LABOUR and the role of dominant GENDER relations (Walby, 1997).   DP
See also CONSUMPTION; LABOUR: DOMESTIC, SWEATED, WAGED, GREEN, SURPLUS.

Colonial discourse Colonial discourse refers to the apparatus of POWER that legitimates colonial conquest over people usually constructed as racially inferior. The critical study of colonial discourse reveals the connections between colonial power and the production of knowledge about colonised people and places as 'OTHER' to a powerful, usually Western, 'SELF'. In this way, colonial DISCOURSE analysis examines the colonial politics of REPRESENTATION. Edward Said's work on 'ORIENTALISM' represents a landmark in this field (Said, 1978). According to Said, British and French Orientalist discourses in the eighteenth and nineteenth centuries were closely tied to IMAGINATIVE GEOGRAPHIES of the Middle East. These discourses served to produce a distant, marginal 'other', which in turn helped to produce and to reinforce a centred, powerful 'self'. Following Said, other writers have been keen to destabilise metanarratives of 'otherness', as shown by Lisa Lowe's work on the gendered production of colonial discourses and the ways in which exoticisation is often closely tied to erotic representations of difference (Lowe, 1992). Other important work in this area includes Homi Bhabha's account of colonial stereotypes as complex and ambivalent (Bhabha, 1994) and Gayatri Chakravorty Spivak's work on the objectification of SUBALTERN subjects in colonial discourse (Spivak, 1988).   AMB
See also WEST, THE/WESTERNISATION.

Colonialism A term closely related to, but different from, IMPERIALISM. Colonialism refers to the government of a place or 'colony' that is at a distance from the ruling, colonial power and to the associated movement and settlement of

people from the colonial power to the colony through a process of colonisation. In the modern period the first stage of colonial expansion dates from the sixteenth century, when Spanish and Portuguese trade led to settlement in South America. From the eighteenth century onwards, many other European countries emerged as colonial powers. By the nineteenth century, the British Empire included settler colonies such as Canada, Australia, New Zealand, Kenya and South Africa, and many emigration societies were established to encourage settlement by British men and women overseas. The twentieth century, particularly post-1945, has been a period of decolonisation as many settler colonies have gained independence. In recent years, the study of colonialism has developed from a largely economic and political focus to one that also addresses cultural concerns (Dirks, 1992; Thomas, 1994). In line with this, an increasing amount of attention has been paid to colonial gender relations. While some accounts describe the colonies as a space of MASCULINE adventure far away from the feminised domesticity of life at HOME (Dawson, 1994) other accounts have focused on white women who travelled to and lived in the colonies, setting up homes and helping to domesticate little-known LANDSCAPES and places (Chaudhuri and Strobel, 1992; Schaffer, 1990). Other work has focused on the impact of colonialism on indigenous peoples, revealing the exoticisation of people and places as 'OTHER' to a WHITE, Western sense of self and the importance of rewriting histories that focus on SUBALTERN experiences (Guha and Spivak, 1988). AMB
See also TRAVEL/TRAVELLERS/TRAVEL WRITING.

Colour, women of A broad category that encompasses women who are defined by themselves and others as being racially distinct from white people of European descent. Ideally, it groups together women of different classes and geographies, including minority women in predominantly white societies and THIRD WORLD women living in their own countries. In practice, however, most women who identify themselves as women of colour in their writing have lived as a minority in the USA or Canada.

'Women of colour' is an inherently political term; it is most often used by non-white women when speaking about the interconnected processes of RACISM and SEXISM. The term politicises identity by assuming that hierarchies of GENDER and RACE are such significant markers of IDENTITY and POWER that there is a shared experience among Latinas, Africanas, and Asian women. By demarking a category composed of women who share only the experience of being non-white in a white-dominated society, the term itself asserts that there is some common subject position that results from being in a subordinate position in terms of gender and race. It is also a political term in that it imagines the potential for COMMUNITY and SISTERHOOD even where none yet exists.

The term is often found in the writing of feminists, at times to point out the racism and exclusion of white feminisms, and at times to facilitate conversations and connections among non-white feminists. The research, writing and experiences of women of colour have challenged feminist scholarship to rethink the relationship between race and gender and to address seriously the barriers to achieving sisterhood.   DM

See also BLACK; SUBALTERN; ALTERITY; POST-COLONIALITY/POST-
COLONIALISM; MULTICULTURAL/MULTICULTURALISM.

Communitarianism A form of political thinking which has as its focus the idea of COMMUNITY. A more precise definition is hard to find, for several reasons: first, many of the ideas coalescing around 'communitarianism' are not new; second, the term community is in itself notoriously difficult to interpret, entangled as it is with both sociological and geographical concepts; and, third, the views expressed by communitarians are seen by many as vague, complex and contradictory. Despite – or, as many people would argue, in a large part because of – its elusiveness, communitarianism has in recent years gathered widespread political support. For example, the work of Amatai Etzioni has influenced political thinking in both the United Kingdom and the USA. A central theme argued by such proponents is the need to counter dominant neo-liberal individualism by appealing to concepts such as 'virtue', 'responsibility' and 'reciprocity' (Etzioni, 1993). Undergirding these general themes is the assertion that society is (or should be) composed of social relations between people. As a result, certain institutions such as the FAMILY, the voluntary sector and neighbourhood organisations are strongly promoted. For feminists, the core concern is that many of these proposals are gender-blind and lack an adequate theorisation of POWER. Indeed, many of the institutions that communitarians seek to restore (such as 'respectable family life') have historically marginalised and oppressed women (Fraser and Lacey, 1993). But, rather than fall back into dualistic choices between neo-liberalism or communitarianism, feminists have sought to garner ONTOLOGICAL insights (such as critique of the logic of identity in favour of notions of relational identity) which can shed light on the possibilities of emancipatory political change.   FG

Community Raymond Williams famously described community as a 'warmly persuasive' word (Williams, 1976). Many other commentators describe its plastic nature and the ways it seems to mean all things to all people. Fraser and Lacey pin down two main ways in which community is interpreted: first, as an entity and, second, as an ideal (Fraser and Lacey 1993). In terms of an entity, they identify four possible meanings. First, community can refer to people who dwell within a geographical area. Second, community can mean a social group held together by ties of KINSHIP, where its members are usually assumed to live and work within the same locality. Third, it can describe collective entities bound together not by place but by common values or shared histories. One example would be the Jewish DIASPORA who are members of a Jewish community regardless of where they live. Fourth, community can refer to people who are engaged in specific shared practices, such as, for example, a linguistic community or the 'business community'. As mentioned above, community operates not only as a descriptive term but also as an ideal whose imaginative force can be politically mobilised at different spatial SCALES. For example, advocates of communitarianism focus on the 'LOCAL', while Anderson traces this mobilising force weaving through movements for national identity (Anderson, B., 1991).

Geographers have addressed this semantic tangle, focusing on the role of SPACE and PLACE in determining different people's perceptions and experiences of 'community'. The implications of the fact that 'community' always implies some kind of exclusive membership has been critically examined, as has the 'slippage' identified above between its descriptive and prescriptive powers. For feminists, these questions are central, as are numerous others, including: the assumed causal relationship between a person's subjective identity and wider communities; the persistent failure to theorise gender and power relations within what are often presented as 'homogeneous' communities; the political, economic, social and cultural connections between 'communities' operating in different places, at different scales and extending across space; and, ways in which notions of community function ideologically, reproducing socio-spatial processes which continue to marginalise and oppress women. Indeed, some feminists have argued that the emphasis on exclusive membership renders the idea of 'community' inherently problematic and that it should therefore be rejected. For example, Iris Young favours notions of RELATIONAL identity and the idea of the social group (Young, 1990c). It seems, however, difficult to reject a term that continues to exercise great emotive and political force across the political spectrum and at different spatial scales. The challenge, therefore, is to acknowledge this force whilst remaining both aware of its complex theoretical and epistemological foundations as well as sceptical of different loads of ideological assumptions the term 'community' can carry. This requires a sensitivity to the complex ways in which people reinvent their identities by drawing on different interpretations of 'community' and how this process of negotiation influences their ability to exercise self-determination.   FG

Community care In a Western context, this refers to the policy and practice of deinstitutionalising people identified as having physical, mental or behavioural differences (including some drug offenders, criminals, etc.). It is a term most popularly associated with the deinstitutionalisation of people with mental health problems, although it can be applied to the transfers of physically impaired and elderly people from institutional and clinical settings to smaller, less medicalised places within everyday neighbourhoods and streets. In Britain, community care refers to a complex set of legal and practical arrangements, introduced in the late 1980s, to be implemented and monitored by health and social welfare agencies, with the intention of facilitating the relative independence of an individual within COMMUNITY settings (Depart-

ment of Health, 1990). Although the possibility of community care for certain institutionalised people was debated in the nineteenth century (Scull, 1977) it only began to materialise as a result of certain practical and therapeutic steps (which were initially locally based: see Parr and Philo, 1996) in the 1940s and 1950s in Britain. Other versions of community care can be found in other European and North American societies, with a notable example being the long-term history of community care at Gheel in Belgium. In contemporary Britain, national programmes of community care have only been legislated for recently, with the 1990 NHS and Community Care Act formalising the need for complex sets of arrangements between local care agencies. In critical human geography notions of community and CARE can be contested, and questions have been asked about who really carries the burden of a seemingly more tolerant society. It has been suggested that women bear the unequal burden of caring as they are traditionally associated with such roles (nurturing) and spaces (the HOME). Community care, then, potentially reinscribes a gendered 'care role' into domestic spaces (WGSG, 1984).   HP
See also WELFARE STATE.

Comparable worth A form of job evaluation which seeks to achieve PAY EQUITY through the construction of '"neutral" measures of skill which make possible the direct comparison of men's and women's jobs' (Pringle, 1988: 22). The term is most commonly used in the American and Australian contexts: in Canada and the United Kingdom women's organisations and TRADE UNIONS have sought measures which would establish 'equal pay', 'equal value' or 'pay equity' (Cuneo, 1990). However, some feminists have questioned the extent to which skill can ever be *objectively* measured: in the case of secretarial work, for example, 'it remains difficult to separate out skills from more stereotypical expectations around personal service' (Pringle, 1988: 22).   SR
See also RIGHTS/EQUAL RIGHTS/EQUAL OPPORTUNITIES.

Compulsory heterosexuality In 1980 Adrienne Rich, poet and writer, and long-time co-editor of the lesbian/feminist journal *Sinister Wisdom*, elaborated and theorised the phrase 'compulsory heterosexuality', drawing upon a 1976 decision by the Brussels Tribunal on Crimes against Women; the Tribunal named 'compulsory heterosexuality' as one of the 'crimes against women' (Rich, 1980: 653). Rich's elaboration was published as an article in the feminist journal *Signs* and was entitled, 'Compulsory heterosexuality and lesbian existence'. The phrase is not dissimilar to that of Gayle Rubin in her classic article, 'The traffic in women: notes on the "political economy" of sex' (1975) which speaks of 'obligatory heterosexuality'.

In the *Signs* article, Rich asks a complex of questions: Why is heterosexuality not seen as a choice but only as biologistic fact (633)? Why is lesbian existence frenetically erased from psychoanalytic theory and popular registers of social life, including historical archives? Is HETEROSEXUALITY, after all, a choice or is it a social and political imposition (648)? How might we see heterosexuality, like MOTHERHOOD, as an overdetermined 'political institution' (637)? And why is it that heterosexual romance is represented as 'the great female adventure, duty, fulfillment' (654)?

Rich begins to address these questions theoretically with a rebuttal of PSYCHOANALYSIS generally: If the primary identificatory bond between mother and daughter is same-SEX, she says, it seems strange that psychoanalysts do not centre their analyses on the processes through which the bond is ruptured and realigned. Psychoanalysis problematises lesbian existence when it is, in fact, many women's heterosexuality that is psychically odd:

> If women are the earliest sources of emotional caring and physical nurture for both female and male children, it would seem logical, from a feminist perspective at least, to pose the following questions: whether the search for love and tenderness in both sexes does not originally lead toward women; *why in fact women would ever redirect that search;* why species-survival, the means of impregnation, and emotional/erotic relationships should ever have become so rigidly identified with each other; and why such violent strictures should be found necessary to enforce women's total EMOTION, erotic loyalty and subservience to men (Rich, 1980: 638, original emphasis).

Rich elaborates a wide range of cross-culturally variable means for rupturing the mother–daughter bond and compelling women to assume a heterosexual IDENTITY, if only for the sake of appearance. These range from the physical control of women's bodies to male control of state institutions and hegemonic culture.

Rich also defines two additional terms – lesbian existence and LESBIAN CONTINUUM. By the latter term, Rich means that same-sex mother–daughter bonds are continually reproduced and/or displaced on to other same-sex relationships and friendships throughout women's lives. Necessarily, then, *all* women lie somewhere along a lesbian continuum. What this means in the context of compulsory heterosexuality is that women lead a 'double life,' a concept perhaps drawn from DuBois's theory that racism produces a 'double consciousness' in those racially oppressed – they need to know the oppressor's culture to survive, a culture from which they are simultaneously excluded. Thus, all women love other women in various ways, but this love must be systematically suppressed and repressed in heterosexist societies. Lesbians might therefore marry men to avoid social ostracism; obversely, heterosexual women may use women to sustain them in key emotional and sensual ways.

Rich argues forcefully for scholars to acknowledge the unnaturalness of heterosexuality and to document politically, culturally, economically and socially how and why it is made NORMAL and/or enforced. Nevertheless, the argument teeters on biologism: 'I perceive the lesbian experience as being, like motherhood, a profoundly female experience' (650).

Although possibilities for analysing how compulsory heterosexuality registers spatially are immense, relatively few geographers use Rich's work. Most geographers examining the compulsoriness of heterosexuality do so through

focusing on how other-than-heterosexuals are oppressed; heterosexuality itself often remains unexamined (but see Nast, 1998b). In contrast, social theorists outside geography have reworked the term in creative ways. Judith Butler (1990: 151–6), for example, uses Monique Wittig's formulation of the 'heterosexual contract' and Rich's 'compulsory heterosexuality' to theorise what she calls the 'heterosexual matrix': 'that grid of cultural intelligibility through which bodies, genders, and DESIRES are naturalized'. Connell (1995) also draws upon Rich's work, but to show how men are also forcefully impelled into heterosexuality, extending Rich's analysis.   HN
See also GAY; DYKE.

Consumption First used by geographers to refer to the purchase of goods and services and mainly used in the context of retailing. More recently it has been applied to the use of goods in a much wider sense. The study of consumption is a relatively recent development within geography; it was not until the 1980s that geographers began to recognise fully the economic, social, political and cultural importance of the problematic of consumption. Early geographical work tended to conceptualise consumption as a BINARY opposite to production and, in the tradition of classical economics, simply important as an end purpose of production. More recently, however, the complexity of the relationship between production and consumption has been acknowledged and studies have started to recognise the significance of consumption in its own right.

For feminist geographers a recognition of the importance of consumption more broadly within the discipline has helped to legitimise research on aspects of consumption, for example, SHOPPING and retail spaces, previously marginalised as being more relevant to the daily lives and experiences of women. Moreover, a concern with both the patterns and the social relations of consumption has encouraged a focus on GENDER. Work has sought to demonstrate, for example, how patterns of consumption vary between women and men, how male and female identities are constituted and reconstituted in relation to consumption practices and how power over consumption is gendered.

The 'NEW' CULTURAL GEOGRAPHY of the 1990s has seen a huge expansion of interest in the 'culture of consumption'. With this interest has come a focus on advertising and the media and on the power of REPRESENTATION in 'selling' goods and services (Jackson, 1993; Miller, 1987; Philo and Kearns, 1993; Wrigley and Lowe, 1996). Attention has also been paid to the ways in which different groups (e.g. young people and women) have been targeted and manipulated in relation to the consumption of particular items or images (in for example music, fashion or food). Geographers have been particularly interested in the PLACES and SPACES of consumption. From a concentration on the physical design and layout of shops and shopping malls, studies have broadened to include an examination of, for example, the atmosphere and themes of clubs and restaurants. Included here has been a consideration of the domestic spaces in which consumption takes place (e.g. work on GENTRI-FICATION and the ways in which consumption practices and preferences have transformed residential spaces). Emphasis has also been placed, in such work, on the PERFORMANCE of consumption and on the importance of consumers as both audiences and performers.

Another important theme in geographical work has been the politics of consumption. Studies at both the global and the local level have looked at the control of commodity chains, the regulation of consumption practices and the shaping of consumer choice. They have also considered the role of ethical consumerist movements and the moral economies surrounding consumption. Such work has focused, in part, on the gendered nature and implications of ethical forms of consumerism.   JL

See also COLLECTIVE CONSUMPTION.

Contraception See BIRTH CONTROL

Cooperatives/cooperative housekeeping Cooperative organisational forms (alternatively called 'collectives') can be seen as strategies to pool resources and create alternative POLITICS and spaces of everyday living, whether in housing, production, consumption or campaigning. In feminist terms, they constitute a rejection of PATRIARCHAL, hierarchical organisation and stress more equal participation. However, cooperative forms are in no way restricted to feminist philosophies, as Soviet collective farms or Israeli kibbutzim illustrate.

'Collective' strategies in HOUSEHOLD organisation, such as multigenerational householding, or coordination among neighbourhood or KINSHIP groups, which involve productive and reproductive labour are often articulated in, and (re)constituted through, spatial divisions in the 'HOME' space, such as family group yards or collective living spaces, but these in turn can function to reflect and sustain patriarchal social relations through a strict gendering of spaces and their functions (Katz, C., 1993; Pulsipher, 1993; Spain, 1992). For many women, adopting collective arrangements for combining productive and reproductive tasks can be a deliberate 'coping strategy' to deal with the problems of being a female-headed household in a patriarchal society (Chant, 1997), or of family POVERTY in times of economic crisis (Oberhauser, 1995). Furthermore collective strategies around domestic tasks such as cooking or food growing form the bases for many 'popular feminisms'. Though none of these *necessarily* disrupt the dominant gendering of household tasks, gendered identities may be refigured (Wilson, 1993).

Cooperative householding as a deliberate challenge to the privatised and atomised bounds of the nuclear FAMILY includes both feminist and non-feminist variants. Women's houses, all-women housing collectives and/or communes remain a significant element in feminist praxis (Wekerle, 1988), and close ties and tensions exist between the ideals of COMMUNITARIANISM and feminist critiques (Rigby, 1974). In recognition of the ways built form, and in particular the 'single family house' (Franck, 1995), expresses the ideological dominance of, and constrains alternative practices to, the patriarchal, HETEROSEXUAL nuclear family (Hayden, 1984; Valentine, 1993b), architectural and design projects from both feminist and other 'alternative' positions (ecological, community formation, anticapitalist) seek to balance the desire for family privacy with the provision of collective spaces for such activities as eating, leisure, community creation, CHILDCARE and 'productive' activities (Franck, 1995; Ley, 1993; Wekerle, 1988). Feminist-inspired approaches in particular seek to overcome gendered divisions of domestic tasks (Spain, 1992) and to disrupt strict divisions of PRIVATE and PUBLIC, productive and reproductive activities. It is, however, worth noting that for many women it is the *lack* of privacy and seclusion which is problematic – in shared hostel accommodation when homeless, or in the openness to public SURVEILLANCE gazes mentioned by LESBIAN women (Smailes, 1995).

Whether design-based strategies (which range from changing interior house layouts to incorporating shared collective spaces and integrating services such as childcare or even productive activities; Franck, 1995) in themselves reinforce or disrupt gender identities depends on specific contexts. Experiments with collective kitchens, laundries and childcare in the USSR, for example, 'liberated' women from household tasks, allowed women greater fulfilment in their new role as workers and challenged bourgeois notions of the household to create a socialist collective (Andrusz, 1984), but such experiments were nevertheless premised on highly gendered assumptions about the divisions of household tasks. Likewise, while design-based strategies may ease the problem of combining a range of roles for some men and women, these benefits often intersect with CLASS and RACE inequalities. Forms of ownership and the everyday politics of the organisation of and participation in collective tasks and collective spaces all influence the meaning and effects of such practices for those involved in them (Franck, 1995; Ley, 1993). FS
See also DIVISION OF LABOUR: DOMESTIC, INTERNATIONAL, GENDER, SPATIAL.

Corporate culture The conventional view of organisations is one of institutions that are dominated by bureaucratic rational assumptions in which decisions are made on the basis of profit-maximisation criteria, either, in the case of the private sector, or, in the case of the public sector, to achieve stated ends and to serve the 'public good'. However, a great deal of feminist research on organisations has documented the ways in which the culture of organisations both reflects and influences societal views about GENDER relations; thus in a patriarchal society is not surprising that organisational cultures are also masculinist in nature, constructing women as an inferior 'OTHER' in the workplace in opposition to a disembodied idealised worker who is in fact, usually, a bourgeois male. One of the earliest and most influential pieces of work on organisational culture is that by Rosabeth Kanter (1977), *Men and Women of the Organisation,* which inspired a number of more recent analyses of corporate cultures in which greater attention is paid

to the ways in which ideas about DESIRE and SEXUALITY also structure organisational practices as well as more material bases of discrimination in hiring and promotion practices, for example, and in the assumptions that lie behind job appraisal and evaluation schemes (e.g. see Halford *et al.*, 1997; McDowell, 1997b; Pringle, 1989). While some theorists argue that to be successful in male-dominated GENDER CULTURES, women have to act as surrogate men (Acker, 1990), it seems that in the shift towards economies dominated by service sector firms and organisations, a new gender culture may be emerging in which 'feminised' values of, for example, non-hierarchical decision-making and less aggressive management styles are more highly valued.   LM
See also BUREAUCRACY.

Corporeality Elizabeth Grosz (1987) has described corporeal feminists as those who treat the female BODY as crucial to understanding women's psychical and social existence. They are concerned not with the biological body but the lived body. For them the body is interwoven with and constitutive of systems of meaning, signification and representation. This group distinguishes itself from ESSENTIALISTS and SOCIAL CONSTRUCTIONISTS by its rejection of the mind–body DUALISM. The body cannot be understood as a neutral screen, a biological *tabula rasa* on to which MASCULINE or FEMININE can be indifferently projected. As sexually specific, the body codes the meanings projected on to it in sexually determinate ways.

A key influence on this group has been the work of Luce Irigaray, who problematised representational systems marking women's bodies as lacking, dependent and orientated towards the PHALLUS. Women have been commodities, signs of exchange between men. The goods have to go to the market and start speaking for themselves. For Irigaray there needs to be a positive re-evaluation of femaleness, which starts from taking women as their own norm. She argues that this requires a return to the pre-oedipal. Women lost the capacity for an auto-eroticism, for a positive identification with their own sex. The MOTHER was our first love-object, but women were forced into accepting her and their own castration. Women now have to lay claim to a body and a SEXUALITY which lacks nothing, which has its own morphology and anatomy and which is not the complement or 'other' to male anatomy. Irigaray (1977) rejects the idea that feminine pleasure is singular, unified or subordinated to a single organ. It is neither vaginal, nor clitoral, but both and more. Her account stresses the multiplicity, ambiguity, fluidity and excessiveness of female sexuality.

This group has often been accused of essentialism but it would be more accurate to see it as working to DECONSTRUCT prevailing notions of the body and of mind–body dualism. Women's bodies have been reduced to biology, which in turn presumes them to be passive. Against this the group argues that BIOLOGY is only one form of knowledge about the body and that biology is also a cultural construction; that the animation and interiority of the body are important; and that the body must be understood not simply as an external object but in terms of how it is lived and experienced by the subject.

Grosz has reminded feminists that, in the 'system of corporeal production', men and women are equally subject to the moulding of gender difference. In a related discussion, Judith Butler (1990) suggests that corporeal style comprises the ways in which the (gendered, sexed) body is made from repetitive stylised PERFORMING acts.

From these theorists, geographers have taken the idea that gender *and* sex are historically and geographically variable categories in which social relations regulate and channel gender attributes. Differentiated corporeality further differentiates experiences of place, flows, and 'spaces of opportunity' (Laws, 1997: 49). Geographical analysis of workplace cultures, urban geographies, sexuality, MIGRATION and mobility have all been influenced by notions of corporeality. Glenda Laws's discussion (1997) of female mobility relates the spatial and social patterning of women's movement between US cities to differential corporeal production by the STATE through its benefit system.   RP, SAR

Cottage industry Industrial activity performed in domestic premises, although organised centrally. In Western economies, cottage industry is usually referred to as an historical phenomenon associated with protoindustrialisation, the period prior to industrialisation (Berg *et al.*, 1983; Houston and Snell, 1984). In contemporary less economically developed regions and nations, such as India, cottage industry is currently evident, although gradually declining as these areas modernise (see Baishya, 1989).

Since production takes place in the HOME, cottage industry has parallels with HOMEWORK. A number of features, however, render these two activities distinct. For example, both historical and contemporary accounts of cottage industry recognise that most household members are involved (Baishya, 1989; Tilly and Scott, 1987). Within the household there is a strict DIVISION OF LABOUR. For instance, in cotton manufacturing, the women and children spin, while the men weave. PATRIARCHAL control characterises production: the intensity and organisation of work is specified by the 'master of the family', normally the eldest male (Mathias, 1983). In contrast to homework, cottage industry tends to emerge mainly in rural economies, especially in pastoral or upland areas, where agrarian conditions are incapable of providing full employment (Thirsk, 1961). Those involved in cottage industry normally also participate in farming activities in order to satisfy simple subsistence needs (Vanneste, 1997).   RK

Crime Offences punishable by law, though official definitions of which behaviours are criminal often reflect interests of powerful groups. Crime statistics are notoriously misleading and support the perception that crime is largely committed by men against men. This is challenged by feminist research on women's and men's experiences. While men are more at risk in PUBLIC spaces, violence against women is largely hidden in the HOME (Painter, 1992).

However, public discourses and fear of crime are highly spatialised, with criminals generally perceived as 'other' people from other places. Such stereotypes are often racialised (Smith, S.J., 1986). Crime and risk are also structured by SEXUALITY, age and ability. Women and men generally commit different types of crime (Walklate, 1995). The female offender has been constructed based upon ESSENTIALIST notions of sexual difference, as women 'fallen' from norms of gender and sexuality, or, more recently, as victims of POVERTY. Either way, women are theorised against a norm of MASCULINITY, only recently explored (Newburn and Stanko, 1994). Criminality is one way of 'doing gender' in particular spatial and cultural contexts: to Campbell (1993), crime amongst young men is contiguous with older forms of masculinity, in expressing difference and maintaining spatial control over street and home. As FEMININITIES transform and diverge, violence and drugs-related offences committed by women appear to be rising. Feminist theory has significant implications for mainstream work on geographies of crime, which has tended to be gender-blind, POSITIVISTIC and to present a specific, spatialised account of offence patterning (Pain, 1997a, b and c).   RHP

See also ABUSE; FEAR OF CRIME; RAPE; STRANGER DANGER; VIOLENCE.

Critical theory Originally formulated by the early Frankfurt School of Social Research, and further developed by Jürgen Habermas, critical theory referred to the systematic study of the relationships between human agency and social STRUCTURES in capitalist societies. The various models developed to explain these relationships were brought into active dialogue with geography in the 1970s by Derek Gregory (1978). The term is still used in this more formal way by feminist political theorists such as Nancy Fraser (1989) and Seyla Benhabib (1986). However, critical theory has also come to function today more as a broad catch-all category for the diverse theoretical arguments emanating from feminist, Marxist, anti-racist, POST-COLONIAL and QUEER THEORY. What unites these heterogeneous theoretical productions is precisely their *critical* stance *vis-à-vis* contemporary cultural, economic and political relations, and their resulting commitment to changing such relations for the better. In this way it can be argued that today's diverse forms of critical theory trace their anti-*status-quo* political engagement back past the Frankfurt School to Marx's own famous call for praxis, in his *Eleventh Thesis on Feuerbach*: 'The philosophers have only interpreted the world in various ways; the point, however, is to change it' (1976: 620).

It is well worth remembering this call for praxis insofar as it reminds contemporary critical theorists of their radical responsibility to the future. This ethical responsibility is another way of describing the *anticipatory* moment of critique which Benhabib (1986) argues was absolutely vital to the notion of critical theory that emerged from the work of the Frankfurt School scholars themselves. It is also an ethical responsibility that has been reasserted by recent attempts to articulate a new, more inclusive and more collaborative critical theory in and of geography (IICCG, 1997). Some theorists of POSTMODERNISM and IDENTITY POLITICS nevertheless depart from critical theory on this point insofar as they tend to view all attempts to anticipate better futures as complicit with various heroic forms of historicist, foundationalist and Western-centric notions of progress, and, as such, doomed to failure. However, this kind of argument misses the rather different POST-STRUCTURALIST point made by feminist theorists such as Gayatri Chakravorty Spivak (1990a) and Judith Butler (1993) that such forms of EPISTEMIC VIOLENCE are open to performative displacement through renegotiation in new more open-ended critical contexts. In addition to missing this possibility of renegotiation, Wendy Brown (1995) argues that certain theorists of identity politics fall prey to what she calls, following Nietzsche, a 'slave morality'. She argues that this is a morality in which the project of critique and social change is replaced by a rhetoric of RESISTANCE as *ressentiment*. Against such *ressentiment* that fetishises individuated identities and personal injury as the twin bases for a moralistic politics, Brown recommends that critical theory in all its contemporary diversity needs to engage once again in actually critically thinking about better futures, doing so as a collaborative and collective project rather than as a slavishly individualistic project of moral self-perfection.   MS

See also ETHICS; MORALS/MORAL REASONING.

Cross dressing The donning of attire, accessories or other accoutrements which disrupt gender NORMS or the sex-gender system; most commonly, the wearing by individuals constructed by dominant Western cultures as either 'men' or 'women' of clothing deemed by the same dominant cultures to be appropriate only for the 'opposite sex'.   LK

Cultural capital Developed by French sociologist Pierre Bourdieu, the concept of cultural capital is used to describe knowledge, taste, manners and patterns of CONSUMPTION as a source of social distinction (Bourdieu, 1984). The concept is developed by analogy with economic capital and plays an important role in the theorisation of POWER and social REPRODUCTION. Bourdieu's theory of domination through culture (symbolic violence) is similar in many respects to Gramsci's concept of HEGEMONY wherein the dominant culture is admired and seen as legitimate by the dominated classes to the extent that the members of the ruling elite need not resort to coercion for the maintenance of a system within which they are privileged. Thus culture can be seen to obscure and add its legitimacy to power relations. Bourdieu argues that ruling-class cultures are arbitrary but inculcated in subtle ways such that cultural capital and competence in making AESTHETIC judgements are only partially achieved through the education system. Cultural capital thus remains unevenly distributed among classes. Families play an important role in socialising children into the more subtle aspects of aesthetic dispositions. Bourdieu argues that one of the important preconditions for the aesthetic attitude and the accumulation of cultural capital is freedom from economic necessity such that practicality need not be the overriding necessity. This, Bourdieu (1984: 55) argues, allows one to concentrate on lifestyle, home decoration, tourism, exercise for the sake of exercise and the reduction of the world to the status of a designed LANDSCAPE. Bourdieu writes of bourgeois women who being 'partially excluded from economic activity, find fulfilment in stage-managing the decor of bourgeois existence when they are not seeking refuge or revenge in aesthetics'. Because taste is seen as natural, it is an important aspect of the HABITUS. Bourdieu states, 'Aesthetic intolerance can be terribly violent. Aversion to different life-styles is perhaps one of the strongest barriers between the classes: class endogamy is evidence of this' (1984: 56). A FEMINIST perspective on cultural capital might aim to question the types of knowledge and cultural competence most highly valued, to denaturalise taste and to focus critical attention on the aestheticisation of power relations, especially gender relations.   ND

Cultural geography See NEW CULTURAL GEOGRAPHY; CULTURAL STUDIES; CULTURE

Cultural imperialism Social and cultural relations, including GENDER relations, are transformed through the experiences of IMPERIALISM and neo-imperialism. Imperialism and culture were and are linked together in many different ways. The imperial project depended upon cultural practices to a large extent, creating the idea in members of imperial nations that empire was feasible and good, establishing many of the practical forms of imperial POWER, and popularising empire (Said, 1993). Empire also stimulated the transformation of CULTURES, both in the metropole and in the colonised places. Understanding of these transformations in colonial contexts has varied. On the one hand they have been seen as the eradication of indigenous culture in the face of the advance of the imperialist intentions of the dominant nation's culture, contributing to the impoverishment of countries by eroding local knowledge and weakening survival strategies. Such erosions could also be understood as necessary to the advancement of poorer nations, as in modernisation strategies of development where the eradication of traditional culture was seen as a prerequisite for economic growth. On the other hand, and in a postcolonial vein, the cultural aspects of imperialism can be understood as involving the SUBALTERN in renegotiating the dominant culture, producing HYBRID forms of cultural practice and IDENTITY. In Homi Bhabha's (1994) formulation, the process of mimicry, whereby the colonised subject reproduced imperial cultures, or the process of repetition, whereby the colonial project was enacted, were never perfectly achieved. The gap, or slippage, involved can be seen as a site of subversion and RESISTANCE. Cultural resources – even those of the coloniser – can therefore potentially be mobilised in a variety of subtle and ingenious ways as a basis for opposition. Insofar as cultural imperialism has substantial impacts upon GENDER relations and particular forms of

FEMININITY/MASCULINITY, these judgements pose significant questions for feminist practices. In some colonial contexts certain practices were outlawed by the colonisers in the interests of humanitarianism – such as *sati*, or self-immolation, by widows in India – and in others existing gendered power relations were reinforced through colonial administrative practices. In a contemporary world, the transnationalisation of culture means that these questions concerning the meaning of transformations in indigenous cultures have immediate import. Most commentators seem to be enthusiastic for the emergence of hybrid and creative new cultural forms as a result of the encounter between dominant world cultures and more local, indigenous cultural practices (see Appadurai, 1996), although some see even that theoretical response as a symptom of global capitalism's cultural logic (Dirlik, 1994). JR

Cultural studies Pioneers of cultural studies in Britain, such as Richard Hoggart and Raymond Williams, set out to apply the methods of literary analysis to more popular cultural forms such as magazines and soap operas, television and advertising. The political edge to their work was further developed by members of the Centre for Contemporary Cultural Studies at Birmingham who made critical use of a range of theoretical ideas from MARXISM and PSYCHOANALYSIS to SEMIOTICS and DECONSTRUCTION. Their study of youth subcultures via Gramscian notions of RESISTANCE (Hall and Henderson, 1977) was particularly influential, though feminist critics charged that there was an undue emphasis on the spectacular, sometimes violent, subcultures of young working-class men. Studies of women's magazines and the readers of romantic fiction partly redressed the balance (McRobbie, 1991; Radway, 1985) while recent work in 'audience studies' (e.g. Morley, 1992) is characteristic of a turn to more grounded empirical studies.

The institutionalisation of cultural studies as an academic discipline provided an intellectual 'home' for many of those whose work had previously been marginalised, including feminists and those associated with GAY and LESBIAN studies. As a result, cultural studies has attracted criticism from those in more established disciplines who question its radical commitments, its often high level of theoretical sophistication and apparent fascination with abstruse terminology. Particularly in North America, cultural studies has been demonised by right-wing critics, who have associated it with the alleged excesses of 'POLITICAL CORRECTNESS'. Despite these attempts at vilification, the subject has brought together some of the most innovative work from across the humanities and social sciences (Grossberg *et al.*, 1992).

Geographers were attracted to cultural studies through their interest in the media and popular culture, leading to new work on the politics of REPRESENTATION and to a general broadening of the agenda of cultural geography. The 'CULTURAL TURN' in human geography has begun to be reciprocated within cultural studies which now recognises questions of SPACE and PLACE as one of its central problematics (Carter *et al.*, 1993) Despite its growing

institutionalisation, cultural studies has retained its critical edge through continued engagements with FEMINISM and POST-COLONIAL studies.   PJ
See also NEW CULTURAL GEOGRAPHY; TEXT/TEXTUAL/TEXTUALITY/INTERTEXTUALITY.

Cultural turn See CULTURE; NEW CULTURAL GEOGRAPHY

Culture Notoriously hard to define; nineteenth-century anthropologists such as Edward Tylor identified 'culture' as 'that most complex whole which includes knowledge, belief, art, morals, law, custom, and any other capabilities acquired by man [*sic*] as a member of society' (1871: 1). Despite the archaic language, Tylor's definition usefully draws attention to the complexity of culture, to its holistic qualities and to the fact that it is a social accomplishment. Etymologically, 'culture' has a number of associations which Raymond Williams (1976) used to highlight the way the word's meaning has shifted with changes in society. Through its relation to 'cultivation', for example, come the associations with plant and animal husbandry (as in 'monoculture' and 'agriculture') as well as the class-based antagonisms of elite and popular culture (as in being a highly cultivated or cultured person). These associations persist in geographical applications of the culture concept, whether in relation to the classic works of cultural ecology associated with Carl Sauer and the 'Berkeley School' on the material transformations of nature and the development of the 'cultural landscape' (Leighly, 1967) or in more recent work on geography, the media and popular culture (Gold and Burgess, 1985).

Culture can also be defined as a signifying system, from which derives a long tradition of approaching LANDSCAPE as a 'way of seeing', interpreting its symbolism and ICONOGRAPHY (Cosgrove and Daniels, 1988). Others have defined culture more broadly as encompassing all of the ways in which we make sense of the world, attaching value and meaning to the material environment. Building on the work of Stuart Hall, 'culture' can be likened to the *maps of meaning* through which the world is made intelligible, the codes through which meaning is constructed, conveyed and understood (Jackson, 1989). This approach requires a pluralisation of 'culture' to include all of the cultures and subcultures that vie for POWER in terms of material resources but also in terms of the politics of REPRESENTATION. From this perspective, culture can be understood as involving the construction and negotiation of different systems of meaning.

Like other social sciences, geography has experienced a 'CULTURAL TURN' in recent years, leading to a greater reflexivity regarding the interpretation of other cultures and drawing particular attention to the significance of different modes of representation. This has led geographers to question their adherence to older 'superorganic' theories of culture (Duncan, 1980) and to broaden the horizons of cultural geography, drawing on recent developments in social and cultural theory. The 'new directions' being taken by cultural geography during the 1980s were summarised as being:

contemporary as well as historical (but always contextual and theoretically informed); social as well as spatial (but not confined exclusively to narrowly-defined landscape issues); urban as well as rural; and interested in the contingent nature of culture, in dominant ideologies and in forms of resistance to them (Cosgrove and Jackson, 1987: 95).

Since then, cultural geography has experienced something of a backlash, accused of distorting its intellectual history and 'reinventing' cultural geography (Price and Lewis, 1993). Others have claimed that 'there's no such thing as culture' (Mitchell, 1995), arguing that the term has no analytical value and that a critical approach would involve an interrogation of those who make ideological claims to 'culture' (as in the formation of 'cultural industry quarters' and other culture-based approaches to economic development, for example).

These debates highlight the extent to which 'culture' is a highly contested term, giving rise to complex issues of 'cultural politics', whereby even the most rarefied AESTHETIC judgments are freighted with political and economic significance. In France, for example, Pierre Bourdieu has highlighted the extent to which cultural markers of taste serve to highlight social distinctions along class and gender lines, as economic capital is exchanged for various kinds of CULTURAL CAPITAL (Bourdieu, 1984). In Britain, a critical approach to culture was developed by members of the Centre for Contemporary Cultural Studies at Birmingham whose work included studies of working-class and youth subcultures (Hall and Henderson, 1977) as well as innovative studies of the cultural politics of 'RACE' and RACISM.

Feminists have criticised this work for its concentration on the spectacular subcultural styles of young working-class men (McRobbie, 1981). As well as highlighting the neglect of GENDER as a central feature in much of this work, feminists have also emphasised issues of embodiment and CORPOREALITY (Longhurst, 1995). They have highlighted the extent to which all forms of cultural interpretation should be reflexively located in terms of a politics of position. Feminist cultural theorists such as Donna Haraway (1992) have also offered an exciting alternative to the sterile opposition between NATURE and culture.   PJ

See also CYBORG; NEW CULTURAL GEOGRAPHY.

Cyberspace A generic concept associated with developments in information and communication technologies that may be existent, emergent or fictional, referring to a computer-mediated space within which humans can interact. The term gained increasing currency in the 1980s after being coined by science fiction writer William Gibson (1984: 51), who envisioned it as a 'consensual hallucination experienced daily by billions of legitimate operators' and a 'graphic representation of data abstracted from the bank of every computer in the human system'. It was, he said later, a cut-up word or 'Neologic spasm' that preceded a concrete referent (1991: 28), but his notion of an imaginal PUBLIC sphere crystallised for some a new sense of COMMUNITY (Stone, 1991), and subsequent use has been fuelled by developments in virtual reality systems as well as considerable hype. Writings on the subject frequently blur categories of fiction, PERFORMANCE and social theory, leading to interest in the recursive relationships between these fields (Burrows, 1997). Research on existing spheres such as the Internet has raised questions about the DOMINATION of network culture by men and masculinist ideologies, and inequalities of access socially and geographically. But, while wary of dominant mythologies of cyberspace, some feminists are considering the potential EMPOWERMENT for women offered by new communication technologies (Light, 1995) and, drawing on POST-STRUCTURALIST theory and cyberpunk, are developing critical discourses exploring new possibilities. Arguing that 'there is more to cyberspace than meets the male GAZE', Sadie Plant (1996: 170, 183), for example, writes of a 'cyberfeminist virus' that is disrupting the symbolic from within, through the proliferation of self-organising systems that escape PATRIARCHAL control, and through the activities of a digital underground that is 'perverting the codes, corrupting the transmissions, multiplying zeros, and teasing open new holes in the world' (see also Haraway, 1991; Lykke and Braidotti, 1996; Marsden, 1996; Plant, 1997).   DP

Cyborg A term used by feminist theorist Donna Haraway (1991) in her critique of the naturalisation of categories, especially SEX, RACE and CLASS and her assessment of the impact of technological change on GENDER relations, particularly the changing links between women and NATURE; originally a composite term used in the defence industry and derived from the fusion of cybernetic organisations: HYBRID systems which embrace organic and technological components.   LM

D

Death of the author See AUTHOR, DEATH OF/AUTHORSHIP

Decentre/decentredness Represents an attempt to destabilise a position and/or a SUBJECT that might be holistic, taken for granted and AUTHORITATIVE. Destabilising a centred subject or position can reveal fractures in IDENTITY and can promote a recognition of partiality and DIFFERENCE that undermines unquestioned claims to AUTHORITY. These terms can be traced to early modernist attempts to destabilise and fracture a holistic, universal SELF, as shown by surrealist art and literature (Lunn, 1985). More recently, notions of decentredness have been important in Derrida's work on DECONSTRUCTION and have influenced feminist attempts to destabilise a unified, ESSENTIALIST subject. Such attempts have been associated with the production of partial and SITUATED KNOWLEDGES and the recognition of difference among as well as between women and men. Other work has shown the importance of reclaiming voices and experiences from places constructed as MARGINAL to a Western, centred perspective. Such work has shown the important links between identities, power relations, and their material and metaphorical SPATIALITY. For example, Gloria Anzaldúa writes of the BORDERLANDS between Mexico and the USA (Anzaldúa, 1987), and bell hooks recovers the HOME as a decentred place of RESISTANCE (hooks, 1991a).

AMB

Decentred subject See SUBJECT/SUBJECTIVITY

Deconstruction This methodological stance seeks to challenge the naïve assumption that a TEXT has 'a' meaning which can be uncovered through careful reading. The term is most closely associated with philosopher Jacques Derrida, especially his 1967 work *Of Grammatology*. Gayatri C. Spivak, the translator of this text, insists that deconstruction is not another term for demystification of IDEOLOGY, as the word is commonly used, but is rather a method to destabilise the TRUTH claims of a text by identifying its unarticulated presuppositions and demonstrating their failure to provide a secure foundation for the statements being made: other meanings can be taken away from the text. It is the use of BINARIES in Western metaphysics that has been the most significant focus of deconstruction: HIERARCHICAL binary oppositions run throughout Western thought and KNOWLEDGE, but by finding traces of one element within the other, deconstruction can refute an opposition between the terms and thus challenge the foundation of the text.

Deconstruction has been used by POST-COLONIAL theorists to undermine the binary of self–other in Eurocentric imagined geographies, and also by feminist theorists to denaturalise the use of the binary male–female in Western thought. Rather than simply reversing the valorisation of the terms (as a POLITICS of opposition would), deconstruction involves a reversal, but then a subversion of the notion of the opposition itself (for an example from geography, see Rose, 1993a).

The RELATIVISTIC nature of deconstruction (the dissolving of concepts of meaning and truth into a 'play of signifiers') has worried some feminists. 'Deconstruction raises questions not only about the social construction of sexual identity and gender roles but challenges the fixity of the very notion of "WOMAN"' (McDowell, 1991a: 124). This can problematise not only PATRIARCHAL constructions of GENDER, but also exposes the use of gender identity by feminists as a politically arbitrary construct. Linda Alcoff (1988: 420) asks, 'How can we demand legal ABORTIONS, adequate CHILDCARE, or WAGES based on COMPARABLE WORTH without invoking a concept of "woman"?'   JS
See also POSTMODERNISM; POST-STRUCTURALISM; RELATIVISM.

De-industrialisation Refers to the closure of industrial plants and the resulting decline of dependent communities. Although most manifest at times of economic crisis, de-industrialisation is an ongoing process whereby capitalists seek the greatest returns on investment, relocating capital from one place to another. Much scholarship has highlighted the impact of such disinvestment on local communities and, in recent years, women have often played a political role in the fight to keep capital rooted in place (as illustrated by the Women Against Pit Closures and the Women of the Waterfront organisations in the United Kingdom (Lavalette and Kennedy, 1996)). Moreover, de-industrialisation has been central to changing GENDER relations as men have found it more difficult to find WORK and the women have often taken over the mantle of breadwinning (see Campbell, 1993; Harris, 1987; Morris, L., 1995).   JW

See also LOCAL/LOCALITY.

Democracy Traditional theories of democracy (*kratos* meaning rule, of *demos*, the citizen body) originated in classical Greek city-states. Nominally, all CITIZENS have the right to decide matters of general concern (although in the Greek system, women and slaves were not included in this universal SUFFRAGE). The modern NATION-STATE has become too large for direct democracy, so indirect, or representative, democracy is the replacement. Representative politicians decide for their constituents. A society is considered democratic if there are free elections, with sufficient choice to bodies with power to enact policy change.

The apparently straightforward UNIVERSALISM of democracy is, however, misleading. Feminists have asked who 'the people' are in democratic formations (Marsden, 1990). It was only in the twentieth century that women won the right to vote. Today, in most democratic states, women hold only a fraction of the positions of POWER than do men, suggesting a POLITICAL culture which is still based around the gendered notion of PUBLIC and PRIVATE spheres, where it is men who dominate the public space of POLITICS.

From the work of Ernesto Laclau and Chantal Mouffe (1985), the concept of radical democracy challenges conventional models of society as a totality, to facilitate the creation of a space for genuinely pluralistic politics through the promotion of radical SOCIAL MOVEMENTS. Radical democracy is primarily a post-MARXIST conceptualisation of society in that it recognises a complex mixture of social positions, groups and struggles and understands politics to work through temporary strategic alliances rather than static political positions. It challenges the limited liberal definition of the political to recognise a much wider variety of issues and actions as occupying the political realm.   JS

Demographic policy Government policy aimed at explicitly controlling and restricting the level of population growth and its composition. Note that many other policies implicitly impact on population growth but cannot be described as 'demographic policy'. The three factors that this policy has to address are FERTILITY, mortality and MIGRATION. Mortality is generally targeted through health-care programmes, and migration through employment and settlement policy, hence this definition will concentrate on controlling fertility. Depending on how the problem is defined, different governments or agencies will adopt different solutions for the perceived problem of overpopulation. Todaro (1997) outlines six different ways in which governments can control population growth, namely persuasion through media and education, through the provision of family planning and health-care, through population redistribution, through coercion using penalties and legislation, through the manipulation of economic incentives and disincentives and, finally, through the raising of the social and economic status of women.

International agencies and donors have historically supported programmes that aim to reduce fertility directly, and the results of these programmes have been mixed. The coercive and manipulative techniques have proved to be highly disadvantageous for women in terms of human RIGHTS, self-determination and personal safety. Men too have been negatively affected. These techniques have largely failed as solutions. Examples of these include the reduction of maternity benefits, the loss of tax exemptions and financial incentives, the subsidisation of smaller families, forced sterilisation and fines. Demographic policy has had severe implications for female children. For example in China where the 'one child per family' policy was adopted in 1982 female infanticide has increased alarmingly as a result of the preference for male offspring.

There is much evidence that the sixth method (described above), that of raising women's social and economic status, is a far more effective way in which to manage population growth. If overpopulation is understood from a feminist perspective it is clear that high fertility rates are related to the economic (and non-economic, such as cultural) value of children in poor countries – that is, it makes sense to have larger families. More specifically, the socio-economic and cultural status of women has a great impact on fertility so that education, employment and independence are crucial in reducing birth rates. It is important, however, to acknowledge the importance of power relations between men and women which often undermine women's ability to make choices about families and employment. The cultural and social importance of children for women must also be recognised. The EMPOWERMENT of women is therefore crucial so that they are able to individually control their reproduction rather than having policies imposed on them that control population.   PM

See also BIRTH CONTROL.

Dependency ratio The ratio of children (0–15 years) and elderly people (over 65 years) to the number of adults in a given population. This indicator is useful in identifying the numbers of people that the adult population, assumed to be economically active, has to support, but is problematic because it assumes full and normal employment, overlooks women's unpaid work, child and elderly labour, and is spatially insensitive.   PM

Deprivation/multiple deprivation At its simplest, the absence or lesser degree of one or more material or social states relative to the general population. The term is problematic and has been widely contested. One of the main difficulties is that it has several near-synonyms, such as poverty, need and want (Bradford *et al.*, 1995; Townsend, 1987). This has led academics to replace the term with more fashionable concepts, such as underclass, disadvantage and social exclusion (Katz, M., 1993; Murray, 1994). The term is still used by governments (e.g. in Britain), although many acknowledge its complex nature (see Coombes *et al.*, 1995).

Deprivation is polymorphous (Berthoud, 1976; Bradford *et al.*, 1995; Brown and Madge, 1982). Researchers have highlighted the importance of distinguishing between material and social deprivation. Individuals experience material deprivation if their diet, clothing or housing is below the customary minimum standard available in a society. Social deprivation is a less developed concept and is more difficult to measure (Bulmer, 1984). Individuals who suffer this type of deprivation may be excluded from the work force and/or educational, recreational and social activities. While some people experience a single deprivation, others face a series of deprivations, usually termed multiple deprivation.

The degree of deprivation and its impact on people varies spatially and temporally. The explanations geographers have given for its incidence include global economic restructuring and ineffective spatial policies (Pacione, 1995). Although it is individuals that experience deprivation, geographers have tended to focus on deprived areas because individual-level information is difficult to handle, data are more widely available for areas, policy responses that tackle deprivation tend to be targeted at areas and arguably the problems of deprivation are intensified when there is a concentration of deprived people in a particular space. Feminists have challenged notions of deprivation because they fail to theorise GENDER, distinguish between men's and women's different experiences of deprivation or provide any causal explanation of deprivation or connection between different forms of deprivation. RK

Desire For Freud desire relates to the experience of satisfaction, which gives rise to a MEMORY trace in the form of a mental image. A link is established, so that the next time a similar need arises it will give rise to a psychical impulse which will seek to recathect or to re-energise the image which evoked the feeling of satisfaction. Wishes or desires are thus linked to memory traces of previous satisfaction and are fulfilled through hallucinatory reproductions of the perceptions, which have become signs of the satisfaction. The search for the object of desire is not governed therefore by physiological need but by the relationship to signs or REPRESENTATIONS. The organisation of these representations constitutes FANTASY, the correlate of desire and a principle of its organisation. Desire cannot be a relationship to a real object but is a relationship to a fantasy. Thus PSYCHOANALYSIS relies on a notion of desire as lack, an absence that strives to be filled through the attainment of an impossible object.

Lacan modified Freud's account, arguing that there is no insistent sexual desire which pre-exists the entry into LANGUAGE and CULTURE. It is the symbolic law of the father that constitutes the lack from which desire springs. Desire can instead be seen as what produces, what connects. This too, it has been argued, relies on a notion of 'lack', albeit linguistically created rather than biologically based, which underlies desire. For Foucault (1980) it still has recourse to a representation of POWER that he calls 'juridico-discursive'. It leads either to a promise of 'liberation' if power is conceived as affecting desire only from the outside; or, if it is seen as constitutive of desire itself, to the belief that one can never be free. Others, like Coward (1984), have argued that the idea of the unconscious and subjectivity as produced in language simultaneously presupposes their pluralism, diversity, heterogeneity and contradictoriness. Her subtle evocation of 'female desire' pursues the 'lure of PLEASURE across a multitude of different cultural phenomena, from food to family snapshots, from royalty to nature programmes'. Here desire is both ingratiating and polyvocal, a potentiality for change, for breaking out of the cage of expectations, and for cooption, for sustaining things as they are.

Like Foucault, Deleuze and Guattari (1983) have stressed the productive nature of desire. For them, any acceptance of Oedipus implies an artificial restriction on a field, the unconscious, where everything is in fact infinitely open. Desire does not take for itself a particular object whose attainment it requires but devotes itself to its own proliferation or self-expansion. For them, desire is not a striving for the lost unity of the womb but the core of a reality which is a state of constant flux. There is in this flux no given self, only a cacophony of 'desiring machines'. Desire is about actualisation, and the body a discontinuous series of processes, organs, flows and energies. This approach has been of interest to FEMINISTS because it rejects the BINARY opposition between mind and BODY and the body is understood more in terms of what it can do than as a locus for consciousness.   RP

Deterritorialisation The term deterritorialisation seeks to capture the 'displacement of identities, persons and meanings that is endemic to the POSTMODERN world system' (Kaplan, 1987: 188). Within a POST-STRUCTURALIST tradition, the term has also been used to describe the radical distanciation between signifier and signified characteristic of language itself (Deleuze and Guattari, 1986).

Deleuze and Guattari first establish that the deterritorialisation characteristic of LANGUAGE unites us all. They go on to point out, however, that, within this field of power, we have very different privileges (Kaplan, 1987). These insights lead Caren Kaplan to espouse a search for humility and self-discovery in which the critique of one's 'home location' – an imaginative deterritorialisation – provides the starting point for critical reflection. In this context, Kaplan highlights the experiences of bell hooks, Michelle Cliff and

Minnie Bruce Pratt. The example of Pratt's widely cited autobiographical text (Pratt, M.B., 1984) is particularly instructive. Pratt describes the process by which she became conscious of her own complicity with racism, heterosexism and antisemitism. At the end of the book, Pratt is living as a solitary WHITE woman in a BLACK neighbourhood in Washington, DC. This location was chosen, in part, as a vehicle for continually destabilising her sense of identity (Pratt, G., 1992).

As Kaplan acknowledges, without fully working through the implications of the statement, there is no 'pure space' of deterritorialisation (Kaplan, 1987: 195). The connections to a space which the autobiographies of women such as M.B. Pratt assume, may in fact may be problematic. Gerry Pratt, for example, seems oblivious to her locatedness and her potential role as first-wave GENTRIFIER (Pratt, G., 1992: 243).   CJ
See also NOMAD/NOMADISM/NOMADIC SPACE.

Detraditionalisation A complex term used by analysts of social and cultural changes associated with late MODERNITY, which refers broadly to a social order in which tradition changes its status (Beck *et al.*, 1994). Detraditionalisation can be seen when tradition is separated from its guardians and its context, a process associated with the increasing complexity of flows and linkages at diverse spatial scales. Detraditionalisation arguably occurs in two directions, each with distinct implications for women and GENDER relations. First, a growing reflexivity about traditions and their meanings in a late modern world may liberate women from constraining relationships, practices and REPRESENTATIONS. Nuclear FAMILY forms, HETEROSEXUAL relationships as well as sex–gender distinctions are all 'traditions' that have been increasingly questioned within FEMINIST GEOGRAPHY, as women have written their own NARRATIVES and histories (e.g. Rose, 1993a).

Second, detraditionalisation may lead to a more rigid interpretation and policing of tradition, such as in FUNDAMENTALISM (Beck *et al.*, 1994). In fundamentalism, the increased contingency of beliefs and their geographical fluidity are firmly rejected in favour of adherence to an apparently threatened tradition. As female subjects become the embodiment of traditional values under fundamentalism, women experience heightened levels of control of their spatial mobility and SEXUALITY from PATRIARCHAL families and institutions, especially religious authorities and, in some cases, the state (Women Against Fundamentalism, 1992). In 1990s Afghanistan, the Taliban rulers exemplify a fundamentalist policy regarding women, as they have closed women's hospitals and schools and have severely restricted women's movement outside the patriarchal family home. Feminist critics of fundamentalism call for the development of social policy that addresses women's needs free from racist or culturalist assumptions about how they are expected to behave.   SAR

Development A highly contested term broadly referring to economic, social and cultural change, particularly among developing countries. Early conceptualisations of development within the modernisation paradigm in the 1950s and 1960s assumed that 'backward' countries in the 'South' would progress along an evolutionary path based on the experiences of the WEST. Despite criticisms from the Marxist Dependency school highlighting the exploitative links between the 'FIRST WORLD' (core) and the 'Third World' (satellite), definitions of development continue to be dominated by BINARY classifications of developed–underdeveloped which inherently assumed superiority–inferiority. These notions have been challenged from an 'anti-development' perspective (Escobar, 1995), and by feminists; the latter have argued that women were initially either ignored in the development process or seen as impediments to development (Parpart, 1993). While early measures of development concentrated on economic dimensions in terms of gross national product (GNP), neglecting women's unpaid LABOUR, more recent indices have included non-material factors such as educational attainment, life expectancy, as well as gender inequalities; these are reflected in the United Nations Development Programme's (UNDP's) Human Development Index (HDI), Gender-Disparity Index (GDI) and Gender Empowerment Measure (GEM) (UNDP, 1995).

Critiques of 'gender-blind' development discourses drew on Ester Boserup's (1970) seminal work arguing that development projects marginalised, rather than improved women's lives. Although criticised (Benería and Sen, 1982), Boserup's work formed the basis for the WOMEN IN DEVELOPMENT (WID) conceptual framework in the 1970s. Rooted in modernisation theory and influenced by Western liberal feminist thinking, WID approaches sought to integrate women into the development process, emphasising their roles as workers and producers. Although successful in increasing the profile of women in development circles, WID policies failed to challenge male-dominated social structures and reasons for gender inequalities, preferring legal and administrative changes. While there are various strands of WID identified by Moser (1989), all are characterised by non-confrontation, treating all women as an undifferentiated group. In arguing that women have always been part of the development process, the Women and Development (WAD) approach assumed that changing international structures would improve women's position. The failure of both WID and WAD to address women's reproductive roles and challenge the causes of gender inequalities led to the emergence of the GENDER AND DEVELOPMENT (GAD) approach in the mid-1980s. Influenced by SOCIALIST FEMINIST thinking, GAD approaches were more holistic, including productive and reproductive spheres, as well as incorporating men in their analyses.

While GAD frameworks claimed to include the voices of Third World feminists, later critiques argued that WID, WAD and GAD were all formulated from a Western feminist perspective. Arising from the call for EMPOWERMENT (Sen and Grown, 1987), postmodern theorising has criticised ETHNOCENTRIC biases of previous frameworks, reflected in the use of ESSENTIALIST and UNIVERSALIST categories such as the gender division of labour

(Mohanty *et al.*, 1991). As well as emphasising the importance of race and class, POSTMODERN views also highlight the place-specific context of gender relations (Momsen and Kinnaird, 1993). Although some consider the postmodern lens the most useful for analysing gender and development issues (Parpart, 1993), others view it as too removed from grass-roots FEMINISM to have any relevance for initiating social change among women and men in developing countries (Udayagiri, 1996). CM

Diaspora Meaning literally the scattering of a population, the word diaspora was originally associated with the dispersal of the Jews following the Roman conquest of Palestine and the destruction of Jerusalem in AD 70 (Keller, 1971). The term now has a much wider currency, applied to other population dispersals (both forced and unforced), for example the Black Atlantic diaspora that resulted from the slave trade (Gilroy, P., 1993). While distinct histories of diasporas can be described, diaspora can also be evoked as a theoretical concept to (re)frame questions of PLACE, CULTURE and IDENTITY. Diasporas involve transnational connections which cut across national boundaries, making linkages which transect the borders of national communities. These linkages intersect and unsettle fixed or 'rooted' conceptions of place, culture and identity. As Hall (1995: 206) argues: 'These are people who belong to more than one world, speak more than one language (literally and metaphorically), inhabit more than one identity, have more than one HOME.' Through these 'extroverted' or 'stretched out' (Massey, 1994b) geographies of flows and connections new kinds of 'open' cultures and cultural identities can be formed which emphasise 'routes' rather than 'roots'. One example of these are the different forms of music of the Black Atlantic diaspora – blues, reggae, jazz, soul, rap – which are produced through a fusion of different influences across different places (Gilroy, P., 1993).

Diaspora experiences are always gendered, and feminists have criticised the ways in which the celebration of theories of TRAVEL or displacement often reflect masculine experiences (Wolff, 1993). Clifford (1997) emphasises that diaspora involves histories of both displacement and dwelling, suggesting that women's experiences of diaspora may be ambivalent. While diaspora interactions may open up new spaces for women (through independent migration, for example), the insecurities of diaspora may involve a tightening of PATRIARCHAL control as women are required to embody the stabilities of the 'home' culture or TRADITION (Ali, 1992). CD

Dichotomy See BINARY/BINARY OPPOSITIONS; CULTURE; DEVELOPMENT; FEMININITY/FEMININE; GENDER; MASCULINITY/MASCULINITIES/MASCULINISM; NATURE

Différance A term that refers to the relational production of meaning through signification. Jacques Derrida has written about *différance* to reveal the divided, incomplete process of signification through which meanings (signifiers) are ascribed to an object of meaning (signified) (Derrida, 1974). In French, *différence* and *différance* are pronounced the same, resulting in an ambiguity between 'differing' and 'defering.' This ambiguity reflects the unstable nature of signification, whereby meanings are produced in terms of both difference and deferral. The process of signification thus does not imply transcendental meanings but rather reveals the complex instability of meanings. As Toril Moi writes, 'meaning is never truly present, but is only constructed through the potentially endless process of referring to other, absent signifiers' (Moi, 1985: 106) In feminist writings, these ideas have been helpful in destabilising BINARY oppositions such as MALE–FEMALE and MASCU-LINE–FEMININE and in critiquing the PHALLOCENTRIC production of meaning more broadly.   AMB

Difference Refers to the relational production of meaning in terms of non-identification. The notion of difference underpins feminist writings in two main ways (Barrett, 1989). First, feminist writers have stressed the differences *between* men and women. In LIBERAL FEMINIST terms, the differential access to education and paid employment by men and women have been central concerns and notions of gender EQUALITY have been paramount in attempt-ing to overcome such differences (Bock and James, 1992; Scott, 1988b). RADICAL FEMINISTS have written about male and female experiences and characteristics as fundamentally different from each other, as shown by some strands of ECOFEMINIST work and representations of MOTHERHOOD. Second, feminist writers have been increasingly keen to stress the differences *among* as well as between women and men. Both the first and SECOND WAVES OF FEMINISM have been criticised for their white, middle-class, HETEROSEXUAL focus (Collins, 1989; Probyn, 1993). For example, Hazel Carby has written about slavery in the USA and the ways in which black women were dehuman-ised in contrast to WHITE women. As she writes, white feminists in the nineteenth century in the USA were often *married* to white men, while black feminists were *owned* by white men (Carby, 1982). Work that has highlighted the differences among women and men has explored the different axes of POWER relations that shape identities along the lines of, for example, CLASS, race, age and SEXUALITY as well as gender. Often informed by POST-STRUCTURALIST, PSYCHOANALYTIC and/or POST-COLONIAL theories, such repre-sentations of difference complicate an ESSENTIALIST, unified SUBJECT. The spatial constitution of subjectivity has been seen as paramount, contextualis-ing difference in material ways. As Pratt and Hanson write, geography reveals 'the ways in which gendered, racialised, and classed identities are fluid and constituted in PLACE – and therefore in different ways in different places' (1994: 6). Some commentators are critical of a focus on 'difference', suggest-ing that it leads to fragmentation and can be politically disabling. In reply, other commentators argue that a gender politics based on an essentialist subject denies the diversity of human experiences and the power relations positioning subjects in multiple rather than singular ways.   AMB

Disability This is a contested term. Traditionally it has been understood as indicating the physical or mental difficulties faced by individuals who have particular 'conditions', 'illnesses', behavioural characteristics or bodily forms. There have been two main ways in which to conceptualise disability: through medical or social understandings. The former has been the predominant way in which 'disabled people' have been understood: here, individual bodies (primarily) have been identified as malfunctioning, and hence individuals suffer from a personal medical 'tragedy' which prevents them from being able to participate in normal social interactions and spaces. The alternative, social oppression, theories advocated by, for example, Oliver (1990) have emphasised societal attitudes to bodily impairment, arguing that it is not the BODY which is fundamentally disabled but rather practices directed at impaired bodies which are disabling. Critiques have been levelled at the latter definition of disability as overlooking the real pain, difficulty and huge challenges that people with impairments live and feel through their body, and some balance needs to exist between critiquing societal attitudes and recognising bodily difference. In recent times geographers have begun to document and theorise body (and mind) differences under the heading of disability, with emphasis on the ways in which society and its spaces may be seen as 'ableist'. Focusing on ableist spaces, on embodied EXPERIENCES and on processes which discriminate against and marginalise people who do not match up to (a mythical) perfect bodily or psychic form may be one way of escaping the constraints of the two 'models' of disability. Such critiques and politics of ablelism link with work by feminist geographers concerned with the idealised 'body spaces' of women, which can also be seen as 'disabling' (Bell and Valentine, 1997; Chouinard and Grant, 1995; McDowell, 1995).   HP

Discourse/discursive practices These terms refer to written and verbal communications but, informed by the work of Michel Foucault, have come to acquire a more distinct meaning in POST-STRUCTURALIST thinking, which refers to 'the ensemble of social practices through which the world is made meaningful and intelligible to oneself and to others.' (Gregory *et al.*, 1994: 136). Particular knowledges of the world are produced and reproduced through REPRESENTATIONS and practices that often come to be naturalised and taken for granted. Foucault has written about the discursive production of, for example, madness and SEXUALITY, the ways in which such discursive formations vary over TIME and SPACE and the ways in which certain discursive practices are institutionalised (Foucault, 1973, 1979). Discourses, and the material inscription of discursive practices, are inseparably bound up with POWER relations, which in post-structuralist terms are diffused throughout society. As Raman Selden writes, 'Discourse determines what it is possible to say, what are the criteria of "TRUTH", who is allowed to speak with authority and where such speech can be spoken' (Selden, 1989b: 76). In terms of GENDER relations, discourses of MASCULINITY and FEMININITY refer to sets of practices and ideas that shape appropriate behaviour, roles, appearance and aspirations for men and women. For example, femininity has been discursively produced in terms of domesticity, maternity, dexterity, nurturing and docility. Many feminists challenge the way in which such discourses and their material effects come to be naturalised, as shown by Cynthia Cockburn's work on the ways in which skills that are seen to be 'naturally' associated with women are devalued in the workplace (Cockburn, 1983). Denise Riley has written about the discursive constitution of the category 'women' over time and suggests that history itself represents a range of discontinuous and disconnected discourses (Riley, 1988). Geographers have increasingly emphasised the ways in which discourses vary over space as well as time and the ways in which spaces are themselves discursively produced. Geography itself is increasingly discussed in discursive as well as disciplinary terms, referring to the production of geographical knowledge beyond as well as within the academy. AMB

Discrimination The action of being treated unfairly, unjustly or unequally. Expressed through a variety of practices and manifested at various spatial scales, discrimination has traditionally been seen to derive from various structural forces – internal COLONIALISM, labour-market segmentation, RACISM, SEXISM, AGEISM, disableism and so on – and expressed through a variety of practices. Most feminist geographical analyses have focused on SEX-based and CLASS-based discrimination through analyses of women's waged WORK, although racial discrimination is now also receiving attention (see Peake, 1997). The term 'positive discrimination' is sometimes used to refer to policies to counter discrimination against women or minority groups. LP
See also DISABILITY; COMPARABLE WORTH; RIGHTS/EQUAL RIGHTS/EQUAL OPPORTUNITIES.

Displacement From PSYCHOANALYSIS. A refusal to accept items of behaviour (slips of the tongue, etc.) as chance occurrences, but instead to look for them as evidence of the unconscious state of the individual. JS
See also FREUDIAN THEORY; LACANIAN THEORY.

Division of labour: domestic, international, gender, spatial The division of labour refers to the arrangement of different groups of workers such that they perform different tasks, both within and outside the formal (waged) LABOUR MARKET. For Sayer and Walker (1992: 13), the division of labour 'is among the most basic social relations governing social life'. Their discussion is primarily concerned with the development of specialisations within waged employment. Additionally, however, feminist accounts have long highlighted the fact that women have performed (and continue to perform) the bulk of WORK within the HOUSEHOLD, which is described as the DOMESTIC DIVISION OF LABOUR. Recent investigations have come to recognise that particular domestic tasks are themselves highly gendered (see Gregson and Lowe, 1994).

61

Gender divisions of labour emerge across TIME and through SPACE as particular forms of waged labour become defined (and redefined) as 'men's work' or 'women's work'. Although the attribution of certain 'natural' characteristics to different groups of workers (e.g. women's 'NIMBLE FINGERS' or their ability to perform CARING work) may be seen to fix a permanent boundary between 'appropriate' work for men or women, it is important to recognise that gender divisions of labour are in fact dynamic. The construction of coal mining as 'men's work', for example, rested on particular historical struggles over the sex of miners (Campbell, 1984; for an account of the printing industry, see Cockburn, 1983). Equally, clerical and secretarial labour was not always gendered as female: until the end of the nineteenth century most secretaries and clerks were men (Walby, 1986; Pringle, 1988).

The concept of a spatial division of labour was first proposed by Massey (1979, 1984) to refer to the structuring of production activities across different regions within the national and international space-economy. Particular localities are positioned within a horizontal structure of investment which differs from place to place and intersects with national and international economies. LOCALITIES are also set within a layered vertical structure, as different 'rounds' of investment reshape economic landscapes over time (see also Gregory, 1989). The term was utilised to understand and explain the gendered construction of labour markets (Massey, 1984, Chapter 5; McDowell and Massey, 1984), although it has been suggested that such accounts tended to subsume the category of gender under that of CLASS. Ultimately, spatial divisions of labour were seen to be formed and reformed by CAPITALISM, rather than by and through patriarchal relations. Nonetheless, the emphasis on the mutual constitution of social and spatial relations highlighted the locally constructed nature of FEMININITIES, MASCULINITIES and gender relations which has informed more recent feminist analyses.

The international division of labour is used to explain the outflow of capital investment from industrialised to industrialising nations, or the internationalisation of production. Here, feminist debates have often centred upon the effects that a new international division of labour has had upon women workers in newly industrialising nations. On the one hand, formal employment might be seen to provide liberation in the form of wages and an escape from the confines of 'TRADITIONAL' village life; on the other, authors have emphasised the new forms of alienation experienced by women workers under industrial capitalist discipline (see Ong, 1987).   SR

Divorce Formal ending of a marriage contract. Divorce involves dissolution of a household unit, physical separation of spouses and unravelling of financial and sexual ties. Feminist research points to financial and labour-market constraints which may 'trap' women in marriage (Delphy and Leonard, 1992) or marginalise divorcees in the housing market (McCarthy and Simpson, 1991).   HC

Domestic labour See LABOUR: DOMESTIC, SWEATED, WAGED, GREEN, SURPLUS

Domestic labour theory An analysis, drawing on Marxist theories of labour and surpus value, which challenges the assumption that tasks done in the private sphere of HOME to enable daily or generational reproduction is unproductive WORK. It stresses the relationship between paid and unpaid work, theorising the low status accorded to women's domestic and related work (Phillips, 1983). It was a cornerstone of SOCIALIST FEMINIST analysis and has underpinned campaigns for 'Wages for Housework' and recent government initiatives in Britain to allocate a monetary value to women's unpaid domestic labour (Gardiner, 1997).   JF
See also CAPITALISM; LABOUR: DOMESTIC, SWEATED, WAGED, GREEN, SURPLUS; PATRIARCHY.

Domestic violence See ABUSE; VIOLENCE

Domestic workers Also known as homeworkers. A category of workers involved in house-based WORK, largely in daily reproductive activities concerned with the maintenance of households, such as cleaning, cooking, shopping, childcare and related tasks. Paid domestic work is mostly carried out by women, with the female share of this job sector reaching over 80 per cent in Latin America, the Carribbean and in Portugal. Straddling the PUBLIC–PRIVATE splitting of space, domestic workers are often marginal socially and politically in that their paid work is made invisible by its location within the domestic space. SEGREGATION of women in this employment sector relates to the varied gender ideologies confirming an association of FEMININITY with the reproductive and domestic arenas (Radcliffe, 1990). Materially too, state regulation and the practices of employers' and enterprises work to shape the domestic service sector. Given its specific characteristics, domestic service has been of interest to geographers examining workplace practices, changing CLASS relations (Gregson and Lowe, 1993), POWER relations and stereotypes. Homeworkers are often characterised by their youth relative to employers, and in some cases their difference in racial or ethnic terms, thereby making the home-based employment relation revealing of the operation of power (England, 1997; Pratt, 1997). The hierarchies ranged across race, class, CITIZENSHIP/nationality and generational lines between homeworkers and employers are revealing of the ways in which these social categories are reproduced, in a number of distinct settings. Homeworkers are frequently disempowered by their employment status, although political movements are infrequent given their spatial isolation.   SAR
See also COTTAGE INDUSTRY; DIVISION OF LABOUR: DOMESTIC, INTERNATIONAL, GENDER, SPATIAL.

Domesticity or domestic ideology This is a concept that feminists have connected to the changes in industrial CAPITALISM in Western societies and the associated spatial division between a PUBLIC and a PRIVATE sphere (MacKenzie and Rose, 1983). In their classic analysis of the rise of the middle class in Britain between 1750 and 1850, feminist historians Leonore Davidoff and Catherine Hall (1987) argued that the origins of the domestic ideology lie in the the rise of the Evangelical movement in England. In this religious movement women were constructed both as spiritually pure and in need of male protection and so it was argued that they should be confined to the home. There, as angels of the house, they would provide rest and respite from the public world for their male partners and bring up their children as moral individuals. Feminist theorists have not only challenged the sexual double standards inherent in this division and its restriction to middleclass households but have also shown that the boundaries between the public and the private sphere are more permeable than contemporary texts assumed. This distinction, however, as feminist scholars have shown (see for example Pateman and Grosz, 1987) has not only influenced gender relations and the construction of the urban built form but was a central theoretical distinction in the development of the social sciences in the eighteenth and nineteenth centuries.   LM

See also BUILT ENVIRONMENT; CITY; LABOUR: DOMESTIC, SWEATED, WAGED, GREEN, SURPLUS.

Domestic division of labour See DIVISION OF LABOUR: DOMESTIC, INTERNATIONAL, GENDER, SPATIAL

Domestic labour See LABOUR: DOMESTIC, SWEATED, WAGED, GREEN, SURPLUS

Domination Implies unequal POWER relations. One party might be said to dominate when they have greater power and control than others. At a societal level, a minority of men have held on to positions of power and authority which can ensure their domination over women.   JW
See also PATRIARCHY.

Dualisms See BINARY/BINARY OPPOSITIONS

Dyke A highly contested term of unknown origin, thought to have first been used to describe LESBIANS in America's black communities of the 1920s. The term was initially used as a simple descriptor – it was neither highly politicised, nor used to denote a particular type of lesbian. Gender and class stereotyping of the word began in the 1950s when it became associated with working-class, 'butch' lesbians – those who adopted a MASCULINIST aesthetic. These lesbians were unfavourably compared with both masculine cultural artefacts such as Mack trucks and with masculine archetypes such as diesel truck drivers. This gave rise to derivations such as 'bull dyke' and 'diesel dyke' which were employed derogatively to describe lesbians deigned to be of a similar size or appearance! Linked implicitly to the POWER relations enacted within and through this masculinist aesthetic, these lesbians and the signifiers used to describe them (dyke, bull dyke and diesel dyke) became increasingly stigmatised, both by HETEROSEXUALS who found such behaviour threatening and by other lesbians who either sought to disassociate themselves from the 'rough trade' or, as Smyth (1992: 37) has suggested, who decided to 'outlaw "butch" and "femme" as a replay of heterosexist power imbalances'.

The emergence of QUEER THEORY highlighted the oppressiveness of prescriptive approaches to sexuality, the need to challenge notions of normativity and to embrace the diversity of deviance. In keeping with the project of redefining past terms of oppression and imbuing them with positive meaning the term 'dyke' is now employed in a celebratory way to refer to any earthy and rebellious lesbian (Lapovsky Kennedy and Davis, 1994). BP

E

Earth as mother A recurring metaphor in human cultures. Merchant (1990) charts several European versions, arguing that this belief imposed cultural constraints on invasive practices such as mining, which required at least ritual propitiation. She quotes Roman warnings that mining caused punitive earthquakes, echoed in medieval debates over whether the earth, which provides so much bounty, would hide anything her children really needed. The notion of the earth as MOTHER gave way to a mechanical conception in the seventeenth century, but has been revived in the modern ecological movement.

Mother Earth is an important conception in ECOFEMINIST spirituality, seen either as a reclaimed and healing cosmic myth, or as spiritual or SCIENTIFIC TRUTH (Diamond and Orenstein, 1990). The wider ecological movement has found inspiration in Native American understandings of the earth, or the land, as mother, and the associated environmental ethic of respect, reciprocity and stewardship (Allen, 1984). However, white American nostalgia for a pre-conquest UTOPIA has met criticism (Weaver, 1996). The Gaia hypothesis has given new support to modern notions of the earth as mother – Gaia being the ancient Greek name for Mother Earth. Scientist Lovelock (1979) argues that the 'highly improbable' relative stability of the earth's climate and chemistry is evidence of homeostatic mechanisms aimed at keeping the earth fit for life. If the biosphere (life considered as a whole) actively adapts the planet, the Earth itself must be understood as a complex organism. Humans cannot destroy the earth our mother, but her self-protective response to our activities may destroy us. CN
See also NATURE.

65

Ecofeminism A range of positions and forms of activism which link the OPPRES-SION of women with the domination and destruction of the natural world. The term was first used by Francoise d'Eaubonne in the early 1970s to protest against the ecology movement's failure to address relations between the sexes. She argued that pollution, overpopulation and planetary degradation result from a male culture which downgrades the needs of women, including the right to choose (d'Eaubonne, 1994). Women's protests against local environmental pollution have been hailed as ecofeminist, for example Swiss women's 1976 protest against the Seveso poisoning, and Namibian women's 1984 campaign against corporate dumping of out-of-date medicines. In such cases women, as MOTHERS trying to protect their children and as the traditional guardians of community health, find themselves in opposition to militarism and the corporate drive for profit. According to Mies and Shiva, women become

> aware of the connection between patriarchal violence against women, other people and nature, and that in defying this ... we are loyal to future generations and to life and this planet itself. We have a deep and particular understanding of this both through our natures and our experience as women (1993: 14).

Ecofeminism can involve the following claims.
1. Women are culturally identified with nature. Both are objectified and constructed as 'other' and are treated instrumentally by men in PATRI-ARCHIAL societies.
2. Women really do have a special affinity with nature, and a special role to play, politically and spiritually, in reclaiming the interconnectedness of all things.
3. Women's liberation and planetary liberation are necessary conditions for each other, and steps towards one are steps towards the other.
4. The mechanisms through which ecologically destructive practices are reproduced are the same as or linked to those which reproduce male DOMINATION.
5. Women's experience is a more reliable source of knowledge than is the patriarchal institution of modern science, which objectifies women and nature.

Ecofeminists do not agree on all these points. For 'cultural' or 'affinity' ecofeminists, such as Collard and Daly, women's reproductive capacity makes them sensitive to the spiritual imperative to bring forth and nourish life, which is key to planetary healing. 'Social' and socialist ecofeminists see the woman–nature link as socially constructed and historically variable. Women's special knowledge and influence originates in their social EXPERIENCE rather than their being, and can and must be extended to all humankind. Hierarchical DUALISMS, such as man–woman, human–nature, human–animal, reason–emotion, power–compassion and so on, are seen as key to a destructive culture. They impoverish and distort the nature of women and men and cannot be overcome simply by reversing the value ordering of the terms in each couple (Plant, 1989; Plumwood, 1993).

Ecofeminists agree in rejecting all forms of domination, recognising the interconnection and interdependence of all things. Patriarchal power structures through which women are oppressed – militarism, industrialism, national state apparatuses and international bodies such as the World Bank – are also those responsible for destroying the planet. Ecofeminists agree that in the long term, women's interests are linked with those of nature, though they disagree about whether women spontaneously recognise this and whether it is also true of men. CN

Economic restructuring An ongoing process whereby companies and economies reconfigure their operations to secure profit. At times of widespread crisis, economic restructuring becomes particularly intense, often affecting whole sectors and spaces of the economy. During the mid-1980s Doreen Massey highlighted the ways in which the UK economy was undergoing uneven processes of economic restructuring, with divergent outcomes in different locations. Economic restructuring changes the landscape of CAPITALISM, and in the twentieth century we have witnessed a new international division of labour (NIDL). Moreover, as Susan Halford and her co-authors (1997) have indicated, restructuring is also a gendered process as the reconfiguration of economic life has differential outcomes for male and female employees. JW
See also DIVISION OF LABOUR: DOMESTIC, INTERNATIONAL, GENDER, SPATIAL; LOCAL/LOCALITY.

Écriture feminine A term that derives from the arguments of FRENCH FEMINIST Hélène Cixous (1981a) where she claims that a subversive and marginal feminine discourse is repressed by the PHALLOCENTRIC symbolic order. It is difficult to define the form of *écriture feminine* as Cixous argues that it does not yet exist, but typically lyrical and poetic forms of writing by women and a celebration of the female BODY and maternity are referred to. It has been criticised as an ESSENTIALIST concept, but other feminists have themselves criticised this criticism and suggest that modes of representation mediate the relationship between a woman and her body and/or that 'WOMAN' herself is a DISCURSIVE construction. LM

Effeminacy First used in the fifteenth century to describe a condition in which men adopted and ostentatiously displayed behaviour or characteristics traditionally associated with women. That these characteristics were defined by men exclusively in terms of weakness, delicacy, softness and FEMININITY – all of which deemed highly undesirable – reveals much about why the term became constructed pejoratively. In a deliberate attempt to subvert this paradigm the term has been reclaimed as a self-identifier by those who wish to celebrate the existence within themselves of those same characteristics.BP

Elder abuse See ABUSE; VIOLENCE

Emancipation A term commonly associated with the liberal tradition of equal rights feminism, especially first wave feminist arguments that women should be freed or emancipated from the legal disqualifications that restricted their RIGHTS compared with those of men. The legal emancipation of women was perhaps first and most famously claimed in Mary Wollstonecraft's *A Vindication of the Rights of Women* (1792). In SECOND WAVE FEMINISM the term liberation was more commonly used than emancipation in recognition that more than legal rights were important in women's struggle against male DOMINATION. LM

See also SUFFRAGE/SUFFRAGE CAMPAIGN.

Embeddedness The term is associated with the New Economic Sociology (NES) (Granovetter, 1985; Granovetter and Swedberg, 1992; Ingham, 1996), although its origins can be traced back to the work of Polanyi (1944) on the social embeddedness of economic processes. It has become widely adopted within economic and social geography, embodying the NES's rejection of 'economic RATIONALITY' as a conception of economic action; rather, the NES argues that economic action is embedded within the social context and the specific institutions within which it takes place. Thus, economic decisions are affected, as are all forms of social interaction, by TRADITION, historical precedent, CLASS, GENDER and other social factors (Smelser and Swedberg, 1995). Zukin and DiMaggio (1990) distinguish between four types of embeddedness: individual, structural, cultural and political – these forms are interrelated and in practice are inextricably bound together, although their separation is a useful explanatory device.

Within feminist writing, the concept informs the approach of Schoenberger (1997) in examining what she terms the 'cultural crisis of the firm'. More specifically, the approach has been adopted by McDowell (1997b) in theorising the way in which the gender of the workplace influences the processes of business. Using investment banks in the City of London as a case study, McDowell argues that some of the business practices of these companies are imbued with MASCULINE attributes – in terms of the social practices of 'doing business'. Women do not fit the hegemonic masculinities of the investment banks as well as men, and this represents a strong explanatory factor in accounting for the relatively few women at the top of these organisations. In this sense, investment banking is 'embedded' in the masculine business cultures of the City. AJ

See also CORPORATE CULTURE.

Emotion Emotions combine mental, social, cultural and bodily dimensions in a way that makes them of great interest to social scientists attempting to theorise the BODY. There have been long-standing debates about how many emotions there actually are, about their relation to each other, whether they are UNIVERSAL or culturally specific, and how they evolved historically. Analogies are often drawn with primary and secondary colours. Some have argued, for example, that there may be as few as four 'basic' emotions (anger, fear, love, sorrow) out of which the rainbow of subtler shades is constructed. While some stress the SOCIAL CONSTRUCTION and cultural diversity of emotions, others emphasise their close links with bodily states. This linkage is evident in the LANGUAGE used to describe emotions: boiling over with anger, shivering with excitement or fear, blushing with embarrassment, burning with passion.

Emotions differ from feelings in that they exist relative to human social acts. They are also cultural acquisitions determined by the circumstances and concepts of a particular CULTURE. It is possible to write the history of emerging emotions, while different emotions can be said to characterise different historical periods. What we now call emotions are constructs of an age of psychological and therapeutic knowledge and practice; they are inconceivable apart from these institutions, social relations and forms of thought. We have become 'emotion conscious' to the extent that we spend time and money 'working on' our emotional conflicts. Emotions have become necessary props for the creation of subjectivity. Our interest in emotion may also be observed at the level of popular culture where we have become consumers of emotion and passion.

Emotion is often assumed to be the opposite of REASON but the two are not so easily separated. For social construction, emotions are neither substances nor states, but rather part of the conscious relations, actions and experiences of selves. It is possible to demonstrate both a RATIONAL dimension to emotions and an emotional dimension to reason. Emotion is also often designated as the realm of the feminine in opposition to the masculine realm of reason. It has been argued both that women are 'more' emotional than men and they are 'better' at managing emotional matters. Feminist sociologists such as Hochschild (1983) have demonstrated that although men and women share the same emotional range, 'gender' itself might be understood as an emotional package. MASCULINITY is normatively understood to include large components of anger, aggression and guilt, whereas FEMININITY is more strongly associated with modesty, shame and fear.   RP

Empire An extensive territory, especially an aggregate of many STATES ruled over by a sovereign state which in its independence and POWER ranks as an empire. A state exercising imperium or dominion over its colonies and other dependencies. The advocacy of imperial interests is concerned with power and AUTHORITY over others. It encompasses the range of state-centred relations of military, political, economic and cultural DOMINATION with attributes which are commanding or domineering. Whilst the historical range of empires is extensive, much recent literature on geography and IMPERIALISM focuses on the past two centuries. It is during this period that the scale, intensity and range of imperial intervention in colonised societies has been unparalled and it has been accompanied by the establishment of GEOGRAPHY as an academic discipline with its associated ideas, practices and institutions (Bell *et al.*, 1995; Godlewska and Smith, 1994; Livingstone, 1992). Recent literature has drawn a distinction between the IMAGINATIVE GEOGRAPHIES of empire associated with the work of Edward Said (1978, 1993) which convey an abstract sensibility (Driver, 1992), and the more tangible geographies created and experienced by women and men both in and of the various European empires (Pratt, M.L., 1992). The gendered nature of imperial practices and the complex interweaving of RACE, GENDER and CLASS relations in particular imperial territories and at specific times are the focus of increasing study. Gendered constructions of geography and empire have also been raised in recent debates on feminist historiographies of geography (Domosh, 1991; Stoddart, 1991).   MB

See also COLONIALISM; CULTURAL IMPERIALISM.

Empirical/empiricism Empirical phenomena are those whose existence is known through human observation – that is, ultimately, through human sensory EXPERIENCE, although this may be mediated through measuring devices. Empiricism can refer both to the idea (i) that all concepts are derived from empirical knowledge and to the idea (ii) that statements of knowledge can only be verified through empirical observation. Thus empiricist approaches, such as POSITIVISM, assume that no theoretical statement can be verified without reference to empirical observation and that a clear separation can be made between theoretical statements and empirical observations. Furthermore, empiricist approaches assume that empirical observations can be replicated – that is, that if empirical observations are made by different people under the same circumstances the same empirical 'facts' will be observed. In order to achieve this replication social scientists must separate their empirical observation from personal judgements and emotion. Such a separation is assumed to be both necessary and possible. The assumptions that theoretical statements and empirical statements can be separated, that empirical state-ments are the basis for theory and that empirical observation of 'facts' can be separated from the 'values' of the social scientist are challenged by MARXIST, REALIST and POSTMODERN philosophies of science. Feminist scholars have been prominent in these challenges to empiricist approaches and in arguing that theoretical knowledge and empirical observation are not independent of one another. In particular, some feminists have argued that the dominant systems of knowledge in the WEST are 'male-biased' – that is, that the interests of men as a group have influenced the structuring and nature of knowledge and have ignored the experiences and knowledge of women.   SB

Empowerment To be empowered is to be invested with POWER. As a process, feminist interpretations have focused on the 'inputs', that is, women's EXPERI-ENCES, and 'throughputs', that is, women's realisations of their own power. The somewhat deterministic assumption was that the 'outputs', that is, women's emancipation through engagement in the WOMEN'S MOVEMENT, would automatically follow. The term was popularised in SECOND WAVE

FEMINISM as an feminist alternative to traditional models of therapy and education, although as early as 1792 Mary Wollstonecraft was setting forward the case for female emancipation. In the 1970s the call for emancipation was based on a designation of women as victims, living under structurally determined forces of OPPRESSION. In the 1990s, with women increasingly refusing to be positioned as victims, feminist notions of empowerment have moved away from the belief of any collective celebration of female AGENCY leading to women's self-determination. While some feminists argue that it is politically imperative to hold on to notions of women's agency, others claim the 'SELF' and, hence, self-determination, is a humanist fallacy, rendering identification amongst women and, hence, feminism, impossible.

Empowerment keeps on bubbling to the surface though. Its most recent revival in academic circles is a result of the influence of post-structural theory, especially Foucault's notion of power as being exercised rather than held, as well as the multiplying diversity of experiences of being FEMALE. Empowerment is currently also a major goal of THIRD WORLD planning and policy agencies (see Moser, 1993). For feminist geographers it has increasingly led to study of the practices of individual women. Two such substantive studies are those of Jan Monk (1996) who has uncovered the long-lost contributions of women geographers to the discipline, and Kathy Gibson (1991) who has investigated work on the empowering tactics of the wives of Australian coal miners, describing their subjects' involvement in the discursive process of research and the potential of such engagement for the production of radical subjectivities.   LP

Enlightenment/enlightenment theory The Enlightenment was a period of change associated with the rise of MODERNITY and liberal theory in Europe in the seventeenth and eighteenth centuries. In particular, enlightenment theory is associated with a group of eighteenth-century French philosophers who questioned traditional authority, especially the belief in the divine right of monarchs, asserting instead their belief in human reason or RATIONALITY and in the inevitability of progress. The claim of inevitable progress is now contested; as is the eurocentrism of Enlightenment beliefs. The belief in the superiority of Western notions of science and reason lay behind Europe's imperial expansion and the enforced introduction of Western systems of, for example, government and religion that are now criticised by POST-COLONIAL theorists. The singular view of progress has also been challenged by postmodern theorists, who argue against the METANARRATIVES embodied in Enlightenment thought, especially in LIBERALISM. Feminist critics have pointed out the patriarchal or masculinist assumptions that also imbue Enlightenment notions of reason (see, for example, Pateman and Grosz, 1987).   LM
See also EMPIRE; LIBERALISM; MODERN/MODERNISM/MODERNITY/HIGH MODERNITY; RATIONALITY.

Environment The context of human activities, the source of the resources they use and the recipient of their intended and unintended consequences. The term can be used in specific senses, as in 'the business environment' or 'the urban environment', but when unqualified means the natural environment. This includes the composition and/or characteristics of air, water, soil, minerals and living things on the planet, and their interrelationship in the ecosystems that support human life. Desertification and the destruction of rain forest, the loss of species, marine pollution and the diminution of fish stocks, the dumping of radioactive waste, the accumulation of poisons in the soil, air pollution, the creation of acid rain, global warming and so on are all constructed as elements in an 'environmental crisis' from a human point of view. These changes threaten current human ways of life, result in increasing POVERTY and will reduce quality of life for future generations. While some see environmental risks as unevenly distributed along class lines, others argue that the common danger transcends old divisions in a new sort of 'risk' society and permits new political alliances (Beck, 1994). This understanding of the 'environment' as the natural context of human life informs environmentalism, but has been criticised by ecologists, social ecologists, socialists and feminists.

1. Ecologists prefer to speak of 'ecology' and 'ecosystems', because 'environment' is necessarily an anthropocentric term, which suggests humans are outside the environment rather than embedded in nature. They insist on the intrinsic value of ecosystems and of all forms of life, preferring an anti-humanist, biocentric ETHICS or even a wider ecocentrism that values rocks and rivers to the 'enlightened self-interest' of environmentalism (Hayward, 1994).

2. The term 'environment' brings with it some of the ambiguities of 'NATURE' and the nature–culture divide. Ecosocialist geographer Harvey criticises conceptions of the environment as separate from social (especially urban) life (Harvey, 1996: 186). Cities are a key part of our environment and can themselves be seen as 'created ecosystems' (186). Any notion of environment that does not recognise 'the production of nature' through commodity exchange and capital accumulation is static and misleadingly dualistic (see also Weston, 1986).

3. 'Social ecologists' also prefer the term 'ecology', distinguishing between 'first nature', which is prehuman, and 'second nature' which includes humanity and all its works (Bookchin, 1989). Neither of these correspond to the term 'environment'.

4. Feminists point out that notions of environment which see it as separate from human life minimise the effects of human agency and the gendered nature of the destructive activities in question (Rose *et al.*, 1997). Seager describes moving as a feminist geographer from an initial view of the environment as a physical system under stress, essentially separate from the organisation of social life, to the position that male culture and interests are responsible for the disruption of the biosphere: 'The task for feminists is to unravel the ways in which gender operates as a structuring condition within the institutions that hold the balance of power on environmental issues' (Seager, 1993: 6).   CN

Environmental movement/Green movement A 'new SOCIAL MOVEMENT' which emerged in the 1960s to oppose environmentally destructive practices. To some extent united by a focus on participative politics and decentralisation, its methods range from parliamentary tactics and traditional lobbying to symbolic protests and non-violent direct action. The movement is both international and local, diverse and internally conflictual. It includes Green parliamentary parties, environmental pressure groups, anti-nuclear campaigners, 'green consumerism', anti-roads protestors, conservationists, not-for-profit providers of recycling, insulation, etc., animal liberationists, ecofeminist networks, organic farmers, indigenous peoples struggling to defend ways of life, and urban ethnic minorities against environmental racism.   CN

Episteme *Episteme* has come to refer to the tacit frameworks in which both theory and empirical work are nested. It is primarily a word associated with Michel Foucault's efforts to distinguish distinct historical regimes of ordering knowledge formation. In *The Order of Things* (1970) Foucault describes three distinct *epistemes*: the premodern, characterised by a relational logic in which earthly arrangements were seen as evoking a heavenly order; the classical, characterised by an abstract tabular or grid approach to ordering knowledge; and the modern, characterised by a constant privileging of human life as the simultaneous centre and judge of knowledge. This approach to historicising different *epistemes* has certainly functioned to unsettle the taken-for-granted universality of modern orthodoxies (and therefore served Foucault well in his polemic at the end of *The Order of Things* against Sartre's human-centred existentialist philosophy as well as in his later battles against what he called Freud's 'repressive hypothesis' (1979)). For this reason, it may be also said to have enabled feminist writers such as Donna Haraway (1991) to replace what she depicts as the cognitive cannon law of formal EPISTEMOLOGIES with less masterful accounts of SITUATED KNOWLEDGES. Nonetheless, Haraway and others object to the finality of the philosophical metaphors such as 'death of the SUBJECT' that characterise Foucault's argument about the end of the modern *episteme* and its faith in representation. Gayatri Chakravorty Spivak (1988a and b), in particular, argues that Foucault's arguments about the limits of REPRESENTATION lead to an abdication of the responsibility to represent the world from a critical perspective and that they remain in this very way complicit with certain structures of modern EPISTEMIC VIOLENCE that have marginalised non-Western and non-masculinist knowledge from modern WESTERN thought.   MS

Epistemic violence A term for describing the world-changing and often oppressive or exploitative implications implicit in imperial systems of knowledge. While it is conceptually underpinned by Foucault's (1980) arguments about the interarticulation of POWER relations and knowledge formation, the notion of epistemic violence has been made popular as a term of critique by the feminist, Marxist and POST-STRUCTURALIST arguments of Gayatri Chakravorty Spivak. Spivak's critical, POST-COLONIAL point is that certain Western systems of knowledge have been established as the enlightened TRUTH and thereby had such a hegemonic effect in the colonised world that they actually displace and destroy native self-understandings.

In her essays on the question of SUBALTERN agency, Spivak (1988a and b) uses the term epistemic violence to describe the diverse ways COLONIAL administrators, IMPERIALIST historians and even would-be radical post-colonial historians have constructed knowledge claims about subaltern women (which is to say the most marginalised and oppressed women of colonial and neo-colonial societies) that effectively do violence to the histories of such people's struggles either to survive or transform their surroundings for the better. In one sense this is just another way of saying that such commentators have made interpretatively mistaken and ideological arguments. However, Spivak's additional and more critical point is that these were and are ideological mistakes that are successfully established as the truth and which, as such, often have violent and world-transformative effects, including being internalised as the truth by colonised people themselves. They comprise thus what she calls 'successful cognitive failures'.

A notable geographical example of such successful cognitive failure is the cartographic epistemic violence implicit in mapping colonial lands as empty space and thus as space not already mapped and interpreted by native agents of knowledge production. Clearly, such CARTOGRAPHY was historically mistaken in the sense that the land was not empty; it was also ideological in the sense that it directly enabled territorial imperialism. Nevertheless, it also succeeded insofar as maps of *Newfoundland, Terra Nullius* and all the rest of the colonised world became the new truth of the supposedly ENLIGHTENED world (Harley, 1989). Indeed, even if the land was not new and empty from the perspective of the colonised, the epistemic violence of colonial maps was such that it effectively anticipated the actual emptying and resettlement of these same lands through colonial conquest. Spivak's DECONSTRUCTION of such successful cognitive failure thus carries an ethical imperative for feminist and other radical scholars to go back and re-examine how any particular narrative or mapping may have effectively eclipsed the agency or knowledge of those resisting colonialism and neo-colonialism (Sparke, 1995).

Spivak's warning about epistemic violence also applies to other forms of geographical knowledge production closer to the contemporary research concerns of feminist geography. Most notable amongst these is FIELDWORK, an old form of geographical research historically connected to imperialism, epistemologically evocative of the battlefield, and psychologically related to a masculinist desire to penetrate and master what were often coded as virgin lands (Rose, 1993a). Nevertheless, and partly because of feminist arguments about the EPISTEMOLOGY of SITUATED KNOWLEDGES, in-depth fieldwork interviews with informants have now become a vital component of feminist geographical research because they offer an opportunity to document some of the power relations, including emotional relations, hidden behind strictly numerical data (e.g. Pratt and Hanson, 1995). In the context of such feminist

ETHNOGRAPHY, the epistemic violence implicit in the imperial and masculinist notion of fieldwork has been questioned through reflexive reconsideration (England, 1994; Gilbert, 1994; Kobayashi, 1994; Staeheli and Lawson, 1994; Nast, 1994) and this has led to reformulations such as Katz's concept of fieldwork in 'the space of betweenness' (Katz, 1994). Such a space evokes a radical reconceptualisation of fieldwork in which both researchers and the people they research can together construct possibilities for dialogue and mutual learning. Critics argue that such attempts to rework structures of epistemic violence are doomed because of their ongoing dependence on older, and thus seemingly tainted, forms of knowledge production including the gazing introspection of self-reflexivity itself (Rose, 1997a). However, even if one accepts this defeatist and somewhat tendentious view of feminist self-reflexivity, the criticism still misses how actual ongoing performances of fieldwork by feminist researchers might open up the transformative possibilities that Spivak (1990c) suggests are available in all negotiations with structures of epistemic violence. To put this in the rather different register of Judith Butler's post-structuralism (1993), it misses the promise of the PERFORMATIVITY of feminist negotiations with the epistemic violence of fieldwork, the promise of displacing the masculinist field in fieldwork (see Sparke, 1996).   MS

Epistemology Epistemology comprises the diverse set of philosophical arguments used to answer the question 'How do I know what I know is true?' Used at its most general, epistemology refers to the different ways of describing and thereby legislating the philosophical commitments that underpin our knowledge of the world. For a generation of contemporary Anglo-American geographers who have at some point or other been trained in the vocabulary of critical REALISM, epistemology is usually understood as the counterpart of ONTOLOGY which, according to the critical realist philosopher Roy Bhaskar (1975), refers to what the world must be like for us to have knowledge about it (but which is more generally understood as the study of being or existing). By contrast to such definitions of ontology, epistemology is generally used to describe the different approaches theorists take to legislating what counts as legitimate knowledge about an ontologically complex world. Such approaches are varied but three notably distinct epistemologies stand out. First of all there is the commitment in the broader philosophy of POSITIVISM to an *empiricist* epistemology of 'letting the facts speak for themselves' (which famously ignores the complex translation problems of interpreting the different languages such facts are supposedly speaking). Empiricists generally presume that truth is guaranteed by direct observation. Second, there is the commitment shared amongst some HUMANIST scholars to an epistemology derived from *phenomenology*, *pragmatism* or *hermeneutics* that asserts the interpretative and local quality of all knowledge claims. What tends to unite this kind of epistemological commitment is an understanding that truths are underwritten by our common experiences of being and communicating as humans. And, third, there is the commitment shared amongst many Marxist,

FEMINIST and critical realist scholars to an epistemology that acknowledges the interpretative quality of knowledge production but which is also attentive to the *systematicity* and often far from local POWER relations in which knowledge claims are made. These scholars have developed an epistemological awareness of processes that are often not directly observable and which are thus often discounted by empiricists.

The critical epistemological position nevertheless asserts that such unseen processes do exist and have causal capacities that create documentable and often systematic effects. As such, it is an epistemology that is open to *empirical* as opposed to *empiricist* research: indeed it enables a form of empirical research that actually helps elucidate power relations while simultaneously attending to the structures of social life through which such critical knowledge is itself produced. However, in contrast to a more STRUCTURALIST (and, in geography, at least, a rarely upheld) Marxist position focused on the ideological superstructures of interpretation in capitalist societies, many *POST*-STRUCTURALIST feminist scholars have attempted to come to terms with how class combines with gendering and other diverse but chronic processes of social ordering-cum-OPPRESSION to inflect both the production and the interpretation of knowledge. Not only has this led to critiques of the EPISTEMIC VIOLENCE implicit in impersonally abstract and geographically UNIVERSALIST affirmations of modern Man's RATIONALITY, it has also enabled feminist theorists such as Donna Haraway (1991) to develop a form of feminist epistemology based around the notion of both empirically and socially accountable SITUATED KNOWLEDGES. MS

Equality/equal opportunities See RIGHTS/EQUAL RIGHTS/EQUAL OPPORTUNITIES

Essentialism This is one of those words with accusatory overtones. It refers to the position that attributes gender inequalities to an essential, BIOLOGICAL difference between men and women. Essentialism, in its simplest form, is, therefore, a biologically determinist argument. There are clear parallels here with the debates about 'race' and racialised inequalities which are also assumed by some to have a biological basis. A number of positions that have been important in the history of the development of feminist approaches have more recently been accused of essentialism. The arguments of radical feminists that were important in the late 1960s and the 1970s are a common example here. RADICAL FEMINISTS, in a move close to, but preceding, that of BLACK and QUEER activists, suggested that 'feminine' characteristics of, for example, nurturance and care-giving, of pacificity and self-denial, were superior rather than inferior to oppositional masculinised attributes such as RATIONALITY and lack of EMOTION. These feminised attributes were not, at least in their earliest versions, seen as socially constructed and so varying over time and space but instead the inevitable consequences of women's biological ability to give birth and nurture children and lay behind the claim of global SISTERHOOD. A strong essentialist thread also runs through, for example, the ECOFEMINIST movement in which women's supposed closer connection to NATURE results in their claims as protectors of the environment.

A version of FRENCH FEMINIST THEORY that draws on but contests Lacanian arguments about women as 'absence' or 'lack' has also been labelled essentialist. The work of feminist theorists such as Julia Kristeva (1987) and Luce Irigaray (1977) (and also see Grosz, 1989; Flax, 1990) who have attempted to define and reclaim a specifically feminine way of being that is impossible within PATRIARCHAL societies is often quoted, and their emphasis on the importance of the female BODY and maternity is seen as a link to the essentialism of earlier radical feminist arguments.

Diana Fuss (1989) and Teresa de Lauretis (1986), however, have suggested that feminists must hold on to the notion of 'STRATEGIC ESSENTIALISM' – that is, the belief that women are united because of their struggles against common experiences, most notably the discrimination that they face in male-dominated societies. In this sense strategic essentialism is a political concept, a way of holding on to the need for feminist POLITICS and change without the necessity of a belief in biological essentialism or universalism. In a helpful extension of these arguments, Avtar Brah has suggested that a threefold distinction might lead to conceptual clarity when defining the term essentialism. She distinguished between:

1. essentialism as referring to a notion of ultimate essence that transcends historical and cultural boundaries;
2. 'UNIVERSALISM' as a commonalty derived from historically variable experience and as such remaining subject to historical change; and
3. the historical specificity of a particular cultural formation (1996: 92).

In this way, she suggests,

> it should be possible to recognise cultural difference in the sense of (3), and to acknowledge commonalties that acquire a 'universal' status through the accumulation of similar (but not identical) experiences in different contexts as in (2), without resorting to essentialism. It is evident that, as women, we can identify many commonalties of experience across cultures which nonetheless retain their particularity. In other words, historical specificity and 'UNIVERSALISM' need not be counterposed to each other (92).

This useful claim has parallels with the notion of strategic essentialism and allows women to point out their similar structural position across different geographical regions without falling into the errors of what perhaps might be termed essential essentialism, or Brah's first definition of the term.   LM

Essex girl Perhaps the only geographically specific term used to refer to women in 1990s Britain; it is, however, a derogatory term implying a dim blonde whose primary interest is in her own appearance and in instant gratification. It may also be used in a parodic sense, however, to open a space for a certain form of transgressive behaviour.   LM

Ethics Ethics can be conceived of as moral philosophy 'involv[ing] systematic intellectual reflection on morality in general ... or specific moral concerns in particular' (Proctor, 1998: 9). MORALITY can be characterised as concerning individual and collective judgements of what is good or bad, right or wrong. Although distinctions can arguably be made between theoretical and applied ethics, both involve thinking about values, beliefs and responsibilities (in academic work these elements can be theorised descriptions of such values (as held by researchers) or conceived more dynamically as calls to action (as new values to inform research activity)). Some writers believe that there are fundamental, underlying and transcendental ethics involved with human existence (e.g. a shared sense of the value of humanness and humankind), while others point to the constantly evolving nature of ethics, as values and beliefs are constituted and regulated by differing sets of social and cultural relations (ones fundamentally infused with power). In geography these questions have been approached from very different perspectives, with some scholars advocating a geographical assessment of current economic and social structures with a view to theorising and pursuing new forms of social JUSTICE (Harvey, 1973). Other scholars in geography have discussed self-reflexive possibilities in the field of ethics, critiquing work which claims to be value-free, detached and scientifically-based geographical work (much of SPATIAL SCIENCE obviously advanced or presupposed such a claim). Humanistic geographers in the 1970s argued that personal values and meanings influence and structure the ways in which researchers relate to and conceptualise phenomena in the world (Buttimer, 1974), maintaining that value-free social science is an unattainable goal. Yet other geographers have conceived of a geography of ethics as connected to the various and continuing calls for the discipline to be socially relevant. In more recent times, POSTMODERN analysis has reinvigorated debates about ethics and geography, with many scholars becoming disillusioned about a perceived relativist stance on issues of social inequality and injustice. A crisis of REPRESENTATION has been debated within such writings as scholars have clashed over the ways in which they portray and theorise different people and their multilayered geographies. This crisis is arguably infused with concerns about ethics, understood here as recognising the multiple possibilities for misrepresentation rather than proposing a search for the 'right' representations.

Feminist geography has also been infused with questions of ethics, if we understand ethics here as values which demand certain actions on behalf of the proponents. Feminist geographers have often stressed the need for EMANCIPATORY outcomes to writing and research, indicating the presence of pervasive values concerned to improve the everyday lives of women. A recent text written by a collective of feminist geographers indicates this sense of shared values: 'feminist methodology is committed to challenging oppressive aspects of socially constructed GENDER relations' (WGSG, 1997: 87). There are other ways in which feminist geography can be argued to be infused with ethics. The call for 'SITUATED KNOWLEDGES' (Haraway, 1991) can be conceived as a move away from myths about value-free knowledge construction, and in consequence the social, cultural and political 'POSITIONALITIES' of authors are increasingly highlighted within research writings as ways in which to contextualise (and perhaps in some ways to justify) any (mis)representations of women's lives. Such strategies of situating knowledge are inherently ethical because they involve a recognition that academic knowledge is partial (and therefore partially constituted by cultural values), but they still aim to uncover (and potentially to expose and even to change) the oppressive conditions which feature in many different sorts of women in many different places. HP See also HUMAN NATURE.

Ethnic/ethnicity Complex words which are commonly used in Britain and the USA to refer to people of colour, or to minority groups who are migrants into a society and who are assumed to be different from their 'hosts' on the basis of a range of social characteristics which usually include their 'race', or skin colour, and their cultural identity. Until recently, the ethnicity of white people has either been denied or taken for granted, although a number of feminist theorists have begun to challenge this incorrect and racist assumption. If ethnic is used with reference to 'white' groups it tends to carry with it associations of homespun traditions or folklore. Thus the black feminist author bell hooks has discussed the representation of WHITENESS as superiority by black writers (hooks, 1992b), and white feminists such as Vron Ware (1992) in the United Kingdom and Ruth Frankenberg (1993a and b) in the USA have begun to document the social construction and spatial/temporal specificity of whiteness.

Ethnicity is often distinguished from the term 'race' to emphasise the social constructivist nature of difference between people of different origins and identities. Whereas 'race' is a disputed term in which the ESSENTIALISM of an assumed biological basis to differences has been hotly contested, ethnicity is often used in the sense of a self-defined and chosen cultural identity (Anthias and Yuval-Davis, 1983, 1992). Whereas race is a label or a stigmatised identity that is forced on the less powerful or disadvantaged in some way by a more powerful group, ethnicity is a term that has more positive connotations. Thus the adoption by Black British people of the slogans 'black power' or 'black is beautiful', first coined in the USA, is an example of an ethnic identity established in opposition to racist stereotyping. The racist label 'BLACK' is taken and reversed, used positively by the 'inferior' group to identify themselves as different and to celebrate this fact. There are parallels here with the way in which gay men have redefined the previously derogatory term 'QUEER'.

Both the term race and that of ethnicity, however, are defined through power relations and categories of exclusion and inclusion. They imply boundary conditions, and membership of an ethnic group is defined by the relative material conditions and social practices of the groups that fall inside and outside the category and by their symbolic representations. Ethnicity is thus a RELATIONAL rather than a CATEGORICAL concept.

Despite the element of self-definition and choice implied in the term ethnicity, in 'white' societies, the superiority of whites is still perceived as 'natural' and there is a dominant or hegemonic discourse evident in which ethnicity, colour, cultural difference and 'tradition' (for which read 'less sophisticated') are closely associated with each other to define 'ethnic minorities' as different from the 'norm': an inferior 'OTHER' in comparison with the white majority. This BINARY distinction is another example of the dichotomous structure of Western ENLIGHTENMENT thought and has similarities to the social construction of FEMININITY as inferior to MASCULINITY. Ethnicity, or ethnic identity, is, however, itself also intercut by GENDER divisions in which particular relations between the sexes are defined by the interactions between the social institutions and practices of an ethnic group itself and the social practices and discourse of 'wider society'. The PATRIARCHAL control of 'dominant' Asian fathers is, for example, manifest as the 'problem of arranged marriages' in hegemonic discourse in the United Kingdom (Brah, 1996). A parallel example in both the USA and in the United Kingdom is the social construction of a 'rampant' sexuality that supposedly characterises BLACK MASCULINITY (Segal, 1990). Thus gender attributes are a key element in the specification of ethnic difference and in racialised stereotyping. Brah has emphasised the current moves to understand the intersections between 'race', gender, class, SEXUALITY, ethnicity and so on are vital, 'precisely because these relationships were rarely addressed together' (1996: 10) and she suggests a move to the analysis of what she terms 'cartographies of intersectionality'.   LM

Ethnocentric This describes ways of thinking which accept our own society as the NORM, without recognising that we speak from particular cultural and historical perspectives. Such universalisation is common to all ethnicities but may be particularly apparent amongst Western academics who privilege a powerful viewpoint without locating it as a specific instance or exploring its limitations for theorisation and practice.

Ethnocentrism may be manifested on a number of geographical scales. Indeed the term is often given geographical specificity in words such as Anglocentric, Eurocentric and even 'westocentric'. Globally, it is reflected in transferring FIRST WORLD perspectives uncritically to other locations. Within First World societies it is evident where WHITENESS is treated as the norm, though differences amongst white ethnicities remain largely unexamined.

Academics outside the centre were the first to draw attention to the problems of ethnocentricity for feminist theorising (Carby, 1982). Thus First World emphasis on the ideology of domesticity has been challenged, since for WOMEN OF COLOUR in the USA the home may be a place of resistance (hooks, 1991b). Women in Eastern Europe have also challenged the assumptions of many in the West that paid work is necessarily emancipatory (Sharp, 1996). Moreover, the priority accorded to family issues and abortion by white feminists overlooks the different reproductive concerns of Black and Asian women, for example.

FEMINIST GEOGRAPHIC METHODOLOGIES are being developed to challenge ethnocentricity. The practice of situating knowledge claims by self-critically identifying our own's political, social and historical position helps to bring such biases into the open (Duncan and Sharp, 1993). BW

Ethnography Contemporary ethnography is a multimethod research tool that includes observation, participation, archival analysis and interviewing in order to understand the inner life and consciousness of a particular group (Jackson, 1985). Traditionally, it has not included testing or large-scale survey methods which would be identified with a POSITIVIST perspective in the social sciences. As many feminist researchers suggest, qualitative methods seem to fit especially well with FEMINIST goals. Feminist ethnographers argue that FIELD-WORK has a special role in upholding a non-positivist perspective and producing new knowledge related to women. Others believe that to place feminist research in opposition to quantitative methods is misguided. Feminists generally have not claimed that fieldwork is inherently feminist or that qualitative methods such as ethnography are the only research methods that feminists should utilise. If this were so, field-based disciplines would have been successful in avoiding patriarchal bias decades ago (Reinharz, 1992). LD
See also FEMINIST METHODOLOGY.

Exclusion See SPACES OF EXCLUSION

Experience Early second wave feminists attacked both academic and dominant common-sense knowledge for ignoring the experience of women. In this claim they were referring not only to lack of attention to the daily pattern of women's lives – the everyday activities of housework, childcare, and shopping which dominated much of their time – but also to lack of attention to women's subjective understanding and interpretation of the meaning of these activities. Some feminists argued that both the KNOWLEDGE of the academy and of common-sense were male-defined knowledges and that feminist knowledge should be constructed on the basis of women's own experience. Conciousness-raising women's groups attempted to arrive at collective knowledge that acknowledged the validity of emotion, intuition and spiritual forms of 'knowing'. However, the notion that there is sufficient commonality between women to allow the development of a specifically female knowledge came under attack as ideas of DIFFERENCE became increasingly significant in feminist politics and it was realised that different women attached different understandings and meanings to their everyday lives. Furthermore, the idea that individual or collectively produced experience can be an adequate basis for knowledge was questioned on the grounds that all knowledge is mediated through the understandings embedded in language, social position and everyday social practices. In response to these difficulties some feminists have explored the notions of SITUATED and SUBJUGATED KNOWLEDGES while others have developed the ideas of STANDPOINT THEORY (Hartsock, 1983). Others have explored critically the significance of social mediation of the experience of embodiment (Braidotti, 1994a and b; Haraway, 1991; Young, I., 1990b and c). SB

Experiential Experiential methodologies rely on attempts to recover and analyse the EXPERIENCE of the researcher in the research situation, of the social interactions involved in research or of individuals whose experiences are the subject of the research. Such methodologies emphasise the importance of emotional and empathetic understanding. Experientially-based knowledge gives epistemological privilege to experience.   SB
See also EPISTEMOLOGY; FEMINIST METHODOLOGY.

Exploitation This is a term that is most commonly used in relation to the capitalist labour process. Following Marx it is argued that workers are exploited through the extraction of their surplus labour at WORK. Exploitation is thus argued to be at the heart of capitalist social relations. Feminists have tended to use the word OPPRESSION to describe the position of women in capitalist society as they are exploited over and above their position in class society. Delphy and Leonard (1992) have taken Marx's argument into the home to suggest that the family depends upon women's exploitation by men, as housework and CHILDCARE are not shared equally between men and women. Moreover, other feminists have used the term exploitation to describe the way women's bodies are exploited for SEX. As Catherine MacKinnon (1982: 516) put it: 'SEXUALITY is to feminism what work is to Marxism: that which is most one's own, yet most taken away'.   JW
See also LABOUR: DOMESTIC, SWEATED, WAGED, GREEN, SURPLUS; PATRIARCHY; SEX/SEXUALITY.

F

False consciousness The gap between the ideas people have and their apparent material interests. The term was originally used by Marx to explain why the working class appears to do things which are not in its best interests but may also be applied to GENDER relations (e.g. women may reinforce gender stereotypes; for a full explanation, see Lukas, 1968). In some ways, it can be argued that women's consciousness-raising is designed to overcome some of the ideological baggage of PATRIARCHAL society, allowing women to develop a FEMINIST consciousness.   JW

Family A term which refers to relations between people of descent through consanguineal or blood ties and/or through affinal or marriage links. Common use of the concept implies shared residence whereby families and HOUSEHOLDS are conflated. In light of the huge diversity of family and household organisation throughout the world whereby not all members of a family share residence, and not all members of a household belong to the same family, most researchers apply stricter definitions which distinguish between the two. In this sense, families transcend spatial boundaries of residence to include wider KINSHIP networks.

Families are therefore seen as institutions providing stability, and encompassing prevailing societal IDEOLOGIES. These familial ideologies are afforded a central role in influencing the social construction of GENDER roles (Barrett, 1980). Through this, the family is viewed as the primary site of women's OPPRESSION, primarily on grounds that families function through exploiting women's unpaid labour and sexuality (Walby, 1990). Underlying this are debates over whether women are 'naturally' embedded within families, through their biological reproductive roles (Moore, 1988); there is now broad agreement that families are seen as social rather than biological constructions (Barrett, 1980). These ideas have also been criticised. First, the notion of the nuclear family as the NORM (comprising a male breadwinner with dependent wife and children) has been challenged; other family structures, such as single-parent or extended units, are ignored and/or perceived as deviant (Chant, 1997). Second, criticisms have been levelled by BLACK feminists on the grounds of ethnocentrism; not only is there great diversity of family forms among black populations, but families may also act as bulwarks against racist society rather than sites of oppression (Phoenix, 1990).   CM
See also PATRIARCHY; DOMESTIC LABOUR THEORY.

Fantasy See DESIRE

Father/fatherhood The term refers in the strict sense to the BIOLOGICAL relationship between a man and his children but is also a social construction in both a particular and in a general sense. In a particular sense the attributes of fatherhood are both socially and legally constructed, usually within a NATION-STATE, and so are spatially variable. In contemporary societies, the attributes of fatherhood usually include social control over children, especially when a man is living in a relationship with the children's mother and the children themselves (often but not only in a nuclear family), legal obligations to provide for these children and control their behaviour as well as certain rights to dispose of their PROPERTY and make decisions about their lives when they are under the age of legal responsibility.

These rights and obligations are often a matter for dispute when a man does not live in the same household as his partner and their children. Thus in Britain in the 1990s there have been legal changes in the financial obligations of men as well as the development of organisations such as Families Need Fathers to press for greater rights of access to children in cases of family breakdown. While the social attributes of fatherhood have been constructed in opposition to those of MOTHERHOOD – fathers typically are absent, distant and not emotionally involved with their children – in recent years there have been moves to encourage greater involvement of fathers in children's everyday lives and emotional development (Burgess, 1997). In Britain there is also a proposal to introduce a 'vow' of commitment by fathers to their children, similar to the marriage vow.

In a more general sense the law of the father or the patriarch is a key notion in the definition of PATRIARCHY. The sets of rights and obligations are also referred to as PATERNALISM – strictly, control over children but, like the term patriarchy, also used to refer to assumptions that women must also be treated as less than adult or as children.   LM

Fatherland See NATION/NATION-STATE/NATIONALISM

Fear of crime Fear of crime covers the wide range of emotional and practical responses to crime of individuals and communities. It is particularly associated with those negative reactions to the prospect of becoming a victim which alter individuals' behaviour and are damaging to well-being, quality of life and opportunity. There is little consensus on how to define or measure fear, but increasingly it is viewed not as a fixed trait which some people have and some do not, but as situational and dynamic. Fear varies according to type of crime, violence provoking more anxiety than PROPERTY crime, and is strongly influenced by spatial, temporal and social context.

Fear is situated in the wider arena of local COMMUNITIES and embedded in the life course and life experiences. It is concentrated in particular places and amongst certain social groups, with GENDER differences being the most prominent dimension. Feminists have provided a strong critique of mainstream explanations for high levels of fear amongst women, which blamed women's physical vulnerability. Instead, feminists relate fear to high rates of hidden violence and harassment, women's 'tacit understanding of risk' and general insecurity (Stanko, 1987). For geographers, women's fear of rape in urban spaces is symbolic of their exclusion from PUBLIC life. Valentine (1989) highlighted the 'spatial paradox' of fear, arguing that women's fear constitutes a 'spatial expression of PATRIARCHY', effectively controlling women through behavioural restrictions imposed by self and others. Fear is also a feature of PRIVATE space for some women (Pain, 1997a); in this setting vulnerability is reinforced by lack of protection from wider society and the state.

Other forms of difference such as CLASS, 'RACE', sexual orientation, age and ability also structure vulnerability to fear (Pain, 1997b and c). Increasingly, parents' fears affect children's geographies, over which parental control is justified by the cultural image of children as 'vulnerable and incompetent in public space' (Valentine, 1996). Formal fear-reduction strategies for women and older people are equally paternalistic. Recent research has questioned men's fear, ostensibly much less widespread than women's. The strong culture of HETEROSEXUAL MASCULINITY shapes men's risk of victimisation, their fearfulness, associated spatial constraints and the likelihood of reporting any of these three (Goodey, 1997).

Sexist and ageist assumptions affect constructions of the fearful. There are concerns that emphasis on women's fear and women as fearful personalities reinforces rather than challenges their 'victim status'. A recent approach stresses the many positive ways in which women and older people deal with the threat of crime, emphasising spatial confidence as well as intimidation (Koskela, 1997; Pain, 1997c). RHP
See also CRIME; STRANGER DANGER; VIOLENCE.

Female English-speaking feminists of the 1970s confidently proclaimed that 'female' refers to the purely biological aspects of sexual difference in contrast to 'FEMININITY', which is a social and cultural construction. The distinction was intended to challenge the deliberate merging of these terms in PATRIARCHAL thought. In French the distinction has always been more difficult to make since there are not separate words for 'female' and 'feminine'. The term 'female' has not functioned as a neutral description of sexual DIFFERENCE. Although the term has no etymological connection with male, it has typically been treated as the BINARY OPPOSITE of male.

Feminists have pointed out that it is in fact 'femaleness' rather than femininity that has been consistently degraded in our culture. Battersby (1994), for example, demonstrated that femininity, when attached to males, may contribute to the spark of genius. She suggests that the pairs male/masculinity and female/femininity are not analogous for, whereas 'femaleness' may refer purely to biological characteristics, 'maleness' does not. It already refers in addition to qualities such as autonomy and independence.

'Female' therefore refers to much more than the raw material on which GENDER as a social and cultural construction is based. The notion of there being two sexes, of opposite polarity, is of recent historical origin, and the social and political significance of anatomical difference has its origins in the social, not the natural. Lacquer (1990) has charted the social contexts in which these taken-for-granted categories emerged. While genital differences have never been ignored, self-definition as male or female did not always rest on possession of a PENIS or vagina. In the medieval period, males and females were seen as possessing the same fundamental constituents. The only difference was that in males these organs were external, and in females internal. The lack of externality of female sex organs was not an indicator of distinction but of incompleteness.

Recognising that 'female' cannot be a simple referent, feminists have increasingly turned their attention back from femininity to the exploration of alternative meanings for femaleness. This takes a number of forms, ranging from the celebration of the female body and its capacities, to deconstructing the current meanings of 'female' in patriarchal thought. It has been suggested, for example, suggests that the social construction of 'femaleness' has contributed to the reading of female bodies in ways that are hidden by the LANGUAGE of AUTHORITY and seem to be irrelevant to it. Grosz (1987) has argued that sexual difference will always have to be represented in any

CULTURE and that the female body is not a *tabula rasa* on which anything can be written. The body cannot be approached merely as an external object, or an object of BIOLOGY, but must be understood from the point of view of its being lived or experienced by the SUBJECT. 'Female' processes like ovulation, menstruation, childbirth and lactation have always been inscribed in patriarchal terms and analysed only according to men's interests. Women's bodies are reduced to biology, which in turn presumes them to be passive. To be 'female' is not merely to be of the sex that produces ova or offspring, and neither is it to be the binary opposite of 'male'. Rather it is to inhabit a certain kind of BODY whose potentialities still need to be explored.

The term 'female' is therefore used not so much to differentiate the biological from the social but to insist that the body itself is an extra-biological phenomenon.   RP

See also BODY; CORPOREALITY.

Femininity/feminine That set of social characteristics that is associated with those who are biologically defined as women, as female or who identify themselves as such. Femininity is usually defined in opposition and as inferior to masculinity and as such is a BINARY distinction which is assumed by some to have an ESSENTIAL basis in a CATEGORICAL distinction between men and women based on biological differences. Feminist scholarship has long challenged this essentialist assumption, as well as documenting the socially and spatially variable construction of feminine and masculine traits. However, the assumption of a categorical difference between women and men is deeply embedded in our sense of ourselves as individuals, in daily interactions, in institutional structures and in Western intellectual thought. Despite a growing recognition of the plurality and diversity of social experiences, the belief that a distinctive version of femininity for women and MASCULINITY for men is appropriate remains extremely powerful. As Doreen Massey (1995) has argued

> deeply internalised dualisms ... structure personal identities and daily lives, which have effects upon the lives of others through structuring the operation of social relations and social dynamics, and which derive their masculine/feminine coding from the deep socio-philosophical underpinnings of western society (1995b: 492).

The belief in categorical difference also influences the production of spatial divisions, assumptions about the 'natural' and built environments and sets of regulations about who should occupy which spaces and who should be excluded.

A list, like the one below, of binary distinctions which are gendered is no doubt a familiar one:

The masculine	The feminine
Public	Private (domestic and subjective)
Outside	Inside
Work	Leisure

Rational	Emotional
Earning	Spending
Production	Consumption
Empowered	Disempowered
Freedom	Constraint

As the list shows, femininity is associated with the PRIVATE sphere or domestic arena. In urban industrial societies, the development of a DOMESTIC IDEOLOGY, in which it was assumed that women's role in life was to become what Virginia Woolf termed 'the angel in the house', has been a potent force in confining middle-class women to the household and constructing the workplace and the wider PUBLIC arena as a masculine sphere. One of the specific aims of a FEMINIST GEOGRAPHY is to investigate, make visible and challenge these relationships between masculinity and femininity, gender divisions and spatial differentiation and behaviour, to uncover their mutual constitution and problematise their apparent naturalness. Thus the ways in which spaces are coded as masculine or feminine and how this affects and reflects their occupation have been investigated in a wide range of circumstances.

As well as arguments connecting the construction of femininity with material divisions, feminists influenced by psychoanalytic theory argue that femininity is also constructed through the UNCONSCIOUS, in which women enter the SYMBOLIC ORDER. Whereas Lacanian-influenced psychoanalyis regards 'normal' femininity as lack, this is challenged by FRENCH FEMINISTS who seek to inscribe a more positive version of femininity through the practice of what is termed ÉCRITURE FEMININE. Other feminists argue that certain aspects of FREUDIAN ANALYSIS offer valuable insights. Jacqueline Rose, for example, while disputing Freud's claim that women achieve entry into the PHALLIC order, believes that his notion may be used to explain the contradictions women face in contemporary societies and suggests that 'femininity is neither simply achieved not is it ever complete' (1987: 7). LM

Feminisation of poverty See POVERTY

Feminisation of the work force The way in which increasing numbers of women have entered the labour market and in which jobs themselves are becoming more feminised (see Jenson *et al.*, 1988). In the United Kingdom women now make up almost half of the labour force although the majority of these women work part time. Although there are significant variations between nation-states the increase in the numbers of women entering waged employment seems to be a widespread phenomenon, with the possible exception of some of the former communist states where women have withdrawn from the labour market in the 1990s.   JW

See also CASUALISATION; LABOUR: DOMESTIC, SWEATED, WAGED, GREEN, SURPLUS; WORK/WORK FORCE/EMPLOYMENT.

Feminism Feminism is both a theoretical and a political movement. The key or general aim of feminist scholarship is to demonstrate the construction and significance of sexual differentiation as a key organising principle and axis of social POWER, as well as a crucial part of the constitution of subjectivity, of an individual's sense of their self-identity as a sexed and gendered person. The aim of feminism as a political movement is to challenge and dismantle the inequalities between men and women and, as such, it has a long history, dating from the early nineteenth century in Europe when individuals, influenced, in part, by Mary Wollstonecraft's (1792) powerful plea for the *Vindication of the Rights of Woman*, began to advocate EQUALITY for women. As a political movement, feminism has had to struggle from this time with the contradictions that lie in claims for equal treatment and claims for the recognition of women's DIFFERENCE from men and their special needs in, for example, the rights to maternity provisions (see Philips, 1987). The claim for equality is related to the emergence of the European ENLIGHTENMENT and, because of this association, at various times in different places, women struggling for recognition have refused the term feminism. As Andermahr *et al.* (1997) have noted Alexandra Kollontai, who headed the women's section of the Soviet Communist Party in the early 1920s and was responsible for introducing wide-ranging reforms aimed at women's EMANCIPATION, rejected the term because of what she considered to be its contamination by bourgeois liberal women's movements in Europe and the USA (Kollontai, 1977). Some members of the Indian women's movement that emerged as part of Indian NATIONALISM disputed the label on the grounds of *anti*-imperialism (Forbes, 1990) and this argument still carries some weight among some THIRD WORLD and BLACK women (Kishwar, 1990: 93).

In spite of these reservations, feminism is a widespread and growing movement in almost all countries of the world, under which label a wide variety of struggles and social and cultural movements have been grouped. In a an edited collection, the feminist art critic, Griselda Pollock defined feminism as follows:

> Feminism here stands for a political commitment to women and to changes that women desire for themselves and for the world. Feminism stands for a commitment to the full appreciation of what women inscribe, articulate, and image in cultural forms: interventions in the field of meaning and IDENTITY from the place called 'WOMAN' or the 'FEMININE'. Feminism also refers to a theoretical revolution in the ways in which terms such as art, culture, woman, subjectivity, politics and so forth are understood (1996: xv).

Pollock's definition, not surprisingly, reflects her training as an art historian. For geographers and social scientists a phrase about the social constitution of everyday behaviours and political actions, as well as the structural or institutional definition of gender – the material as well as representational interventions – might be added to this definition, which otherwise is helpful. Pollock also added a significant rider to her definition: 'feminism does not imply a united field of theory, political position, or perspective' (1996: xv).

Thus, many feminist scholars tend to refer, in the 1990s, to feminisms in the plural, rather than a singular feminism. But even from its resurgence in the 1960s there have been theoretical differences within feminist scholarship and different strands within the WOMEN'S MOVEMENT. Thus RADICAL and SOCIALIST FEMINISTS in Britain in the 1970s disagreed about the basis of women's OPPRESSION, the former emphasising patriarchal dominance as a global phenomenon, the latter linking specific forms of PATRIARCHY to CAPITALISM, in a dual or single systems model. There are also differing emphases on women-only political organisations and on joint struggles against other forms of oppression, especially RACISM and imperialism.

As a theoretical enterprise feminists have engaged in a radical rethinking of the nature of social explanations, not only challenging the GENDER-BLINDNESS of many disciplines but also demonstrating the ways in which knowledge itself, as well as the structure of social institutions, is deeply gendered. Thus the universalist and masculinist assumptions of Western knowledge have been challenged, through, for example, the deconstruction of the dichotomous definition of MASCULINITY and femininity and their asociation with a set of BINARIES such as public–private, mind–body, and culture–nature that construct women as opposite and inferior to men.

There is a difference of opinion at present in the extent to which the arguments of feminism are now widely accepted (Coward, 1998) and the extent to which there is a backlash against them (Faludi, 1992; Oakley and Mitchell, 1997). In Britain at the turn of the twentieth century, young women are gaining eduactional credentials at an accelerating rate and there is a widespread debate about boys who are underachieving. However, there is also evidence of the development of a msyogynistic movement, sometimes referred to as new LADDISM, which continues to denigrate women, especially through sexist REPRESENTATIONS.   LM

See also BLACK FEMINISM; FEMINIST GEOGRAPHY; SECOND WAVE FEMINISM.

Feminist anthropology Because feminist anthropology has significantly contributed to the recent critical and linguistic turn in cultural anthropology, it is often dated to the 1970s when it was institutionally recognised as a distinct subfield (Lamphere, 1996). This recent date, however, underemphasises the role of women in the history of anthropology, FEMINIST METHODOLOGY, and the historical uses of GENDER and SEX as analytical categories. Visweswaran (1997), for example, describes four periods in feminist ETHNOGRAPHY:

- 1880–1920, biological sex was seen as determining social roles (gender was beginning to emerge as an analytical category but was not seen as separable from sex) (Parsons, 1916);
- 1920–60, sex was no longer viewed as determining gender roles, and sex and gender were viewed as distinct analytical categories (Mead, 1935);
- 1960–80, the idea of a sex/gender system developed to describe how different societies gendered biological facts (Rubin, 1975); and

- 1980–97, critics of the sex/gender distinction challenge the idea of 'woman' as a biologically defined, UNIVERSAL category, and redefine sex as a social construct (Butler, 1990) (for an historical overview, see anthropology).

Feminist research of the 1970s and early 1980s either focused on women as a topic of study or criticised the ANDROCENTRIC bias of the discipline (Ardner, 1975; Caplan and Bujra, 1979; Martin and Voorhies, 1975; Reitner, 1975; Rosaldo and Lamphere, 1974). Many of these works continued the long tradition in anthropology of examining human universals across cultures, such as Rosaldo and Lamphere's volume, that explored the universal subordination of women across cultures. More recently, feminist anthropologists have challenged the gender ESSENTIALISM of Western European and English-speaking feminists, the Western hierarchies of NATURE–CULTURE and PRIVATE–PUBLIC and the assumptions that gender is the centre of all women's experiences (Braidotti and Butler, 1995; Martin, 1994). Foucauldians argue that gender is the 'discursive origin of sex', not vice versa, to challenge the earlier focus in the field on sex difference (Morris, R., 1995: 568–9).

By the 1980s and early 1990s, feminist anthropological research focused on production and work (Ong, 1991; Romero and Higgenbotham, 1995); reproduction and SEXUALITY (Ginsburg and Rapp, 1995; Lewin, 1993); gender and the STATE (Gailey, 1987); and the history of marginalisation of women in anthropology (Parezo, 1993). More generally, feminist research methods explore the positionality of the anthropologist as ethnographer and have proposed strategies to deal with multiple positioning (Behar and Gordon, 1995; Visweswaran, 1994). Feminists argue that researchers should promote a dialogue between, rather than an 'objective' distancing from, themselves and the women and peoples they study through community activism, play writing, performance ethnography, and writing multivocal texts (Brettell and Sargeant, 1992; di Leonardo, 1991; Jones, J., 1996; Kondo, 1995; Moore, 1988; Morgen, 1989). Recent work in feminist ethnography also attempts to displace gender as a universal category to engage in strategies of disidentification with and/or as women (Visweswaran, 1997).   KT

Feminist geography Feminist work within geography dates from sometime in the 1970s, influenced by the resurgence of interest in feminist issues by the SECOND WAVE FEMINIST MOVEMENT that developed from the late 1960s onwards. The key aim of feminist scholarship in general is to demonstrate the construction and significance of sexual differentiation as a key organising principle and axis of social power, as well as a crucial part of the constitution of subjectivity. Although the emphases of feminist geographers have changed over the years, it is possible to identify a constant and particular aim or focus of feminist geography (or geographies):

> to demonstrate the ways in which hierarchical GENDER relations are both affected by and reflected in the spatial structure of societies, as well as in the theories that purport to explain the relationships and the methods used to investigate them (McDowell and Sharp, 1997: 4).

In this work a broad shift of emphasis is evident, from a dominant interest in material inequalities between men and women in different parts of the world in the earliest publications to a focus on the significance of LANGUAGE, symbolism, REPRESENTATIONS and meaning in the definition of gender and a greater emphasis on the spatial constitution of subjectivity, IDENTITY and the sexed BODY. However, these two aspects of gender difference – gender as defined by and constructed through material social relations and gendered symbolic meanings – cannot really be separated, because social practices, including social interactions, and ways of thinking about and representing place and gender are interconnected and mutually constituted. People act relative to their intentions and beliefs, which are always culturally shaped and historically and spatially positioned. These actions in turn have an effect on beliefs and future intentions, on people's knowledge of and understanding of the world and their place in it. Thus what people believe to be appropriate behaviour and actions by men and women reflect and affect what they imagine a man or a woman to be. So spatial location matters.

It is argued that:

> spatial relations and layout, the differences within and between places, the nature and form of the BUILT ENVIRONMENT, images and representations of this environment and of the 'natural' world, ways of writing about it, as well as our bodily place within it, are all part and parcel of the social constitution of gendered social relations and the structure and meaning of PLACE. The spaces in which social relations occur affect the nature of those practices, who is 'in place' and who is 'out of place' and even who is allowed to be there at all. But the spaces themselves in turn are constructed and given meaning through the social practices that define men and women as different and unequal (McDowell and Sharp, 1997: 2–3).

Thus physical and social boundaries reinforce gendered social relations that are unequal. It is also possible, however, to challenge hegemonic gender relations through transgressive occupations of spaces and through the construction and maintenance of oppositional spaces; even to create what the feminist writer and critic bell hooks has termed 'spaces of radical openness' (1991a: 148), challenging conventional power relations from marginal locations.

The specific aim of a feminist geography, therefore, is to investigate, make visible and challenge the relationships between gender divisions and spatial divisions, to uncover their mutual constitution and problematise their apparent naturalness. Thus the purpose is to examine the extent to which women and men experience spaces and places differently and to show how these differences themselves are part of the social constitution of gender as well as that of place. In a common-sense way, there is a clear geography to gender relations as there are enormous variations between and within nations in the extent of women's subordination and relative autonomy and, correspondingly, in male power and domination as well as an evident multiplicity in the social construction of gender, gender divisions and in the symbolic meanings associated with FEMININITY and MASCULINITY. Constructing a geography or geographies of gender, as Griselda Pollock has noted 'calls attention to the significance of place, location and cultural diversity, connecting issues of sexuality to those of NATIONALITY, IMPERIALISM, MIGRATION, DIASPORA and genocide' (1996: xii). But gender relations are also of central concern for geographers because of the way in which a spatial division – that between the PUBLIC and the PRIVATE, between inside and outside – plays such a central role in the social construction of gender divisions. The idea that women have a particular place is the basis not only of the social organisation of a whole range of institutions from the family to the workplace, from the mall to political institutions, but also is an essential feature of Western ENLIGHTENMENT thought, the structure and division of knowledge and the subjects that might be studied within these divisions.

Feminist geography is, however, distinguished by another feature: that is, a commitment to change. Not only is it scholarship that focuses on revealing the ways in which gender relations are structural relations of power and inequality but also it is scholarship 'for feminists', work that embodies a progressive and more equal view of the future. Feminism is, therefore, associated with the aims of the WOMEN'S MOVEMENT, which, like the theoretical and methodological shifts since the 1970s, have also changed in emphasis over the years, from, for example, a predominant interest in women's position in the labour force to issues about SEXUALITY and reproduction and representations in different media.

So feminism within geography is the place where the question of gender is posed, and where too the idea that the academy should also address political questions about who is represented within its walls, both as scholars and as subjects of scholarship, and a number of geographers working in the academy in different nation-states have documented women's unequal position. This commitment to change has often been used by critics to accuse feminist geographers of politically motivated work rather than 'unbiased' scholarship, but the wider acceptance of recent arguments about the situated or positioned nature of all academic products has reduced the force of this charge.

It has, nevertheless, been a long struggle to gain recognition within geography as a discipline that gender relations are a central organising feature both of the material and symbolic worlds and of the theoretical basis of the discipline. It is also too often assumed that either feminist scholarship is about or 'for' women, perpetuating the mistake that men are an ungendered 'norm' – One to women's gendered 'OTHER' – that feminist geographers have a sole focus on gender to the exclusion of other axes of subject constitution and of discrimination. This assumption, however, is a misguided one. As Griselda Pollock noted:

> Feminism is not for gender what MARXISM is for class, and POSTCOLONIAL THEORY for race. First, there is a range of feminisms, in varying alliances with all the analyses of what oppresses women. SOCIALIST FEMINISM has always concerned itself with matters of class, and BLACK FEMINISM details the configurations of imperialism,

sexuality, femininity and RACISM. In their breadth, as the plural, feminisms deal with the complex and textured configurations of power around race, class, sexuality, age, physical ability and so forth, but they have of necessity also to be the particular political and theoretical space that names and anatomises sexual difference as an axis of power operating with a specificity that neither gives it priority, exclusivity or predominance over any other nor allows it to be conceptually isolated from the textures of social power and resistance that constitute the social (1996: 3–4).

This shift to analyses of the mutual constitution of social relations has been an important part of more recent feminist geographies, while not denying the continuing centrality of gender both as an axis of power and a key part of subjectivity.

There are many other places to look for a comprehensive definition of feminist geography and for a history of changing perspectives (Bondi, 1990a, 1991; Duncan, 1996a; Jones III, *et al.* 1997; McDowell, 1992 and b, 1998; McDowell and Sharp, 1997; Massey, 1994a; Pratt, 1993; Rose, 1993a; WGSG, 1984, 1997).   LM

Feminist methodology In recent years there has been an animated and somewhat contentious debate among feminist social scientists about whether there is a universally accepted feminist research method. Although there is no *single* feminist method this diverse academic community is searching for techniques which are consistent with their feminist convictions (DeVault, 1996; McDowell, 1992c). For this reason, it important to distinguish between methods, as tools for research, and methodology, as theory about the research process. Feminist innovation regarding the research process has focused on the EPISTEMOLOGICAL realm, rather than on the invention of new methods; however, feminist critiques of research practices have modified established methods such as participant observation and other ETHNOGRAPHIC methods (DeVault, 1996; McDowell, 1992a). DeVault argues that feminist methodology is distinguished by a commitment to three goals:

1. Feminists seek a methodology that will do the work of 'excavation', shifting the focus of standard practice from men's concern in order to reveal the locations and perspectives of (all) women.
2. Feminists seek a science that minimizes harm and control in the research process.
3. Feminists seek a methodology that will support research of value to women, leading to social change or action beneficial to women (1996: 32–3).

These goals might imply that feminist fieldwork is necessarily ethnographic. It is true that historically feminists indicted positivist methods as part of the patriarchal production of knowledge. Consequently, feminists argued for alternative methods – most specifically, opened-ended interviewing and ethnography. These methods are dependent on an 'intersubjective understanding between the researcher and the person studied' (Reinharz, 1992: 46), therefore they are most appropriate for a feminist methodology which seeks out the everyday experiences of women.

Many feminists rely heavily on qualitative research methods; however, feminist fieldwork may also involve archival research, extensive surveys and can also be enhanced with secondary data analysis (Staeheli and Lawson, 1994). Gender relations has become a multidimensional area of research, thereby generating many types of question. Staeheli and Lawson (1994) argue that some questions are more appropriately addressed with qualitative techniques, while others may be better answered with quantitative techniques. The place-specific nature of ethnography can ignore non-local forces that operate at many different scales, such as 'the BODY, HOME, COMMUNITIES, NATIONS' and 'international political economies' (Staeheli and Lawson, 1994: 98). Many feminists now advocate combining qualitative and quantitative research tools, often through collaboration with other researchers. Collaboration both gives breath to a qualitative study and adds the depth of a feminist inquiry to the quantitative tradition.   LD
See also EPISTEMIC VIOLENCE; FIELDWORK.

Fertility The reproduction performance of a woman (or couple, group or population) measured in terms of a fertility rate which is the number of live births per 1000 women aged between 15–44 years in a given year (note that definitions of fecund ages vary between 15–44 and 15–49 years). Fertility is different from fecundity, which is the theoretical capacity to reproduce. Fertility analyses may specifically analyse the 'marital fertility rate' or the 'age-specific fertility rate'. Together with mortality and migration, fertility has a significant impact on the structure and size of a population; however, it is poorly understood and projections are hence problematic. Specific variables that impact on fertility are rates of marriage, ABORTIONS, CONTRACEPTION and post-birth infertility, but it is accepted that fertility is largely affected by socio-economic circumstances such as education, the status of women, POVERTY, employment, urbanisation, increased individual control of fertility decisions and lower infant mortality. In this regard it is obvious that fertility rates differ spatially and are also gendered. Fertility rates are generally higher in Third World countries, but there is much differentiation between these countries. Africa and the Middle East have the highest rates of fertility (both above 5) whereas in the Far East fertility rate declined to 2.9 in the 1980s (as a result of DEMOGRAPHIC POLICIES). Intra-regional spatial variations in fertility are notably discernible between rural and urban areas. Feminists understand that fertility is gendered because of the close association of women's status to fertility rates.   PM

Fetishism Implies the masking of social relations. It is most commonly used with regard to commodities (the term 'commodity fetishism' was orginally used by Marx), whereby the labour embodied in a product and the social relations of which it is a part are masked by the monetary exchange of commodities in the marketplace. Feminists have also used the term to refer to the way in which LESBIAN and GAY iconography is adopted by mainstream culture, with its orgins masked (see Clark, 1993). In addition, the term is used in PSYCHOANALYTIC theory to describe the way in which sexual satisfaction can be realised through particular objects. Such fetishism has always been linked to MASCULINITY, and it is only recently that feminists have begun to consider female fetishism.   JW

Fieldwork A means of collecting primary data by way of observation. Historically, this work included quantitative techniques such as questionnaires and modelling as well as qualitative approaches such as ETHNOGRAPHY and participant observation. In the 1990s feminists have challenged traditional approaches to fieldwork as part of a PATRIARCHAL production of knowledge, consequently favouring qualitative methods over quantitative techniques. However, feminist scholars have recently come to accept that quantitative methods have a place in feminist research and interestingly now argue there is a power relationship inherent in fieldwork. Heidi Nast maintains that we need to 'question how relational qualities between the researched and researcher inform research agendas and knowledge claims' (1994: 54). As part of this critique feminists question the very notion of *the field*. Cindi Katz argues that the boundaries which constitute the field are exclusionary, with the researcher defining who will be ruled in or out. Most importantly, feminist researchers have questioned the dichotomous role of the researcher as insider–outsider and whether they are able to truly represent the experiences of those they have researched. Therefore, feminists now argue that in every academic endeavour researchers need to be aware of their POSITIONALITY. As Cindi Katz contends 'we are always already in the field – multiple positioned actors, aware of the partiality of all our stories and the article of the boundaries drawn in order to tell them' (1994: 67).   LD
See also EPISTEMIC VIOLENCE; FEMINIST METHODOLOGY.

First wave women's movement See SUFFRAGE/SUFFRAGE CAMPAIGNS

First World See WORLD: FIRST, SECOND, THIRD, FOURTH

Flâneur/flâneuse The *flâneur* is associated with activities of strolling and observing, especially in the arcades and streets of nineteenth-century Paris. But the figure has recently roamed further afield to feature prominently in wider discussions of urban experience, the CITY and MODERNITY. References to the *flâneur* appeared as early as 1806, and Baudelaire later celebrated the figure's 'heroism' as a man of the crowd and a spectator. The figure's cultural significance was particularly explored by Walter Benjamin who traced its histories and ambiguities in his investigations of nineteenth-century Paris. Feminist criticism has generated considerable debate about the gendered activity of *flânerie*. It has been argued that the *flâneur's* freedom to wander was essentially MASCULINE since the ideologies and sexual divisions of the nineteenth century restricted the movements of unaccompanied 'respectable' women (Wolff, 1985), and the practice itself was structured around a male GAZE (Pollock, 1988). An equivalent female *flâneuse* was thus rendered impossible. But Elizabeth Wilson (1991) emphasises the increasing emergence of women into PUBLIC spaces towards the end of the nineteenth century and the new opportunities for wandering opened up by department stores and other urban sites. She further stresses the insecurity and MARGIN-ALITY of the *flâneur*, and suggests that he might be reinterpreted as 'a shift-ing projection of angst rather than a solid embodiment of male bourgeois power' (109). Other critics have explored paths of female *flânerie* as regis-tered in fictions of cities by women such as Virginia Woolf and Dorothy Richardson, as suggested in the mobile gaze made available by the develop-ment of early cinema (Bruno, 1993; Gleber, 1997) and by 'machines of virtual transport' (Friedberg, 1993) and as performed in relation to contemporary urban spaces (Munt, 1993).   DP

Flexible/flexibility Notions of flexibility have framed debates about the trajec-tory and nature of contemporary ECONOMIC RESTRUCTURING since the 1980s. There is now a substantial literature on the transition from FORDISM, a liter-ature which has emphasised the reconfiguration of industrial organisation through the implementation of new production techniques, changes to the labour process and the reworking of labour relations. Debates have tended to focus upon either the emergence of a new, more 'flexible' era or (at a different level of analysis) the growth of more 'flexible' working practices (see Macdonald, 1991). Initially, the flexibility literature appeared highly attractive to the extent that it sought to draw out the novelty of changes to production and working arrangements. However, in so doing the imagined coherence of 'flexibility' ultimately has undermined its utility as a conceptual framework: it is a term which often 'has been invoked to explain too much' (Allen and Henry, 1997: 193).

Flexibility is perhaps most problematic when used to describe the chang-ing experience of work and employment in the late twentieth century. Although increases in part-time, temporary, subcontracted and casualised working arrangements may provide flexibility for employers, they certainly do not represent flexibility for employees. Part-time jobs, for example, often involve unpaid overtime working, and employees have little choice about the number or the type (e.g. late-night shifts) of hours that they work. It is also difficult to interpret employers' demands for increasing 'flexibility' from full-time workers in a positive light, particularly in cases where jobs place 'extremely long and unpredictable demands on the time-spatiality of life' (Henry and Massey, 1995: 61). For the partners of elite 'flexible' workers too, the implications of extended working hours are significant: 'the expectation that [male scientist-engineers] should be highly flexible in the organisation of their domestic environments turns itself into a *constraint* upon the lives of female partners' (Massey and Allen, 1995: 129).   SR

Fordism Fordism can be viewed either as a labour process, involving a moving ASSEMBLY LINE and mass-production processes, or (in the language of the French REGULATION SCHOOL) as a mode of capitalist regulation. Fordism as a labour process involved the decomposition of work tasks, the specialisation of tools, and the transformation of craft production into large-scale industrial production (see Beynon, 1984). As a 'total way of life', Fordism can be seen to refer to a particular historical era, to 'the long post-war boom, from 1945 to 1973, [which] was built upon a certain set of labour control practices, technological mixes, consumption habits, and configurations of political–economic power' (Harvey, 1989: 124). It is commonly argued that the Fordist era involved a particular configuration of relations between capital, organised labour and the state. More specifically, Fordism rested upon a particular 'gender order', a 'compact between male workers, industrial capital and the institutions of welfare Keynesianism' (McDowell, 1991b). This is not to say that (*contra* the theoretical arguments of the Regulation School) women's participation in the Fordist LABOUR MARKET was always and everywhere 'secondary', emerging *solely* through their role in the nuclear FAMILY (McDowell, 1991b). In a discussion of women assembly workers in inter-war Britain, for example, Glucksmann argues that '[these] women were the pilots and pioneers of the new relation between capital and labour which was integral to mass production ... they were not marginal, peripheral or secondary but central' (1990: 4). SR

Fourth World See WORLD: FIRST, SECOND, THIRD, FOURTH

French feminist theory This body of work has come to be identified with feminist theorists who have engaged, primarily, with the works of Lacan and Derrida. Although there are many writers who might be included, for anglocentric scholars three key theorists – Hélène Cixous, Luce Irigaray and Julia Kristeva – are most commonly referred to and have tended to come to stand for the entire body of work. One consequence is that the differences between these theorists tend to be ignored, and in general becauses of their reliance on psychoanalytic and linguistic theories their work is often regarded as difficult by anglophone feminists.

Their work was first introduced to British and American feminists by the collection *New French Feminisms* edited by Elaine Marks and Isabel de Courtivron in 1981. Here, the construction of a feminine subject position – that of WOMAN – through language acquisition is a uniting feature of the contributions. Dismissed as ESSENTIALIST by many feminists, especially socialist feminists (see for example Moi, 1987), in the 1990s interest in the poetics of women's writing has led to a revival of interest, especially in the work of Irigaray. Among geographers, the 'cultural turn' to questions about REPRESENTATION and symbolic meaning has led to some exploration of the work of Irigaray, Kristeva and Cixous (Rose, 1995b; 1996a, Shurmer-Smith, 1994). LM
See also JOUISSANCE; LANGUAGE.

Freudian theory Psychoanalytic theory and practice are both widely understood to have been founded by Sigmund Freud (1856–1939). His published writings span a period of nearly 50 years. Freud changed his views in significant ways during his life so that the full body of his work does not provide a coherent body of theory. However, in the decades since his death, his later writings have come to be viewed as constituting 'Freudian theory'. As well as this 'glossing' of Freud's ideas, it is worth remembering that in English-speaking parts of the world most people engage directly or indirectly with the *Standard Edition*, which was translated from the original German in the 1950s and 1960s.

Through his publications, Freud sought to establish PSYCHOANALYSIS within the field of medical science, both as a clinical practice and as a body of scientific theory concerned with the workings of the human psyche, which he framed largely in terms of human development. Freud's writings present an ambivalent view of the relationship between BIOLOGY and psychic health. On the one hand, he asserted that 'biology is destiny', and so propounded the view that psychic health can only be achieved within conventional HETERO-SEXUAL roles and relations. On the other hand, he conveyed the idea that 'NORMAL' gender/sexual identities are difficult psychical achievements, and that biological sex is only very problematically related to cultural norms (also see IDENTITY; GENDER).

Two crucial elements in Freud's thinking and distinguishing features of psychoanalytic theory concern the UNCONSCIOUS and SEXUALITY. Freud did not invent the concept of the unconscious but he did open up ways in which unconscious aspects of the human mind could be explored and theorised. He argued that these aspects could be glimpsed through slips of the tongue, through neurotic symptoms, through the clinical method of 'free association' and, above all, through dreams. He viewed the unconscious as radically different from the conscious mind and detached from ordinary reality. 'Real' time and logic are suspended. Unconscious mental processes are characterised by DISPLACEMENT, exemplified by puns, where one word replaces another, and by condensation, in which one SYMBOL expresses many ideas. According to Freud the unconscious includes a variety of forgotten and instinctual experiences, some of which can easily be retrieved (those constituting the 'pre-conscious') but most of which cannot, including those that are REPRESSED because they are not acceptable to the conscious mind. He argued that psychoanalytic treatment works by enabling elements of the repressed unconscious that produce neurotic symptoms to be brought at least partially into consciousness.

According to Freud, many of the wishes and DESIRES that constitute the repressed unconscious relate to sexuality. The significance he attached to sexuality is connected with his idea that one of the key ways in which human biology makes itself felt is through what he calls the 'drives' (in the German original *Trieb* but in the *Standard Edition* translated as 'instinct'). The principle drive Freud identified was a sexual one, which he termed the LIBIDO. He argued that sexuality is central to infantile experience, and that neurotic symptoms can be understood in terms of the repression of infantile sexual wishes and desires. And so from clinical evidence derived from his adult patients he developed a theory of human psychological development in which sexuality is accorded decisive significance.

For Freud, during the earliest phases of human development sexual life is polymorphous: erotic pleasures may be derived from many different parts of the body. These early phases have been explored more fully by psychoanalytic theorists subsequent to Freud, most especially within OBJECT RELATIONS THEORY. But one of Freud's most influential legacies concerns what he called the OEDIPUS complex. Freud conceptualised the complex in terms of a crucial stage through which the child takes up a position in relation to its mother and its father and during which sexuality becomes organised in a particular way. As well as operating as a crucible in the formation of gender and sexual identity it is the moment at which the child enters into three-person relationships, thereby symbolising a key step towards integration of the individual within the wider society.

Within the Freudian theoretical framework, successful resolution of the oedipus complex is what enables little boys to become, in due course, heterosexual men, and little girls to become heterosexual women. But in naming this process after the tragic myth of Oedipus, Freud drew attention to the psychical costs entailed both in relinquishing incestuous infantile desires and in replacing a disorganised eroticism with the organised structures of conventional gender and sexual identities.

In developing his account, Freud began with what he took to be the MASCULINE experience. He argued that at a crucial point the little boy becomes aware of the (VISUAL) difference between the genitals of his mother and his father. Freud postulates that the little boy 'discovers' that, unlike himself and unlike his father, his mother has no PENIS. And he interprets her state as castrated. He then fears that he too might be castrated. (Freud links certain responses to the initial discovery of the mother's lack of a penis, particularly denial or disavowal, to the development of FETISHISM.) For Freud the boy's CASTRATION ANXIETY is inextricably tied up with the taboo against incest: in realising that, just like his father, he has a penis, whereas his mother does not, the little boy becomes positioned as competitor for his mother's love. But his father has the power to castrate him and so he comes to accept that he must redirect his sexual desires to women other than his mother. The oedipal phase is the period during which competition with the father for the mother is played out. Its resolution is marked by the little boy's acceptance that his identification with his father will in due course enable him to take up the position of adult heterosexual man in relation to a woman other than his mother.

Within the Freudian account it is at this stage in development that the little girl's experience first diverges from that of the little boy. Prior to the oedipal phase she too is a 'little man'. In this way, Freud represented FEMININITY as derived from the masculine position and as a deviation from a supposed 'human' norm. He postulated that when the little girl recognises genital difference she responds by immediately developing PENIS ENVY. In other words, she recognises not only difference but inferiority, and so she internalises something of what Freud took to be a universal contempt for those without penises. Under such circumstances her attachment to her mother gradually shifts to an (initially rivalrous) identification with her: being like her mother holds out the possibility that she might 'acquire' a penis through adult heterosexual relations with a man. But, according to Freud, she also knows that she cannot acquire a penis in this way and instead she comes to long for a baby as a substitute for the penis.

Freud's ideas have circulated very widely indeed. Via their impact on Western intellectual thought, medical practice and the development of the 'psy' disciplines, many aspects of Freudian theory have become a part of everyday talk in many cultural contexts. For example the 'Freudian slip' indicates how Freud's ideas about the unconscious and about the influence of sexuality are taken up in common parlance. As a consequence of this, Freudian theory has inevitably made itself felt within human geography. However, this influence has generally been indirect, attenuated and implicit, direct references to Freud's writings remaining relatively rare (for exceptions see Pile, 1996; Smith, 1980). The same is true of a good deal of feminist geography. But to understand the overall impact of Freudian theory on feminist geography it is important to review its reception within feminism.

Freudian theory has provoked diverse responses among feminists. During the first half of the twentieth century many women took up Freud's ideas and trained as psychoanalysts. Some of these women, notably Karen Horney and Helene Deutsch, engaged with Freud's ideas in ways that indicate a concern with feminist issues. However, the initial response of SECOND WAVE FEMINISM was very hostile: Freudian thinking was condemned for portraying women as passively conforming to patriarchal domination and for over-generalising from a tiny sample of women who were drawn from a very particular social group, while psychoanalytic practice was criticised as a deeply insidious device through which rebellious women were manipulated into accepting the oppressive conditions of 'normal' femininity. A shift Freud made early on in his thinking attracted particularly vehement rebuttals. This shift concerns the status Freud attached to accounts offered by many of his women patients in which they reported that, when they were young girls, adult men (sometimes their fathers, sometimes other male relatives or family friends) had engaged in sexual activity with them. Initially Freud took these accounts to be literally true and he began to theorise about the impact of what we would now call child SEXUAL ABUSE. But he suddenly and dramatically changed tack, arguing that his patients' accounts should be viewed as fantasies or as expressions of repressed wishes. Thus, Freud transformed evidence of child sexual abuse into evidence of girls' incestuous desires for their fathers. Feminist hostility to Freudian theory has probably been a factor in a general absence of interest in psychoanalysis in much feminist geography.

Against the initial negative response of second wave feminists to Freudian theory, Juliet Mitchell (1974) offered a very different and influential perspective through her book *Psychoanalysis and Feminism*. What Mitchell argued was that Freud's ideas provided enormously powerful insights into the oppression of women. Her account suggested that the normative element in Freud's work could be left aside and the broad principles of psychoanalytic thinking put to good use by feminists in understanding and undoing sexual inequalities.

Juliet Mitchell's work engages with Freud's ideas via LACANIAN THEORY, sometimes referred to as the French return to Freud. Since her book was published Lacan's ideas, together with those of post-Lacanian feminist psychoanalytic theorists, including Hélène Cixous, Luce Irigaray and Julia Kristeva, have been taken up quite widely by feminists writing in the English language. In parallel a number of other feminist writers, most notably Nancy Chodorow, have engaged with another strand of post-Freudian psychoanalytic thinking, namely object relations theory, which was initially developed principally in the United Kingdom. Feminist geographers have drawn on both strands of this feminist engagement with post-Freudian thought, typically within a POST-STRUCURALIST framework. This has enabled the UNIVERSALISM of much human geography to be uncovered as fraudulent: it has been shown to be ANDROCENTRIC, MASCULINIST or PHALLOCENTRIC. It has also enabled feminist geographers to open up spaces for the exploration of forms of SUBJECTIVITY unimaginable within Freudian theory itself.   LB

Fundamentalism Historically associated with a neo-Conservative movement in American Protestantism in 1919, but now recognised as a phenomenon across other religions, fundamentalism refers to the strict maintenance of traditional, orthodox, religious beliefs usually defined by the irrefutability of religious texts and literal acceptance of religious creeds. Sahgal and Yuval-Davis (1992: 4) argue that fundamentalist religious movements are characterised by opposition to pluralistic systems of thought and by the use of political means to impose their version of the religion on others. Feminists have been active in resisting fundamentalist movments (see Connolly, 1991; WAF, 1990) because of their focus on patriarchal FAMILY structures and control of women's reproductive rights.   CD
See also DETRADITIONALISATION; TRADITION.

G

Gay Thought to derive from the French word *gai* meaning high-spirited, the term gay has also been used since the sixteenth century to describe 'those addicted to social pleasures'. As this descriptor was thought at the time to describe appropriately the condition of many male HOMOSEXUALS, it is therefore difficult for historians to discern at what point the word 'gay' came to denote sexual orientation rather than simply emotional disposition. Although references to the use of the word 'gay' as an explicit descriptor of homosexuals appear as early as 1935 when it was used as English prison slang (*Oxford English Dictionary*) the term was adopted more widely from the 1970s onwards in a deliberate and reactionary move away from the use of existing identifiers which had become overlain with negative connotations. The term homosexual, for example, had by the 1950s become constructed exclusively in terms of either social deviancy (as a sin or a crime) or clinical disorder (caused by degeneracy or biological anomaly and treatable by aversion therapy, PSYCHOANALYSIS or castration; see Foucault, 1980).

A useful parallel can be drawn here between the use of the words 'gay' and 'camp', which had also became popular as a homosexual signifier during the 1960s. 'Camp' was a term used to describe an aesthetic sensibility that was founded on a celebration of the unnatural, of all things frivolous, artificial, extravagant. As Sontag (1967: 290) has suggested, homosexuals began to adopt and promote this aesthetic as a means of integrating themselves into society. 'Camp', as she argues, 'is a solvent of morality [in that it] neutralises moral indignation and sponsors playfulness'. It achieves this, however, through a process of disengagement rather than confrontation – by blurring the boundary between performed effeminacy and active homosexuality. The term is consequently not strictly denotative of sexual orientation and hence, some would argue, inherently depoliticised.

At the time of the emergence of the gay liberation movement in the late 1960s the term 'gay' was promoted as one which was inclusive of both male and female homosexuals; however, it was later criticised by some LESBIANS as a term which had become gender stereotyped, being popularly understood as referring exclusively to male homosexuals. Although successfully deployed as terms which counteracted negative stereotyping, both 'camp' and 'gay' became closely linked to notions of excessiveness in representation and performed effeminacy, neither of which were deemed to be of particular relevance to the lived experiences or political sensibilities or aspirations of other male homosexuals or lesbians.

A desire to retain a separate, visible IDENTITY led lesbians to lobby for the institution of the double-barrelled moniker 'gay and lesbian'. Despite these concerns 'gay' remains a popular identifier for female homosexuals, many of whom prefer it to the word lesbian which they themselves reject as a consequence of internalised HOMOPHOBIA. With the advent of queer theory, the term 'gay' has taken up what some would argue is its rightful place (amongst lesbian, QUEER, bisexual, transgendered and TRANSSEXUAL) as but one of a host of descriptors of deviant sexuality. **BP**

Gay and lesbian communities An inherently plastic and highly contested term which refers to COMMUNITIES which are at once both real and imagined. The term is used describe actual entities – collectives of gays and lesbians who either reside together in a particular geographic locale or who are linked together across space by a shared SEXUALITY and cultural values – or to refer to an idealised conception of those entities.

A growing awareness of the politics of identity and representations of identity, induced in part by the AIDS crisis, informed a desire to adopt it as a celebratory term which could be employed to counter historically negative representations of gays and lesbians. It is now used symbolically to appeal to, or convey, a notion of a coalition of gay men and lesbians united in resistance against shared injustice.

However, as theorists such as I. Young (1990d) suggest, the notion of 'community' can be an oppressive one which sets up exclusionary boundaries, overrides internal differences, privileges certain geographic sites and makes invisible those who do not conform to the normative paradigm. Lesbians and gay men have criticised, and at times rejected, attempts to subsume them under this term, arguing that it remains an ESSENTIALIST construct which privileges shared sexual orientation whilst remaining insensitive to persistent differences relating to CULTURE, GENDER and lifestyle. Other sexual transgressors such as BISEXUALS and TRANSSEXUALS have also questioned why they are excluded from this IMAGINED COMMUNITY.

A new area of research for geographers, work on gay and lesbian communities began by focusing on white, urban, Western enclaves, but is increasingly exploring the geographies of ethnic, rural and international gay and lesbian communities (see Bell and Valentine, 1995; Knopp, 1990; Peake, 1993). **BP**

Gaze The process by which people and things in the world are made into objects of CONSUMPTION by the vision of a privileged subject. According to Foucault, 'the clinic was probably the first attempt to order a SCIENCE on the exercise and decisions of the gaze ... it was no longer the gaze of any observer, but that of a doctor supported and justified by an institution' (1973: 89). Because this process is unidirectional (the world is not allowed to gaze or look back), implies a viewer in a privileged position of power and involves turning subjects into objects through instruments of visuality, feminists have argued that the gaze is a MASCULINIST form of knowledge/power (Haraway, 1991). The viewer controls what is to be seen and how it is to be seen; space and people are made transparent by the 'objective' gaze of the scientist (Rose, 1993a).

Much feminist analysis of the power of the gaze has been influenced by Lacan (1981) who argues that the gaze does not rely upon vision or the eye of the beholder but rather upon a looking relationship between the SELF and the OTHER. The gaze is not literally seen but imagined by the Self to be 'in the field of the Other' (Lacan, 1981: 84). The self (and Other) sees himself or herself as others see him or her through representational forms such as art, photography, film, maps and ETHNOGRAPHY.

In the social context of PATRIARCHY, the gaze reifies the split between the active Male (the surveyor) and the passive Female (the surveyed), thereby allowing for the construction of femininity by the surveyor (Berger, 1972; Burgin, 1986; Mulvey, 1985). In geography, scholars argue that LANDSCAPES, women and non-Western peoples are represented as FEMININE by the male gaze; looking relationships are defined by sexual DESIRE, PLEASURE and POWER (Rose, 1993a; see also Deutsche, 1991). Other scholars have examined the institutions and social practices of the gaze to describe the construction of POSTMODERN places of consumption by corporate interests (Falconer Al-Hindi and Staddon, 1997; Urry, 1990). The term gaze also has been used in FEMINIST GEOGRAPHY to refer to the method of reflexivity. Pamela Moss (1995), discusses the double reflexive gaze of researchers, who turn inward to examine their own 'identity' and outward to view their relationship to what they research. By thinking about the process of reflexivity in visual terms, however, scholars continue to represent space as knowable through their privileged gaze (Rose, 1997a).

Some scholars question the gaze as essentially masculinist, arguing that race and class – as well as gender – are central to looking relations (de Lauretis, 1987; Gaines, 1988; Jameson, 1984; Lutz and Collins, 1991; Tagg, 1988). These scholars maintain that the social and historical contexts in which looking takes place must be a central component of any analysis of the gaze. POST-COLONIALISTS, for example, suggest an ambivalence, rather than unidirectionality, of colonial looking relations. The viewer may recognise him or herself in the racial Other but, because he or she must deny that recognition, loss or absence may occur (Bhabha, 1983). To disrupt the authority of dominant representations, Anne Williams (1987) advocates a feminine gaze, a kind of oppositional looking in which the spectator can identify with both the surveyed and the surveyor.   KT

Gender A complex word to define, as its meaning has changed over the past two decades or so and is now highly contested. It is, however, an extremely significant term for FEMINISM and feminist theory, and the history of its changing use encapsulated the changing emphases in feminist scholarship. Since the resurgence of feminism in the late 1960s the term has been introduced and redefined and is now used in two different, although interconnected, ways.

In her essay 'Interpreting gender' Linda Nicholson (1995a) outlined the history of its usage, and her work is drawn on here. The first way in which the term gender tends to be used is in contrast to the term SEX. Whereas sex depicts biological differences (male and female), gender describes socially constructed characteristics (MASCULINITY and FEMININITY). In 1949 Simone de Beauvoir, the great French existential thinker and feminist, published *The Second Sex*, in which she claimed that women are not born but made, and so challenged the assumption of biological determinism:

> One is not born but rather becomes a woman. No biological, physiological or economic fate determines the figure that the human being presents in society: it is civilisation as a whole that produces this creative indeterminate between male and eunuch which is described as feminine (de Beauvoir, 1972: 525).

While de Beauvoir's positioning of women between men and eunuchs may have been challenged as well as the ETHNOCENTRISM of her work, her recognition that femininity is socially constructed was extremely important in the advent of SECOND WAVE FEMINISM in Britain and the USA. The 1960s resurgence of feminist thought and action, so called to distinguish it from the 'first phase' or struggle for SUFFRAGE, drew on de Beauvoir's work. Her argument about the social construction of femininity was taken seriously and developed through the adoption of the term gender to refer to the ways in which women were 'made'. Thus the key aim of a great deal of early theoretical work by contemporary feminists was to challenge what seemed like immutable sexual differences based in biology and so to undermine the claims of absolute sexual difference between women and men and, importantly, women's supposed inferiority in matters of physical strength and mental agility. By differentiating sex from gender, gender could be theorised as the cultural or social elaboration of the latter and as such amenable to change.

In one of the best-known and influential early articles of second wave feminism, Gayle Rubin showed how these processes were interrelated through what she termed a sex/gender system. Such a system is, she argued, that 'set of arrangements by which a society transforms biological SEXUALITY into products of human activity and in which these transformed sexual needs are satisfied (1975: 159). Through this transformation and social regulation 'sex' becomes 'gender'. In a vivid analogy, Linda Nicholson suggests that this model or relationship between sex and gender is like a coat rack. Sex or biological difference is the basic frame on to which different societies in different historical periods have hung various coats – the socially defined arrangement of gender characteristics.

The great advantage of the sex/gender distinction was that it enabled feminists to challenge the 'naturalness' of gender divisions and to theorise them as variable. It also enabled both commonalties and differences among women to be postulated and, for geographers, it proved invaluable, as gender relations were theorised as spatially variable phenomena across a range of different scales. Thus it was recognised that gendered characteristics vary not only between countries and over historical time but also in everyday spaces and interactions. So, for example, in bars, clubs, parliamentary buildings, student hostels and offices the use of gender symbols and expectations of gender-appropriate behaviour vary.

In its second, and later, usage, the term gender is not seen as distinguishable from sex; rather, the latter is subsumable within the former. Nicholson (1995) quotes Joan Scott's explanation of this second way of understanding or defining gender:

> It follows then that gender is the social organisation of sexual difference. But this does not mean that gender reflects or implements fixed and natural physical differences between women and men; rather gender is the knowledge that establishes meaning for bodily differences. ... We cannot see sexual differences except as a function of our knowledge about the BODY and that knowledge is not 'pure', it cannot be isolated from its implication in a broad range of discursive contexts.

In this way, the biological foundationalism of the first perspective on gender differences is challenged and the attributes of sexual difference which were assumed to have universal applicability have been revealed for what they are: 'matters specific to Western culture or to specific groups within it (Nicholson, 1995: 42). This recognition means that

> we cannot look to the body to ground cross-cultural claims about the male/female distinction ... differences go 'all the way down' ... tied not just to the limited phenomena many of us associate with gender (i.e. to cultural stereotypes of personality and behaviour), but also to culturally various understandings of the body, and to what it means to be a woman and a man (43).

Thus the body too is open to analysis and theorisation as a variable, not a constant.

In the earliest work undertaken by feminist geographers and others, gender was interpreted in the main in the first sense and the major emphases were, first, on uncovering variations in gender roles and, second, the ways in which material social practices resulted in inequitable gender relations. In this work the idea of PATRIARCHY, which has common features with the notion of a gender order, was important.

The concept of a GENDER ORDER was introduced by Bob Connell, an Australian sociologist who has written widely about gender and especially about masculinity (Connell, 1987, 1995). A gender order or regime, according to Connell, consists of a threefold model 'distinguishing relations of (a) power, (b) production and (c) cathexis (emotional attachment) (1995: 73–4). He defines these relations as follows:

> (a) *Power relations* The main axis of power in the contemporary European/American gender order is the overall subordination of women and the dominance of men – the structure Women's Liberation named 'patriarchy'. This general structure exists despite many local reversals (e.g. woman-headed households, female teachers with male students). It persists despite resistance of many kinds, now articulated in feminism.
>
> (b) *Production relations* GENDER DIVISIONS OF LABOUR are familiar in the form of the allocation of tasks, sometimes reaching extraordinarily fine detail. ... Equal attention should be paid to the economic consequences of gender divisions of labour, the dividend accruing to men from unequal shares of the product of social labour. This

is most often discussed in terms of unequal wage rates, but the gendered character of capital should also be noted. A capitalist economy working through a gender division of labour is, necessarily, a gendered accumulation process. So it is not a statistical accident, but a part of the social construction of masculinity, that men and not women control the major corporations and the great private fortunes. Implausible as it sounds, the accumulation of wealth has become firmly linked to the reproductive arena, through the social relations of gender. [Note that the significance of intermarriage and of legitimate births for inheritance purposes has long interested the capital-owning classes and was, indeed, the basis of Engels's arguments about women's domination in the family well over a century before Connell's work.]

(c) *Cathexis* Sexual DESIRE is so often seen as natural that it is commonly excluded from social theory. Yet when we consider desire in Freudian terms, as emotional energy being attached to an object, its gendered character is clear. This is true both for heterosexual desire and homosexual desire. The practices that shape and realise desire are thus an aspect of the gender order. Accordingly we can ask political questions about the relationships involved: whether they are consensual or coercive, whether pleasure is equally given and received. In feminist analyses of sexuality these have become sharp questions about the connection of HETEROSEXUALITY with men's position of social dominance (Connell, 1995: 74–5).

Connell's arguments are paralleled by the work of feminist theorist Sylvia Walby (1997). She has recently suggested that her earlier definition of patriarchy as consisting of six interrelated structures might be mapped on to specific but changeable GENDER REGIMES. She distinguishes between a primarily DOMESTIC or PRIVATE gender regime or order in which relations of power and dominance in the household are the primary reasons for women's subordination, and a PUBLIC gender regime or order in which women's position in the public arena, especially in the workplace, is the primary cause of their subordination. Both Walby and Connell insist on the complexity and variety of the ways in which gender relations result in unequal relations between women and men and emphasise the interconnections between gender, class position and ethnic origins. In his book *Masculinities*, Connell (1995) shows, for example, how white men's sense of themselves is constructed in relation to an idealised notion of black masculinity as well as in opposition to white femininity and argues that acceptable notions of both manliness and femininity vary by class position, as well as over time and between regions and nations. Thus, the idea of dominant and oppositional gender regimes, which are complex and variable, gives a useful and structured way of investigating the geographic diversity of gender relations. There are many different ways of 'doing gender' (West and Zimmerman, 1987), of being a man and a woman. Multiple and oppositional as well as hegemonic versions of femininity and masculinity exist, which are geographically and historically specific and vary across the range of spatial scales.

Despite this recognition of multiplicity and differences between women, a range of critics, influenced in particular by post-structural and POST-COLONIAL theory, began, from the 1970s, to deconstruct the very notion that gender is a stable construct at all. Their critique built on earlier arguments from black feminists that most feminist scholarship and politics implicitly assumed a white subject. Similarly, lesbians pointed to the implicit heterosexuality of a great deal of feminist writing and scholarship and demanded that questions about 'alternative' sexualities be put on the agenda. Their demands were reinforced by the rapid rise of GAY and QUEER THEORY in the 1980s (Herdt, 1992; Sedgwick, 1990; Simpson, 1994; Weeks, 1986).

These critics forced a recognition that feminist scholarship so far had been written from the particular position of white, often middle-class and predominantly heterosexual women, not as 'Woman'. This exploration of the significance of differences between women was accompanied by a more challenging development as feminist scholars versed in the psychoanalytic literature, as well as influenced by the work of Michel Foucault, began to retheorise the subject as relational and contingent. The subject, it is argued, rather than being a fixed and stable entity which enters into social relations with its gender in place, is always fluid and provisional, in the process of becoming. Gender, it is argued, is constructed and maintained through DISCOURSE and everyday actions. There have been a number of directions in which this work has taken by those that we might group together as deconstructionist feminists. Briefly, this work questions the dichotomous distinction of sexual difference as well as the mapping of gender attributes on to the bipolar division.

A key theorist here is Judith Butler. She argues, somewhat differently from Haraway, that sexed bodies are constructed as such from the standpoint of already dichotomised gender – that is, it is taken for granted that there are two genders, male and female – and that through scientific, medical and other discourses bodies are identified in the same way. Butler (1990, 1993) suggests that this identification and its maintenance over time is constructed through what she terms a gender performance, in which the regulatory fiction of heterosexuality constrains most of us to perform within the hegemonic norms that define bipolar feminine and masculine norms in specific societal contexts. She suggests, however, that the discursive or taken-for-granted construction of bipolar gender may be challenged through subversive performances, and identifies drag as a key subversive act. Butler's work has given rise to an important debate about the materiality or reality of the body.

The retheorisation of gender as a discursive construction and a performative fiction has been extremely exciting for geographers. In studies of the workplace, for example, it allowed a new set of questions to be asked about workplace cultures, how gender identities are constructed through daily interactions at WORK and, importantly, offering ways to think through challenges to unequal gender relations in immediate ways. The recognition of diversity, and possible oppositional strategies, provided a more nuanced way to think though women's domination in the workplace than relying on the overarching concept of patriarchal DOMINATION. It also led to the introduction of new research questions about LANGUAGE, gestures, speaking styles and bodily presentation in the construction of a sexed or gendered SUBJECT.

A further impetus to the theorisation of gender as fluid has come through the impact of new technologies, in the spheres of reproduction and reconstructive surgery. Gender seems nowadays to have escaped the constraints of the body, or more accurately the body has been redefined. Control over reproduction, for example, in a wide range of ways from *in vitro* fertilisation, fertility treatment and surrogacy, has led to gender and sexuality increasingly becoming fluid and malleable, almost something to be chosen as one further aspect of personal IDENTITY. Women are now 'free' to become MOTHERS after the menopause, or to reshape their bodies as they wish, with, in most cases it seems, income level being the sole constraint. We seem to have arrived, in the words of the feminist anthropologist Marilyn Strathern (1992), at a time 'after nature' when conventional bonds of blood, KINSHIP and marriage seem almost irrelevant to the process of reproduction and family life.   LM

Gender and Development (GAD) Programmes introduced to ensure that the differential relations of POWER between men and women are taken into account in the development process. This approach has replaced the earlier emphasis on Women in Development (WID).   LM
See also DEVELOPMENT; UNITED NATIONS; WORLD: FIRST, SECOND, THIRD, FOURTH.

Gender audit See GENDER DEFICIT

Gender-blind This is a straightforward term used to refer to theories or policies which ignore gendered differences; geographical scholarship in previous decades was gender-blind – a good example here is locational analysis in which 'rational economic man' stood in for all humanity.   LM

Gender culture See CORPORATE CULTURE

Gender deficit A term used in the policy literature to refer to the absence or unequal position of women within organisations and institutions; one way of revealing this deficit is through the mechanism of a GENDER AUDIT in which the impact of policies on women, as well as their absence in structures, is documented.   LM

Gender division of labour See DIVISION OF LABOUR: DOMESTIC, INTERNATIONAL, GENDER, SPATIAL

Gender impact analysis A research tool and basis for social policy which systematically examines the differential impacts of planned interventions on varied groups of women and men. Rather than 'adding women in' to policy, gender impact analysis is premised on the belief that gender issues should be integrated into *all* aspects of policy; otherwise, unchallenged assumptions about the distribution of resources minimise women's gains. Devised by feminists interested in overcoming women's economic, social and political marginalisation in both the North and the South, gender impact analysis aims to provide policy-makers and planners with the requisite information for gender-neutral or, at best, pro-women policies. Research methods used in gender impact analysis vary with context and include ethnographic documentation, rapid rural appraisal, time-budget analysis, disaggregation of official statistics and surveys and interviews. Consultation with and listening to women are key elements in the analysis.

Development planners have been at the forefront of gender impact analysis, in order to put gender inequality at the heart of development concerns. Longwe (1991) argues that the collection of gender-differentiated data and then the design of gender-specific policies must be based on the assessment of women's needs at five levels – welfare, access, concientisation, participation and control. Moser (1993) similarly argues that gender planning can be addressed to women's strategic needs for empowerment, alongside practical assistance. The British development organisation Oxfam, with its Gender and Development Unit (GADU), assesses project proposals for their gendered impacts. SAR

See also GENDER AND DEVELOPMENT; WOMEN IN DEVELOPMENT.

Gender order See GENDER

Gender Perspectives on Women (GPOW) A group formed by women geographers in the USA with an aim similar to that of the first aim of the WOMEN AND GEOGRAPHY STUDY GROUP (WGSG) in the United Kingdom – that is, to investigate the ways in which gender is geographically constituted and the differentiation between men's and women's lives in different places. It is one of the speciality groups of the Association of American Geographers, which also sponsors a Commission on the Status of Women. LM

Gender regime See GENDER; PATRIARCHY

Genealogy Traditionally, genealogy is concerned with tracing lines of descent as in the Old Testament sequences of begetting. Since these are MASCULINE lines, feminists have at times tried to explore what female genealogies might look like. However, Nietzsche, and more recently Foucault, have given genealogy a rather different meaning, linking it not with origins but with beginnings. Rather than investigating the chronological process of what happened in time, it examines the historical record itself, the narrative account of what happened and the way people record, narrate and explain their own past. Accounts of origins seek to articulate the basis from which things grow, to find in earlier circumstances a unified pattern that gives birth to what comes after. Traditionally, the task of the historian was to work over the discontinuous elements of history in order to reveal the continuous relations of causality that underlay the surface dispersal of isolated events.

Genealogy, by contrast, celebrates the discontinuities, and shifts the inquiry toward a history of accidents, disconnections and disparities. It 'initiates a mining of fields of meaning for the many possible stories they hold rather than the excavation of the one true story' (Ferguson, 1993: 21).

In his *Genealogy of Morals*, Nietzsche (1956) sought to show that all of man's economic, social and political activities are ultimately determined outside the consciousness of the individual subject. What was presented by historians and philosophers as the 'purest' human morality could in fact be traced back to the most naked of power struggles. For Nietzsche what underlies such narratives is desire, a will to power or to life. Drawing on Nietzsche, Foucault has also emphasised the discontinuities, of practices and ideologies over time, thus challenging their pre-given or monolithic character. Some feminist historians, dissatisfied with the idea that one may unproblematically present 'women's history' have been engaged in a project of charting a genealogy of the category 'WOMAN' or 'women'. From this approach, 'woman' itself is understood to have a history, a genealogy, a 'line of descent'. Rather than assuming that 'woman' has a stable referent across the centuries, or that women have a traceable linear history that is uniquely their own, a genealogical approach asks, how has woman/women functioned as a DISCURSIVE category throughout history? In similar ways, other feminists have taken up such categories as 'dependency' and 'need', asking how such terms have come to be stigmatised.   RP

Generation The creation, maintenance and perpetuation of social divisions between members of one age group and another. Age is often thought of as an individual attribute, but chronological age is frequently used by the state and other institutions to legitimise the exclusion of certain groups from access to resources and opportunities that may benefit others. An example in the United Kingdom is unemployment benefit which is denied to people under the age of 18 years and over retirement age. The most visible age stratification occurs at both ends of this axis and is itself value-laden and conceptually flawed (Biggs, 1993; Jones, 1993).

Feminists consider the social, economic and cultural relations between women (and men) from different generations as they move through the LIFE CYCLE. This research looks to sources of intergenerational conflict within and between household units and draws out their implications for GENDER relations. Common themes include the hierarchical age and gender relations involved in the allocation of HOUSEHOLD resources, such as food (O'Brien, 1995), the contribution that children make to housework (Morrow, 1996) and the organisation of CHILDCARE (Cotterill, 1992). Although feminist geographers have been slow to address questions about age, a growing literature looks to connections between PUBLIC space and generation. This shows that the organisation, use and symbolic control of public space reflect divisions of age, often with deleterious effects for the participation of the elderly (Laws, 1994; Pain, 1997e; Valentine, 1996).   HC

See also AGEISM; HUMAN CAPITAL THEORY.

Gentrification Gentrification can be defined as neighbourhood regeneration, usually in inner-city locations, by relatively affluent incomers who displace lower-income groups. It encompasses a highly varied set of processes whose manifestations range from converted warehouses and renovated housing to large-scale redevelopment, and from 'sweat equity' by individual gentrifiers to commercial development activities (Mills, 1993). Gentrification involves aspects of 'production, reproduction and consumption' (Beauregard, 1986: 41) and has been debated widely between those explaining the availability of 'gentrifiers' (Rose, 1989), those discussing the AESTHETICS of the new 'style' and those stressing the conditions created by capitalist restructuring which enable gentrification (Smith, N., 1986a).

Accounts of the 'gentrifiers' stress the importance of changing GENDER relations (increased participation by women in paid employment, particularly in well-paid career jobs (Rose, 1989), delayed childbearing and alternative HOUSEHOLD constitution). Indeed, Warde (1991) argues gentrification is more about 'household composition and household organisation' than it is about CLASS positions (Warde, 1991). Other work suggests a more complex intersection of gender and class, highlighting the differing involvement by women according to class position, employment status and prospects, family situation (Butler and Hamnett, 1994; Lyons, 1996; Rothenberg, 1995). Furthermore, new place-based lifestyles may offer desirable sources of IDENTITY and identification both for women and for men (Mills, 1993). However, gentrification involves not only new residents but also displacement of previous residents and the disruption of existing COMMUNITIES. While moving to inner-city locations may offer a (temporary) solution to the mismatch between the gendering of labour-market structures and of urban structures for some women and men (Bondi, 1991), it may do little to overcome inequalities between different women and men in terms of income and social status (Beauregard, 1986; Lyons, 1996).

Liz Bondi (1990a, 1992b) criticises the inadequate conceptualisation of the differences and connections between gender *relations* and gender *identities* and highlights a tendency in some analyses to treat gender as a 'given' individual characteristic while treating class as a 'social relation'. She argues one way forward is to 'disentangle the symbolic and sociological aspects of gender, rather than assume direct correspondence' between them (Bondi, 1992b: 157) and to engage with the 'terrain of meaning' (Mills, 1993) of gentrification, working through the tensions between highly gendered imagery of, for example, gentrification's 'urban frontier' as it is contested and re-negotiated through practice. Far from marking a reordering of gender relations, the practices of gentrification may in fact draw on ESSENTIALIST notions of gender and merely 're-present hierarchical gender relations' (Bondi, 1992b: 157), albeit in ironic references to Victoriana or working-class culture. Mills (1993) identifies to some extent a conscious feminisation of marketing and construction among developers. This move repackages for sale other more 'EMANCIPATORY' aspects such as alternative lifestyles and the rejection of the (gendered) strictures of suburban living.

Myths, meanings and practices may be highly contested and reworked by a range of those involved in gentrification, including those resisting displacement. Some analyses of gentrification take seriously the mutual constitution of identities through social and spatial practices and in the creation, contestation and circulation of meaning. Studies of the influence of SEXUALITY (Brown, M., 1995a; Rothenberg, 1995) and the politics of 'RACE' and POST-COLONIALISM (Jacobs, 1996a) emphasise that gentrification is shaped, in part, through contested and highly selective reinterpretations of the past as 'HERITAGE' and the present as selectively defined CONSUMPTION.   FS

GIS See CARTOGRAPHY

Geography Until recently, perhaps one of the less significant of the social sciences geography has in these postmodern, reflexive times, when POSITION-ALITY and situatedness, borderland and margins, STANDPOINT and DIFFER-ENCE, are key words among feminist theorists, suddenly found itself at the forefront of contemporary debates, not only in the social sciences but also in the humanities (McDowell, 1996; Pollock, 1996). A combination of material and theoretical shifts has brought to prominence the concept of location: a place to speak from, a place to claim in a world in which mobility and displacement are increasingly common experiences, as well as the reconceptualisation of 'real' places as fluid and relational networks rather than bounded or CATEGORICAL spaces.

It is common to define geography as a discipline not in terms of substantive subject material but rather as a synthetic point of view addressing both difference and particularity, the causes and consequences of uneven development and the patterns of flows and movement discernible over space, linking places with real and symbolic meanings for their inhabitants and outsiders that differentiate them from elsewhere. It is now axiomatic that spatial relations affect and reflect social relations (albeit less well reflected in the practices of many of the social sciences) and that location makes a difference to the constitution of gendered subjectivities and identities. Thus the terms SPACE (flows or relations) and PLACE (locations) are what define geography. To these two terms must be added a third: that of SCALE. A geographical imagination spans the spatial scale from the body to the globe, and feminist inquiries focus not solely on the small-scale or the local – the BODY, the HOME and the COMMU-NITY – but also on the region and the NATION, indeed the world in its entirety. What holds together this ambitious agenda is the idea of connections: the links between places at a range of spatial scales that produce the particularity of any single place (Massey and Jess, 1995). Thus in a world in which millions of women are wrenched out of place, driven away from their homelands by wars or by famine, by economic hardship or household changes, by global capital flows, economic restructuring and structural adjustment programmes, and are forced to recreate their place in the BORDERLANDS of other nations, a range of social processes and events at different spatial scales combine to reshape the links between their IDENTITY and their location. The results often challenge conventional geographic divisions. As Mohanty (1991a) noted, 'Third World women', defined by their common political interests, are as likely to live in the 'FIRST' as the 'THIRD' WORLD.

Global flows of labour and capital in combination with the development of new technologies for overcoming the friction of distance and permitting an instantaneous transfer of information and capital have resulted in TIME–SPACE COMPRESSION (Harvey, 1989) or distanciation (Giddens, 1991), even, according to some, the end of geography in the sense of marked differences between places. With the growing dominance of capitalist social relations at the global scale and the HEGEMONY of Western culture, it seems that everywhere is increasingly similar, and that the 'non-places' or 'spaces of flows' (Castells, 1996a) of global capital – airports, stock exchanges, the Internet – have replaced fundamental attachments to place and territory on which cultural identities were based and which gave us our sense of belonging. While some POSTMODERN theorists celebrate and others rue this apparent DETERRITORI-ALISATION, the counter claims are ANDROCENTRIC and ETHNOCENTRIC. Money, power and knowledge are needed to escape the bounds of place, yet many women without these assets might approve of reducing the ties that bind them in place. But as feminist critics have pointed out, through a range of examples at different spatial scales, from the ethnic nationalism of the former Yugoslavia to the barrios of the world's poorest cities, the poor, and paradox-ically even the mobile among them, remain trapped in place, in the least desir-able spaces of exclusion in cities and nations, where daily life is spatially bounded but affected by large-scale forces and spatial flows beyond local control. The significance of place has not declined, although it has been rethe-orised to combine local and global processes.

Feminists have also retheorised concepts of place and location in metaphorical, as well as material, ways. The redefinition of notions such as borderlands and MARGINS has become an important part of the broader challenge to the inherent spatiality of Western ENLIGHTENMENT theory, in which the distinction of mind from the body, reason from emotion, the public sphere from the private arena placed men on one side and relegated 'WOMAN', as OTHER, to the other side. Thus all that was 'naturally' FEMALE and FEMININE was located inside, in private, at the smallest spatial scale and so taken for granted and untheorised. The long struggle to unseat these divisions and spatial associations is well known. More interesting and poten-tially disruptive for feminism as a theory and a political practice has been the recognition of difference, and in this the importance of location as a place to speak from. The internationalism of Virginia Woolf's much-quoted claim that 'as a woman I have no country. As a woman I want no country. As a woman my country is the whole world' was challenged by the development of what Rich (1987) termed the politics of location. Such a politics is not necessarily divisive but may lead instead to alliances between women in different locations and speaking positions. This politics marks welcome progress in the displacement of the singular and universal view from 'nowhere' in PHALLO-GOCENTRIC knowledge claims. Feminism too is not a universal perspective. Position and location – geographies – matter in the constitution of subjectivity and identity, in theory construction, in the making and marking of the body, in customs and practices and in political struggles against oppression. The place from which we speak affects the claims we make and recognition difference strengthens these claims. As hooks (1990) insists, the margins are a place of radical openness from which the singular view of phallogocentrism may be undermined and superseded.   LM

Geopolitics An element of the practice and analysis of statecraft and international relations more generally which considers GEOGRAPHY and spatial relations to play a significant role in the constitution of international POLITICS. Certain 'laws' of geography, such as distance, proximity and location are understood to influence political processes. The effect of geography on politics is based upon 'common sense', rather than ideology: the 'facts' of geography are seen to have predictable influences upon political processes. However, recently certain authors have challenged the political innocence of geography to suggest that rather than being timeless concepts, geographical relationships and entities are specific to historical and cultural circumstances, and as such there is always a politics to the use of geographical LANGUAGE (Ó Tuathail and Agnew, 1992).

Some FEMINIST commentators have remarked on the lack of WOMEN in geopolitics and international relations. Cynthia Enloe (1989) suggests that women have been written out of the stories of international politics which have traditionally described the spectacular confrontations between mighty states led by powerful statesmen, of the speeches and heroic acts of the elite, and the specialist knowledge of 'intellectuals of statecraft'. Enloe (1989) refuses to accept this story as covering the full extent of the workings of international relations, and instead focuses on those elements that it excludes and silences: the role of international LABOUR MIGRATION, the availability of cheap female labour for transnational corporation investment, the availability of sex workers for the TOURIST INDUSTRY in South East Asia and so on. Enloe's is a very different account of international politics than the traditional story and certainly one that lacks its glamour. She ties international geopolitics to everyday geographies of GENDER relations. This linking of the PERSONAL and the POLITICAL suggests that these alternative political geographies need to be uncovered because, 'if we employ only the conventional, ungendered compass to chart international politics, we are likely to end up mapping a landscape peopled only by men, most elite men' (Enloe, 1989: 1).

Other feminist critiques of geopolitics, and international relations more generally, concern the techniques of VISUALISATION they employ. The (geo)politician is distanced and all-seeing of the world-as-exhibition. Ó Tuathail has introduced the concept of an 'anti-geopolitical eye' (Ó Tuathail, 1996), an embodied vision from somewhere, which subverts the (geo)politician's GAZE.   JS

Girl power A term which has become popular in the United Kingdom in the 1990s, used in a general sense to refer to the growing confidence which young women demonstrate in their everyday lives and in a particular sense associated with the all-women group, the Spice Girls, popular in the late 1990s, especially with pre-teenage girls.   LM

Global corporation A capitalist business organisation which has activities in many different countries throughout the world economy (Dicken, 1992, 1994). The term is indistinct from multinational company (MNC) and transnational company (TNC) but can perhaps be distinguished by the scope, scale and financing of business operations in numerous national contexts.   AJ
See also GLOBALISATION.

Globalisation The term is an ill-defined (King, 1991; Robertson, 1990), if currently fashionable, 'catch-all' concept which is increasingly widely used in human geography as well as in other social science disciplines and popular discourse. Globalisation should be understood primarily, argues Giddens, as 'the reordering of TIME and distance in social life' (1993: 528). This tends to be linked to the wilting of the idea of a cohesive and sequestered national economy and society (Castells 1996a; Giddens 1989) as the globalisation process produces increasing levels of societal integration across time and SPACE in the contemporary world. Many commentators also refer to the 'dialectic' between the global and LOCAL (e.g. Harvey 1989, 1996) and suggest that the 'local' has been transformed by globalisation (Massey, 1991b). Indeed, much of the 'globalisation debate' within geography is framed in terms of the relationship between the global and the local.

However, as Amin and Thrift (1994) point out, these general arguments are rather unsatisfactory, by no means pointing to 'a finished theory of globalisation'. Globalisation is a complex idea whose meaning is often contradictory in different usages and, as a general 'process', is ambiguous. Partly as a consequence of this, the literature often divides up the debate into discussions of economic, cultural, political or other 'sub-forms' of globalisation – although these sub-divisions too have been criticised for separating integral processes (Pieterse, 1996; Robertson, 1996). Furthermore, recent criticism has been levelled at the 'strong globalisation thesis' (O'Brien, 1991; Ohmae, 1990) where proponents have often presented globalisation as some inevitable, unstoppable process which will erode the significance of place and context (Kofman and Youngs, 1996). Geography in particular has been central in arguing that globalisation redefines the local rather than reducing its significance (Massey, 1994b).

Clearly, the notion of economic globalisation and an increasingly 'global' economy are central issues. Geographers have contributed substantially to this debate in, for example, developing theorisations of the nature of global CAPITALISM, uneven development and the circulation of money and commodities (Harvey, 1982, 1989; Smith, N., 1986b). Economic geography has also concerned itself with the key significance of GLOBAL CORPORATIONS (Dicken, 1992), examining the nature of global manufacturing industry and theories of the post-Fordist 'global factory' (Amin, 1994). However, geographers have also focused on the growth of global financial markets (Corbridge *et al.*, 1994; Peck and Tickell, 1994) and service industries (Beaverstock, 1994). Workers within these fields have argued that the financial services and markets are one of the most strongly 'globalised' of economic sectors, owing primarily to the abstract nature of money as a commodity (see Corbridge *et al.*, 1994) and advances in the sophistication of telecommunications and information technology (IT) in the past 30 years (Castells, 1996a; Talalay *et al.*, 1997). There has also been considerable interest in the implications this might have for the powers and roles of nation-states (Camilleri and Falk, 1992; Dicken, 1994; Rosenau and Czempiel, 1992; Strange, 1994; Thrift and Leyshon, 1994) and financial centres (Martin, 1994; Thrift, 1994).

Social and cultural geographers have been similarly concerned with the globalisation of cultural flows and the relationship with transnational migration and employment movements. Interest in the changing nature of the new INTERNATIONAL DIVISION OF LABOUR (NIDL) has recently turned to consider the way in which post-FORDIST firms utilise cheap labour in the developing world, whilst at the same time supplying increasingly globally branded products (e.g. Levi's jeans, Nike trainers) to an emergent global middle class scattered throughout the spaces of 'comfortable capitalism' (Agnew and Corbridge, 1995). Debates about what might be meant by global culture(s) remain multi-faceted, although most commentators would agree that the notion of any singular 'common global culture' (McDonald's-isation, Coca-Cola-isation) is problematic (Featherstone, 1993). Theorists of cultural globalisation have also considered in depth the problems of articulating the relationship between the global and the local (Featherstone and Lash, 1996; Robertson, 1996) and the problems that ideas of HYBRIDISATION present – embodied in notions of the 'glocal'. (Pieterse, 1996).

Feminist geography has engaged with globalisation in many of these fields of study. For example, workers have pointed to the feminised nature of many of the developing world labour markets which are being used by transnational corporations. Companies producing for the global middle class often rely on a heavily feminised low-skilled labour force to produce brand clothing or electronics goods in the developing world (Allen, 1994; Barff and Austen, 1993; Phizacklea, 1990). Feminised low-skill work forces are also a feature of the core economies, with foreign direct investment in Europe and the USA also seeking out cheap labour (Kochan *et al.*, 1986; Lash and Urry, 1994; Peck, 1996). AJ

Goddess The image of God as male legitimises patriarchal rule and asserts 'man's' dominance over NATURE. To counter this, spiritual feminists have consciously drawn on images of the Goddess as an affirmation of female power and divinity, celebrating the female BODY as part of the living cosmos. CN

Green Labour See LABOUR: DOMESTIC, WAGED, GREEN, SURPLUS

Green politics All who accept the thesis of 'limits to growth' may be seen as politically green, though some reformists hope for a technological 'fix' to solve current environmental problems or believe that a sustainable 'green capitalism' can be brought about by consumer action, recycling and environmental legislation. Such 'light greens', or environmentalists, work in pressure groups, local protest organisations, provide environmental services and form green parties, though all these organisations may also involve more radical greens. Radical greens, or ecologists, insist that energy-saving, 'green' TRANSPORT policies, extensive recycling and so on cannot eliminate the necessity to limit consumption (especially, but not only, in the rich countries). They differ about the sort of society we should aim for and the strategic value of lifestyle politics, protest and revolution (see Dobson, 1995). They include ecosocialists, ecoanarchists, social ecologists, ECOFEMINISTS and 'dark greens' or 'deep ecologists'. The former three approaches believe that CAPITALISM is inevitably expansionist and destructive and must therefore be replaced by new non-exploitative socio-economic forms (for the differences between them, see Dobson, 1995; Merchant, 1992; Pepper, 1992). 'Deep ecology' insists that the biosphere, all ecosystems and the physical features of the planet have intrinsic value and should be respected for their own sakes (Dobson, 1995: 48; Naess, 1995). Humans need to reduce both population and their impact on the planet. Some deep ecologists take an authoritarian position, others work for a spiritual change in human consciousness, while others (such as Earth First!) take part in radical direct action. Ecofeminists agree with deep ecology that a radical change in our relationship with NATURE is necessary, but believe this is impossible unless men's POWER over women is itself addressed – which deep ecology ignores (Salleh, 1984, 1997).   CN

Greenham Common Women's peace camp outside the US base near Newbury, UK. Established in 1981 in opposition to Cruise missiles. In various mass demonstrations in the 1980s, tens of thousands of women 'embraced the base' by joining hands around the perimeter fence and covering it with symbols of women's concern for life on earth. Camps at the various gates developed their own characters and were resourced and visited by a national network of support groups. The women resisted considerable harassment and VIOLENCE from police, soldiers, bailiffs and some local residents. Non-violent direct action included cutting the fence, dancing on the missile silos, blockading the camp and monitoring the movement of missiles along motorways, which may have been a factor in the withdrawal of Cruise. The protest focused on nuclear weapons but many broader feminist issues were raised, including the validity of LESBIANISM as a choice for women. Cruise was withdrawn in 1991, and the peace camp formally disbanded in 1994. (See Roseneil, 1995.)   CN
See also MILITARY/MILITARISM; RESISTANCE; SOCIAL MOVEMENTS.

H

Habitus The term habitus was developed by the French social theorist, Pierre Bourdieu. He defines it as 'a system of shared social dispositions and cognitive structures which generates perceptions, appreciations and actions' (Bourdieu, 1984: 279; see especially Bourdieu, 1977). He argues that it is important to understand social relations (including, importantly, gender relations) in terms of habitual practices. Everyday, taken-for-granted practices arise from habitus which, although embodied in individuals and individual practices, is in fact a widely shared set of dispositions giving rise to relatively uncritical, naturalised behaviour. It is similar to Giddens's (1984: 375) notion of practical consciousness in that it *can* be interrogated as an object of self-reflection, but by definition tends not to be, in large part because it is usually learned in childhood and serves as the basis of social expectations. The idea of habitus is similar also to culture but tends to operate at a smaller scale. Fields or structured systems of social positions within which struggles for rewards and resources (see CULTURAL CAPITAL) take place each tend to have their own habitus. Examples include fields based on employment, formal politics, education, other intellectual, scientific or artistic endeavour, lifestyles (CLASS or COMMUNITY) and major arenas of consumption such as housing, music or art. Unarticulated POWER relations and strategies for manoeuvring into positions of prestige within these fields are based on habitus. Habitus is a useful concept because naturalised, unquestioning behaviour based on socialisation within a group accounts for a large percentage of human behaviour. Nevertheless, Bourdieu has been criticised for overemphasising structured behaviour and underemphasising conscious, rational decision-making. There are good political reasons why feminists and others with progressive political agendas wish to emphasise the fluidity of conscious action and agency, but the idea of habitus as a constraining factor can easily be adapted to a social theory with greater emphasis on AGENCY and negotiation than Bourdieu's. Alternatively, it can be argued that whether or not one places emphasis on naturalised or self-conscious behaviour in a particular study should be an empirical question and that any political agenda should be on the lookout for structures of inequality (e.g. class, PATRIARCHY or COMPULSORY HETEROSEXUALITY) and therefore for constrained or structured behaviour.   ND

Harassment Harassment is behaviour directed at others because of their gender, 'race', sexual orientation, age or ability. Despite the growth of policies to combat harassment, the term is contested ground, but generally involves unwanted and intrusive behaviours which cause a feeling of threat or humiliation, such as verbal ABUSE, touching, flashing, stalking and the display of PORNOGRAPHY. Radical feminists have labelled as harassment all 'the processes of social control enacted by men over women in which the totality of our lives are available to being policed by them' (Wise and Stanley, 1987: 15), but others advocate clearer delimitation for strategic reasons. Harassment is an exertion of POWER. Its content frequently makes verbal or physical reference to bodily difference, objectifying women and minority groups, and threatening or appropriating their bodyspaces. Those whose presence is a threat to the existing social order of space, such as lesbians and gay men negotiating heterosexual space, may be subject to harassment (Valentine, 1993b). Harassment deters people from particular spaces at particular times, influences identity and has a role in the social control of 'others' (Gardner, 1995). In some circumstances it creates fear as a portent of more serious violence, further circumscribing use of SPACE (Pain, 1991). Harassment is an expression of the male-dominated and HETEROSEXUAL nature of many workspaces (MacKinnon, 1979); one of the ways in which female bodies 'out of place' are disciplined (McDowell, 1995). Similar effects have been charted in the home, and in education, against girls, against 'effeminate' boys in schools and against women in academia (McDowell, 1990). While personal resistance strategies are common, few of those who suffer harassment report their experiences because of the stigma attached to the term.   RHP

See also FEAR OF CRIME; HOMOPHOBIA; VIOLENCE.

Health Whilst many geographers have studied health care with emphasis on services, facilities and their locations, others have more explicitly investigated health. Traditionally, within human geography health has been studied (and thus implicitly defined) by medical geographers through the mapping of indices of disease and morbidity and of death and mortality. Such mappings are characterised by an emphasis upon spatial distributions of disease and death, often with the purpose of linking different environmental or spatially delimited social factors into causal or influential relationships. Health has thus been conceptualised as an environmentally determined state, and only usually recorded as 'ill-health'. It is notable that 'health' has thereby been biomedically defined and often unquestioned by geographers seeking to describe and explain clustering and dispersal of various diagnoses. More recently, medical geography has arguably seen what has been termed a 'cultural turn' in which contested geographical concepts such as LANDSCAPE and PLACE have been critically evaluated in relation to therapeutic processes and spaces, notions of health and ill-health and relations of power.

Feminist writings on the BODY have increasingly become influential in such a rewriting of the subdiscipline and have influenced an increasingly critical take on what constitutes health and more especially on how health is embodied. In accordance with other influential rewritings of what health means (such as the World Health Organization, which defines health as a combination of mental and physical well-being), geographical understandings of health have increasingly become informed by qualitative materials, ones which emphasise the contestations and resistances of biomedical inscription and its meanings for embodied, everyday geographies. Women's bodies and experiences of such inscriptions, and associated treatments, have played a key role in this critique of medical geography, as well as in challenging notions of health and illness more generally.   HP

See also AIDS/HIV.

Hegemony The Italian MARXIST theorist Antonio Gramsci defined hegemony as the capacity of a dominant group to control not through visible rule or the deployment of coercion or force but through the manufacture of consent which ensures the willingness of people to accept their subordinate status. This is achieved not through POLITICS, but through social and cultural institutions in civil society. Cultural norms and standards reproduced in institutions (FAMILY, education, church, media) preserve the *status quo* and the sense that citizens have a vested interest in its maintenance (even though the *status quo* is in the best interest only of the dominant classes). Hence, acceptance of things is managed through the detailed scripting of some of the most ordinary and mundane aspects of everyday life rather than through spectacular confrontation between the STATE and its citizens.

Gramsci's notion of POWER and DOMINATION in capitalist social relations was not of an unassailable monolith. Hegemony is always fragile, an 'unstable equilibrium between the CLASSES' (Gramsci, 1971: 245) consisting of the different interests of the institutions making up a hegemonic assemblage. The concept of hegemony is relevant to feminist analysis as it broadens the definition of 'the POLITICAL' beyond the narrow political landscape of state politics to encompass civil society and the contested cultures of family and media. JS

Heritage Described as 'a NOMADIC term, which travels easily, and puts down roots – or bivouacs – in seemingly quite unpromising terrain' (Samuel, 1994: 205), heritage refers to a range of historically valued phenomena, events, LANDSCAPES, material artefacts, in a variety of locations and media, whose status have in recent times become greatly inflated as well as contested. Rather than a unified object, it is an ever-changing process involving diverse practices. Its schemes may conserve landscapes or places but also revalourise them and change their uses, with an accent on 'historic' values and on marking out and staging aspects of the past. Concern about the creation of sanitised, packaged versions of the past that erase histories of OPPRESSION, EXPLOITATION and conflict is at the heart of numerous critiques of the 'heritage industry'. Tracing a huge expansion of 'heritage' in Britain since 1975, critics have connected it to reactionary NOSTALGIA in the context of industrial decline, and to conservative attempts to bolster a sense of national IDENTITY (Hewison, 1987; Wright, 1985). It has also been linked to the commodification of the past and attempts to market the distinctiveness of places. Other commentators have questioned general critiques, however, and emphasised the tangled genealogies of heritage practices and their diverse connections with more 'progressive' politics, popular MEMORY and pedagogy (Samuel, 1994); in addition, they have argued that heritage should be studied in terms of its enactment or 'utterance', with criticism challenging reactionary attempts to reify and organise its field as 'Heritage-*qua*-Thing' (Crang, 1994). These latter approaches allow recognition of the importance of heritage for recovering and giving presence to the interests of SUBALTERN groups for potentially EMANCIPATORY ends, as explored by organisations such as 'The Power of Place' in relation to struggles over GENDER, RACE and ETHNICITY in Los Angeles (Hayden, 1995).   DP

Heterosexual/heterosexuality/heterosexism 'Heterosexuality: not normal – just common.' This definition of heterosexuality, which appeared on the front of a tee shirt in New York, aptly summarises the challenge which has been posed over the past decade to the seemingly unassailable normativity of heterosexuality. The term heterosexual, from the Greek *heteros* meaning 'other' or 'different', is classically defined as those who exclusively desire and/or engage in sexual relations with members of the opposite sex. Heterosexuality is believed to be the most common form of sexuality in human society and has been traditionally privileged on the basis of the central role which it apparently plays in the process of biological reproduction. Consequently, although heterosexuality exists as simply one of a variety of different human sexualities it has become socially constructed as normative.

Challenges to the normativity of heterosexuality have been consistently raised by feminists and QUEER theorists. They have done so by raising a series of questions. Why, as the tee shirt suggests, should prevalence – the fact that most people 'do' it – make heterosexuality normal? Why should biological sex (which is defined solely by the presence or absence of particular anatomical features) determine either gender roles or sexual orientation? Why, as Coote and Campbell (1982: 213) enquire, 'should heterosexuality be defended more vigorously than any other precept on which our society is supposed to be founded – more than religion, equality or freedom of expression?'

Heterosexuality has since been retheorised as a socially constructed rather than an ESSENTIAL or a natural phenomenon. Rather than comprising simply a series of sexual acts, heterosexuality is better understood, as Rich argues, as a 'compulsory' set of relations which are produced not only at the level of the body but also at the level of DISCOURSE and social practice. Some feminists (see Dworkin, 1976) have argued that heterosexuality is responsible for naturalising and institutionalising a set of ideals (the idealisation of heterosexual romance and marriage for example) which perpetuate men's DOMINATION of women. Within this context alternative forms of sexuality become, as Chouinard and Grant (1995: 40) suggest, 'profoundly political because they both resist and threaten the oppressive system of male dominance which requires women to be heterosexual for its very existence'.

Attention has also been given to the role of heterosexism in sustaining the hegemony of heterosexuality. Heterosexism refers to a set of social practices, ideas and behaviours which act to reinforce the belief that heterosexual relations are the only truly 'natural' or 'normal' sexuality and that all other types of sexuality are consequently deviant, sick or 'unnatural'. Also constructed as a belief in the superiority of heterosexuals and heterosexuality it is evidenced by an absence of interest in and/or respect for non-heterosexual persons, practices, events or activities (Sears and Williams, 1997). Heterosexist practices range from the overt and violent (HOMOPHOBIC attacks and verbal abuse) to the subtle (asking lesbians, but never oneself, 'How did you get that way?') and are directed not only at gay people but also at transgendered and bisexual people.

Heterosexuality and homosexuality are classically constructed as BINARY opposites. Queer theorists have recently called this binarism into question suggesting that sexuality might be more usefully constructed as a continuum along which people may be variably positioned. Whilst some people operate at the extremes most are situated at some point along it: they may contemplate heterosexuality but would never act on it, or may contemplate it and possibly act on it, or may contemplate it and actually act on it, but only under certain contingent conditions (e.g. when intoxicated). Sexuality is understood as not necessarily fixed, but mutable and thus subject to change over TIME – and over SPACE.

As heterosexual behaviour is viewed as normative there are few strictures on the way in which it can be performed either in public or in private spaces. Conversely, other 'deviant' sexualities are deemed sufficiently threatening that specific laws have been introduced to regulate the ways in which they may be enacted in the public domain – and, as the case of Operation Spanner in the early 1990s in the United Kingdom demonstrated, even in the privacy of one's own home. Factors such as legal regulation, social SURVEILLANCE, the implicit heterosexual coding of PUBLIC spaces and heterosexist practices affect the PERFORMANCE of sexualities, creating spaces of inclusion and exclusion which combine to form a very evident geography of sexuality which is of increasing interest in geographical praxis (Bell and Valentine, 1995; Knopp, 1992).   BP
See also COMPULSORY HETEROSEXUALITY; GAY; LESBIAN; LESBIAN CONTINUUM.

Hexis A term used by Bourdieu to refer to the ways in which class and gender characteristics, status and CULTURAL CAPITAL are made through the stance, style, gestures and movements of the BODY.   LM
See also HABITUS.

Hierarchy In addition to highlighting the use of binary oppositions in dominant Western thought, feminists have demonstrated the inherent hierarchicalisation of the opposed concepts. BINARIES insist on absolute difference, but never DIFFERENCE with EQUALITY, one term is always privileged.   JS

High-technology industries/information-technology industries Terms that refer to industries in which the application of new scientific techniques and technological innovations are significant, including, for example, computer-based industries and new pharmaceutical industries based on advances in genetic manipulations. These new industries are often located in 'new industrial districts' or parts of a nation-state where older 'smokestack' industries were never significant; they are frequently based in suburban science parks. The high-status jobs in these industries typically are jobs for men, whereas women are concentrated in lower-status, assembly jobs. Doreen Massey (1995) has explored the ways in which MASCULINIST assumptions structure not only the occupational hierarchy also but the organisational culture of high-tech firms. LM

Historical materialism See MATERIALISM

Historicism The argument that all social phenomena are historically determined; in contemporary usage in cultural geography it implies the reading of traces of the past in TEXTS and LANDSCAPES using the methods of the new cultural criticism; a recognition of the ways in which causal or structural influences are mediated through discursive practices. The term cultural poetics is associated with the new historicist school and has been used by the geographers Barnes and Gregory (1996), as the subtitle to their book *Reading Human Geography*, reflecting the growing emphasis on the discursive construction of geographic difference in the 'cultural turn' being experienced in human geography. LM
See also DISCOURSE; NEW CULTURAL GEOGRAPHY.

Home An elusive and complex term. Although scholars have always acknowledged the great importance of ideas of 'home' throughout history, its precise meaning has proved harder to define. The association of home with a house or place of physical shelter is, as Benjamin argues, a 'small part of the story' (Benjamin, 1995: 1). Shelter is certainly a basic need and in societies where this need has been commodified there are HOMELESS people who are left physically vulnerable. Yet, the testimonies of homeless people and people who are institutionalised teach us to look beyond physical need alone: issues of control, personal dignity and privacy seem to be some of the defining features of home (Daly, 1996). Indeed, the word home can be seen as a vessel in which a tangle of abstract, cultural concepts are found. To understand these meanings demands a sensitivity to different historical and geographical contexts. It is also necessary to examine critcally some of the ways in which dominant assumptions and attitudes associated with the home can function IDEOLOGICALLY. For example, during World War II both the US and British governments strategically manufactured symbolic images of home and national HOMELANDS in order to foster patriotic sentiment which fed into the post-conflict commitments to provide 'Homes for Heroes'. These post-war images were of concrete places as well as of particular social constructions of decent and respectable domestic FAMILY life where gender roles were quite literally mapped out by housing design. Home life was organised around a series of DUALISMS such as home versus WORK and PRIVATE versus PUBLIC life – dualisms epitomised by post-war suburban developments. The architecture of the houses also shaped gender relations, signalling, for example, that the woman's place was not only in the home but also carrying out domestic tasks in the kitchen. There is a large feminist literature which critiques the oppressive nature of these dominant ideas and geographies of the 'dream home' (Friedman, 1965; Hayden, 1984). At the same time, authors such as bell hooks have argued powerfully that we should not seek to impose a revised ETHNOCENTRIC notion of what constitutes the home but instead listen to people's different experiences of home life, homelessness and journeys towards and away from home (hooks, 1990). Indeed, the idea that home can be equated with a fixed, safe place of residence and a permanent social support network has been questioned by novelists, poets, feminists and geographers amongst others (Massey, 1994c; Rushdie, 1991). Physical and psychological security of place is instead seen as a rare and privileged fate that many women have never experienced.   FG

See also BUILT ENVIRONMENT; CITY.

Homeland The place to which a NATION or ETHNIC group feels itself to belong. It is a particularly resonant notion for DIASPORIC COMMUNITIES which retain personal links, MEMORIES and myths of their countries of origin. The 'myth of return' can bind together migrant people and their descendants. Often it is strongest for men, paralleling the myths of the HOME as a source of belonging, identity and security through which the masculine SELF is constructed in opposition to threatening others (Massey, 1992). Escape from home may therefore represent freedom for women but a frightening loss of boundaries for men.

However, it is often migrant women who manage links with the homeland through close ties to remaining family members. Those who earn money outside the home send remittances, often more reliably than men (Fitzpatrick, 1986), and ensure that children visit or hear stories about their ancestors and kin (Lennon *et al.*, 1988).

Homelands are explicitly gendered when they become the markers of RACE. They are frequently represented as 'mother' countries, drawing on imagery of nurturing and protection. Cohen (1993) suggests that negative feelings repressed from this idealised maternal embodiment are projected on to outside others, contributing to the intensity of nationalist rhetoric and activity. In a related metaphor, the FAMILY is used to represent the nation, naturalising interconnections between its CITIZENS and disguising power relationships within the social hierarchy (McClintock, 1993).

The notion of homeland also draws on domestic imagery of containment and limited horizons when it is applied to the territory allocated to native peoples in South Africa, for example (Shurmer-Smith and Hannam, 1994). Men who are confined to the homeland are feminised as dependent.   BW

Homeless Traditionally, the term pertains to individuals without accommodation. The definition has recently been extended to include those at risk of losing their accommodation as a result of, for example, DOMESTIC VIOLENCE, eviction or other unsatisfactory living arrangements. Homelessness intersects with GENDER, RACE, CLASS and AGE: particularly vulnerable groups are single mothers, ethnic minorities, people on low income, children, youths and the elderly (see Daly, 1996).

Explanations for homelessness vary according to place and different people's experiences. Those usually given are ECONOMIC RESTRUCTURING, poverty, lack of affordable housing, cutbacks in welfare programmes and deinstitutionalisation (Baer, 1986; Bluestone and Harrison, 1982; Dear and Wolch, 1987; Ralston, 1996; Wolch *et al.*, 1988).   RK

Homework Homework is a global phenomenon defined as work performed in domestic premises, usually for piecework payment. Those engaged in homeworking are economically and technologically dependent on one or more coordinating firms (Meulders *et al.*, 1994). Homework is a feature of certain sectors, particularly those that are highly feminised, such as garment manufacture (Phizacklea and Wolkowitz, 1995).

It has been suggested that women choose flexible forms of employment, such as homework, because it allows them to combine work and family responsibilities (see Hakim, 1996). A more realistic explanation offered by Allen and Wolkowitz (1987) is that homework is a feature of ECONOMIC RESTRUCTURING and a means to cut production costs.   RK
See also COTTAGE INDUSTRY.

Homoeroticism The representation or experience of same-'sex' aesthetic, sensual or especially sexual appreciation or DESIRE; may also be a set of practices engaged in as an element of AESTHETIC, sensual or sexual relations or activity; may be conscious or unconscious, acknowledged or denied, intended or unintended or some combination of these; frequently seen as manifesting itself as a subtext underlying many otherwise heteronormative practices, relations, images or NARRATIVES.

Homoeroticism's role in struggles over heteronormativity and the sex–gender system is much debated. Bell *et. al.*, (1994) suggest that where homoeroticism's crucial role in constructing HETEROSEXUAL MASCULINITIES and FEMININITIES is made explicit (as in hyper-'masculine' GAY male and hyper-'feminine' LESBIAN identity PERFORMANCES), it may heighten these struggles; others (Arens, 1981: 5–8) argue that in more traditional contexts (such as male contact sports) it may serve primarily to reinforce traditional norms and power structures. Still others, including Andrea Dworkin (1981) argue strongly that homoeroticism can reinforce traditional norms and structures even while being fully acknowledged and celebrated, as in much male-orientated PORNOGRAPHY. Mark Blasius (1994), meanwhile, argues that 'erotics' (defined, more narrowly, as the use of bodily and other pleasures in relationships), especially gay and lesbian erotics, may be key to the development of a radical new (and more egalitarian) social ethic.

The distinction between homoeroticism and homosexuality revolves around the more general (and problematic) distinction between eroticism and SEXUALITY.   LK

Homophobia A deceptively complex term whose meaning has evolved considerably over its relatively short 20-year or 30-year lifespan; refers generally to the fear and/or loathing of same-sex sexual activity and/or desire and of people who identify in some way or other with this (especially GAY and LESBIAN people); often used colloquially to refer to HETEROSEXUAL-supremacist policies, practices, attitudes and values. However, the term 'heterosexism' is now increasingly used for this latter purpose, leaving 'homophobia' with a more limited and etymologically literal meaning.

Homophobia's origins are contested. Frequently it is seen as deriving from ignorance about the nature of SEXUALITY and DESIRE generally and from a host of fears about the imagined personal and social consequences of same-sex sexual activity and/or desire in particular. These in turn are viewed as deriving from deep-seated cultural values (and, some argue, anxieties) about gender and sexual norms as well as the nature and meaning of MASCULINITY, FEMININITY, sexuality, reproduction and the various social institutions associated with these (which is to say virtually all social institutions!). Most feminist perspectives on homophobia see it as at least having the effect, if not deriving from a need or intention on the part of dominant social interests (especially those of straight men), of policing sexist gender norms; others, however, see homophobia and heterosexism as producing sexism. Still others regard sexism, homophobia and heterosexism as elements of a single, integrated sex–gender system, none of which is logically or functionally prior to any other.   LK

Homosexual/homosexuality A highly problematic catch-all term referring to same-'sex' sexual or erotic activity, practices, DESIRE or IDENTITY; also refers to individuals, COMMUNITIES and CULTURES which include or are organised, to some degree or other, around the accommodation of the same.

According to many theorists and historians of sexuality (Greenberg, 1988; Katz, 1995) the concept has origins in the scientisation and, in particular, medicalisation of sexual activity and desire generally, and same-sex sexual activity and desire in particular, in the nineteenth century. These scholars argue that prior to the modern era (and especially nineteenth and twentieth centuries) sexual activity and desire were less organised socially around the (socially constructed) gender of individuals' sexual object choices. Indeed, some argue that the sexual itself was a less distinctively conceptualised realm of human experience prior to the modern era. This is not to suggest that sexuality and gender as categories did not exist prior to the modern era nor that sexual objectification was not focused, at times, on people's GENDERS. Rather, it is to suggest that the concepts and language through which sex, gender, sexual activity and erotic desire were all understood included a wider range of possibilities than in the current era and, in particular, were not structured around a homo–hetero BINARISM. Some feminist interpretations of the construction see it as at least having the effect, if not deriving from a need or intention on the part of dominant social interests (especially those of straight men), of policing sexist gender norms.

Homosexuality is usually distinguished from homoeroticism by its specific association with sexual (as opposed to less specifically sexual, sensual and/or aesthetic) activities, practices, appreciation or desire. But this begs the question of what distinguishes the sexual from the erotic, something which most feminists (and many others as well) see as highly problematic.

Another problem with the concept is its grounding in binary conceptions of both sex and gender, and the conflation of these two with each other. In forcing sexuality into a binarism defined by human beings' allegedly self-evident anatomical sexes, and then mapping these uncritically on to the SOCIAL CONSTRUCTION 'gender' (as the supposed 'natural' basis of gender distinctions, which are then similarly seen in binary terms), the socially constructed nature of both, and the POWER relations embedded in these constructions, are obscured. LK

Household A term referring to a residential unit where people share reproductive activities and expenditure. The household is usually defined as a spatial unit which may or may not include members related by blood (Chant, 1997: 5–6). Varying across cultures, households may be male-headed or female-headed and comprise a range of compositions (nuclear, extended, and so on). CM
See also FAMILY.

Housework The work involved in running a household. This includes a diverse range of tasks such as cooking, cleaning, doing laundry and DIY: in short, the WORK that ensures smooth running of the domestic economy. Housework is segregated by sex, with women tending to have overall control and performing most of the tasks.

Housework was the cinderella of the work world until the 1970s when second wave feminists paved the way for a series of research programmes on the social organisation of the household. Here housework was conceptualised as an occupational role (Oakley, 1974) and, under MARXIST THEORY, as a separate mode of production (Delphy and Leonard, 1992). More recently studies have stressed the connections between the gendered organisation of housework and the FEMINISATION OF THE WORK FORCE. A key theme is the entrenched nature of the sexual division of housework: there has been

relatively little change in the housework that women and men do, in spite of the increased participation of women in paid employment and the rise of male unemployment (McKee and Bell, 1985; Morris, L., 1993). Other avenues of enquiry include sequential scheduling (Hanson and Pratt, 1995) and the division of housework in dual-career households (Gregson and Lowe, 1993). Increasingly, the division of housework is thought to offer a window on the nature of relationships, for example between retired adults (Cliff, 1993). Housework is seen as a way by which people indicate what they think of themselves and others, and what they regard as just or fair.   HC
See also DIVISION OF LABOUR: DOMESTIC, INTERNATIONAL, GENDER, SPATIAL.

Housing, allocation of Distribution of HOUSEHOLD and FAMILY types amongst different housing tenures which creates a divided city and perpetuates GENDER inequalities (Morris and Winn, 1990). Research emphasises the economic constraints which disadvantage women in the private market (Gilroy, R., 1994; Munro and Smith, 1989), and connections between the spatial marginalisation of women and housing policy (Heenan and Gray, 1997).   HC

Human capital theory An individual's stock of abilities, skills and knowledge can be defined as their human capital. Human capital theorists explain individuals' educational and labour market choices as decisions about investment in their human capital in relation to labour-market opportunities and explain variation in individuals' wages as resulting from differences in their human capital. The differential between women and men's pay is held to result from women's lower investments in education, training and work experience, linked to their expectations of spending time out of the labour market in domestic activities (Mincer and Polachek, 1974; Polachek, 1975). The approach has been criticised for ignoring gender biases in the valuation of skills and analysis of the PATRIARCHAL power relations in the home and in the labour market which create the situation within which women make their educational and labour-market choices.   SB
See also DIVISION OF LABOUR: DOMESTIC, INTERNATIONAL, GENDER, SPATIAL.

Humanism See HUMAN NATURE

Human nature The characteristics and capacities shared by human beings, by virtue of which we are a distinct species. Theories of human nature vary according to whether they stress human continuity with other animals ('naturalism') or whether they see the human/animal boundary as qualitatively different from the distinctions between other animal species.
- *Strong naturalism*: for Midgely the nature of a species is 'a range of powers and tendencies, a repertoire, inherited and forming a fairly firm characteristic pattern, though conditions after birth may vary the details' (1979: 23). She draws on ethology for evidence of what humans share with other

animals – such as a strong dislike of being stared at. Sociobiologists are also strong naturalists, with their reductionist view of human nature in terms of the 'competition' between genes (Dawkins, 1978).

- *'Non-reductive' naturalism*: Benton describes himself as a 'non-reductive' naturalist, stressing human/animal continuity yet describing humans as bearers of special 'emergent powers' (1993: 17). For him, our language speaking and our complex social life do not set us aside from the rest of nature – they are just what is special to our *human* nature, based in the structure of the human brain (Chomsky, 1996).
- *Religious dualism*: historically this contrasted human beings to supernatural beings on the one hand and animals on the other. Human nature had aspects of each, but while the soul was eternal and essential to human identity the BODY was temporal and destructible (Trigg, 1988).
- *Culturalism*: a secular dualism; for example, Levi-Strauss held that the irreducibility of the cultural is demonstrated by the universal taboo against incest (Levi-Strauss, 1969).

Positions on this issue affect but do not decide questions of the *characteristics* of human nature. Are humans naturally individualist (as Hobbes believed), or COMMUNITARIAN (as Marx held)? Are humans RATIONAL (as Hegel believed), or largely instinctually driven (as Freud argued)? Are humans 'basically good', as humanist psychology believes, Or is hatred and opposition to society a permanent part of human nature? Positions on these questions have moral and political implications (Berry, 1986; New, 1996). If humans are perfectible, we should try to build societies in synchrony with human nature (Maslow, 1973). If humans are incorrigibly individualist, firm government is necessary to enforce cooperation.

In a trivial sense human nature must permit all that humans ever do. A non-trivial theory of human nature must describe the mechanisms underlying human needs, capacities and motivations, in what conditions these become active and their range of likely effects. A capacity could be part of human nature yet usually remain an unrealised potential, or it could be sex-specific or age-specific. Theories of human nature are nowadays most openly expressed through psychological theories and models. These arguably have implications for what is socially possible, for example whether socialism is achievable (New, 1991). Other disciplines, such as politics, ANTHROPOLOGY and economics, often make implicit assumptions about human nature. Thus economic theory assumes individuals will normally pursue their own economic interests and instrumental goals (Elster, 1984).

Assumptions about human nature have frequently been used to 'naturalise' relations of domination. In the Hobbesian view, competitiveness is said to be a universal aspect of human nature, resulting in the domination of weaker individuals or groups. Human nature theories are frequently 'gendered' or 'raced', attributing fixed natures to women or certain ethnic groups from which relations of domination inevitably follow. The use of the concept to legitimise oppression has led to its rejection by many feminists and social constructionists, on the grounds that human actions, customs and inclinations vary and change and are best understood as effects of discourse. Opposition to the idea of human nature is part of the more general POSTMODERN critique of humanism and of the 'subject' (Rosenau, 1992). In extreme versions of this culturalist, antinaturalist position features of bodies, such as sex, are seen as constructed within CULTURE rather than given (Butler, 1993).

From a REALIST point of view the concept of human nature does have value. Human embodiment together with human mental capacities establish the possibilities of human culture – including that of creatively transcending some bodily limitations. Humans have physical and psychological needs at all stages of development, which can be seen as part of our nature. Unloved babies fail to thrive; mistreated adults respond in various determinate ways. Knowledge of human needs has ethical implications, and human-nature arguments can be used to oppose relations of domination rather than to legitimise them (Doyal and Gough, 1991). Human nature need not be seen as fixed and static but as potentialities which could be realised in many ways in different social contexts.   CN

Hybridity Of seventeenth-century origin in the English language, hybridity refers to a condition resulting from the interbreeding of two different species or varieties of plants or animals. More generally it applies to a condition derived from combining characteristics from heterogeneous sources or incongruous elements. It is associated with a hybrid, meaning the offspring of human groups who transgress established social categories, including differences of CLASS, RACE or ETHNICITY. The term cultural hybrid is used also to refer to groups as a mixture of local and non-local influences. Their character and cultural attributes are a product of contact with the world beyond a local place. The concept therefore challenges the coherence implied in structured social and spatial categories such as SELF/OTHER, insider/outsider, local/global. Young (1995) suggests that heterogeneity and diversity have become the self-conscious IDENTITY of MODERN societies and that the hybrid nature of cultural exchange reinforces the difficulties of separating the local from the global. Notwithstanding this evidence, claims to specific places based on adherence to an exclusive identity (e.g. national, ethnic, class), with its implications of purity, continue to contribute to the maintenance of BINARY divisions and such ESSENTIALIST categories as outsiders and OTHERNESS (Hall, 1991).

Within the Western social sciences numerous models have been developed to analyse the nature of cultural interchange. These included diffusionism and evolutionism which, within the context of nineteenth-century European IMPERIALISM, conceptualised cultural encounters as deculturation of the less powerful society and its transformation towards the norms of the West. More recently, certain strands of POST-COLONIAL criticism (which has itself been interpreted as a form of intellectual hybridity) have stressed separateness rather than acculturation through intercultural exchange. The construction of antithetical groups, self and Other, coloniser and colonised, have followed, with the seconds of each group knowable onlt through a necessarily false REPRESENTATION. In consequence, the essentialist categories which it seeks to undo have been reproduced. Alternative interpretations of cultural hybridity have been suggested by Homi K. Bhabha (1991, 1994) who refers to the 'hybrid displacing space' which develops through interactions between dominant and subordinate cultures and which has the effect of challenging and resisting the political authority and authenticity of the dominant culture.

Numerous applications of hybridity find common ground with feminist approaches to gendered identities and ideas of the subject as multiple and fragmented (Duncan, N., 1996a). An appeal to hybridity as a description of new cultures and subjects is associated with the BODY, for example, as an active mode of making new connections. Donna Haraway's (1991) notion of CYBORG, a hybrid of animal and machine, challenges Western evolutionary, technological and biological narratives by demonstrating the unstable nature of social identities and otherness. Further illustrations exist in an expansion of what is regarded as 'the political', the meeting of different peoples in CITIES, the intermixing of bodies in sexual activity and the varied outcomes of imperial and post-imperial contact zones (Pile and Thrift, 1995). Applications of the concept occur also in attempts to replace the DUALITIES implicit in NATURE–CULTURE or nature–society by new hybrid representations and ethical considerations in which the human is decentred and no longer in opposition to the non-human.   MB

See also HUMAN NATURE.

I

Icon/feminist icon The term icon literally means an image or a statue; it is used in a specific sense to refer to a woman who is a conspicuous symbol of success in ways that challenge conventional assumptions of FEMININITY; often used in association with Madonna – the singer, not the Virgin Mary, who is, of course, an iconic image of conventional femininity.   LM

Identity One of the keywords of the 1990s, 'identity' is a complex and highly contested term, central to the recognition and articulation of DIFFERENCE. Notions of identity are invoked in defining our sense of SELF and in marking ourselves off from various culturally constructed OTHERS (Pile and Thrift, 1995). Whereas older theories of identity posited a stable and core sense of self, often closely tied to differences of social class, recent theories have asserted the possibilities and problems associated with a more 'HYBRID' (unstable, mixed and multiple) notion of identity, often conceptualised in highly voluntaristic terms as part of an individual 'lifestyle' choice. Between these two extremes, the term has also come to be associated with a range of SOCIAL MOVEMENTS (including those around gender and generation, ethnicity and NATION, NATURE and ENVIRONMENT) that has complicated the allegedly universal basis of class-based politics, giving rise to an often heated debate about the 'sectional' nature of IDENTITY POLITICS which 'hardly ever mobilizes more than a minority' (Hobsbawm, 1996: 44).

Several theorists have related the process of identity formation to the social conditions of late-modernity or POSTMODERNITY. According to Anthony Giddens (1991), for example, contemporary identities should be theorised as a reflexive project, shaped by the institutions of late-modernity and sustained through narratives of the self that are continually monitored and constantly revised. Others adopt a more RELATIONAL approach, theorising identity as a contingently defined social process. Margaret Somers's work is representative of this latter trend, approaching identity as a DISCURSIVELY constituted social relation, articulated through NARRATIVES of the self:

> Narrative identities are constituted by a person's temporally and spatially variable place in culturally constructed stories composed of (breakable) rules, (variable) practices, binding (and unbinding) institutions, and the multiple plots of family, nation, or economic life. Most importantly, however, narratives are not incorporated into the self in any direct way; rather, they are mediated through the enormous spectrum of social and political institutions and practices that constitute our social world (1994: 635).

While identities may be plural and dynamic, they remain subject to social regulation through cultural norms and expectations. Our sense of *who we are* is related in quite fundamental ways to *where we are*: identity is spatially as well as socially constituted (Keith and Pile, 1993). Thus, the multiple forms that Frank Mort (1989: 169) describes for contemporary MASCULINITIES ('we are not in any simple sense "black" or "gay" or "upwardly mobile"') depends on where we are ('at work, on the high street and the spaces we are moving between'). What it means to be a GAY man in London revolves around particular places (such as Soho) and what those places have come to represent.

The spatial constitution of identity through the social recognition of *symbolic boundaries* may even have gained in strength as the *literal* boundaries of COMMUNITY have loosened and as the interconnections between places have proliferated: 'The presumed certainties of cultural identity, firmly located in particular places which housed stable cohesive communities of shared TRADITION and perspective, though never a reality for some, [have been] increasingly disrupted for all' (Carter *et al.*, 1993: vii). Geographers have therefore been searching for new theories of SPACE and PLACE which are more consistent with the increasingly hybrid character of contemporary identities. Doreen Massey's (1994b) argument for a 'progressive' or global sense of place can be read in this light. Rather than searching for a lost

'AUTHENTICITY' based on distorted MEMORIES and a NOSTALGIC search for roots, the identity of particular places can be understood as a distinctive articulation of social relations, linking a variety of scales from the GLOBAL to the LOCAL. According to Massey, places are not bounded areas but are porous networks of social relations.

These ideas are consistent with recent feminist theories of identity which seek to repudiate the false certainties of ESSENTIALISM by arguing that all social identities are culturally constructed, changing from time to time and from place to place (though many feminists would also recognise that groups and individuals may wish to declare a strategic sense of 'fixity' in order to make specific political gains). Judith Butler's work on the permeable boundaries of gender can be seen in this light, arguing that our identities are a PERFORMATIVE creation but that they are made *culturally intelligible* through regulatory grids such as those associated with an idealised and COMPULSORY HETEROSEXUALITY (1990: 135). Identities, she argues, are a 'regulatory fiction', actively forged within a regulatory framework of socially sanctioned norms and expectations.

The cultural politics of identity are also central to the work of feminist artists such as Cindi Sherman whose photographs present a series of images of her own 'identity', shaped through the visual conventions of the cinema and subject to the MASCULINIST GAZE. The choice of 'image' may be voluntary on Sherman's part but its interpretation depends on the culturally constructed knowledge that the viewer brings to the photograph (Williamson, 1986). Similarly, the identity of 'lipstick lesbian' or 'gay skinhead' may be intended in a knowing or ironic way but is always subject to opposing interpretations with sometimes unpredictable consequences for the individuals who adopt such a 'look' (Bell *et al.*, 1994). Questions of identity are therefore hard to divorce from a wider cultural politics of REPRESENTATION.   PJ

Identity politics See IDENTITY

Ideology A complex term defying simple explanation. Eagleton (1991) highlights two broad traditions of thought; the first, rooted in Marx, 'has been much preoccupied with ideas of true and false cognition, with ideology as illusion, distortion and mystification' (see FALSE CONSCIOUSNESS); the second is a more sociological approach concerned with the function of ideas in society. Post-structuralists have largely rejected the term because of its associations with a universal assumption of a singular TRUTH, and instead prefer to talk about POWER relations and DISCOURSE. Althusserian thought has been important in socialist feminist discussions of ideology (Althusser, 1971 [1985]) shaping understanding of the way in which PATRIARCHY operates, at all levels of society, to reinforce women's oppression (Barrett, 1980).   JW

Imagined community This notion derives from the term coined by Benedict Anderson (1983, 1991) to define the special character of nations whereby even though individual members will never 'know most of their fellow-members, meet them or even hear of them, yet in the minds of each lives the image of their communion' (1991: 6). It has quickly been accepted as a fruitful way to conceptualise the ties of identity between collectivities. Although primarily used to describe relationships at the national scale, Anderson goes on to argue that all communities are imagined and can be distinguished 'by the style in which they are imagined'.

In elaborating the concept, Anderson exposed the masculinist thinking which lay behind it, for example using willingness to die in battle for the collectivity as evidence of its importance. He does not examine the reverse process of the construction of the national citizen, which in fact involves its gendering as male (Sharp, 1996). An inclusive definition would acknowledge the diversity which must be addressed in creating equality, with openly negotiated priorities (Young, 1989b).

The mythical unity of national 'imagined communities' is maintained and reproduced by symbolic 'border guards' (Armstrong, 1982). These are linked to specific cultural codes of dress, behaviour, customs, religion and language. Gendered bodies and sexuality play pivotal roles as markers, for example in rules governing the 'proper' behaviour and clothing of women. Such codes usually keep women in inferior positions in power relations (Yuval-Davis, 1997).

Notions of DIASPORA construct global imagined communities, which draw on different sets of identities from those at the national scale. Indeed diasporic identities 'often involve an explicit refusal to play the assimilation game' (Cohen, 1993). Diasporic peoples are 'at home' locally, but retain significant ties globally with their countries of origin (Clifford, 1994). Although diaspora discourse is unmarked, it is implicitly also masculine, paralleling other metaphors of TRAVEL (Wolff, 1993) and yet women play central roles in sustaining and transforming diasporic communities (Walter, 1997).   BW

See also MIGRANT/MIGRATION.

Imagined geography A term used by Edward Said in his work to refer to the social construction of a discursive or imagined distinction between East and WEST; it is also sometimes used in a similar way to Benedict Anderson's term IMAGINED COMMUNITY.   LM

See also NATION/NATION-STATE/NATIONALISM; ORIENTALISM.

Imperialism Edward Said defines imperialism as 'The practice, the theory and the attitudes of a dominating metropolitan centre ruling a distant territory' (1993: 8). Colonialism, which generally involves settlement and direct rule of a TERRITORY, has by and large ended, but cultural, economic and political aspects of imperialism arguably remain significant in contemporary society, commonly referred to as 'neo-imperialism'. Imperialism has historically involved a wide variety of different agents, and the forms and practices of imperialism have also varied across national and cultural contexts and over time. Traders, missionaries, educationalists, government agents, military and industrialists have all shaped the complex networks which bind people in different places into the unequal relationships with one another which combine to produce the practices and experiences of imperialism. Not only are all these imperial practices themselves gendered but also they have had considerable impacts upon the nature of gender relations in both the metropole and those areas within the imperial ambit. The often close relationship between GEOGRAPHY as a discipline and imperialism has attracted some attention (see Driver, 1992; Godleskwa and Smith, 1994), perhaps most famously in Edward Said's (1978) reference to the role of 'IMAGINATIVE GEOGRAPHIES' which in an epistemic sense enabled the practice of imperialism. The importance of international politics in shaping the form of contemporary imperialism means that geography in the form of geopolitics is also a significant element of neo-imperialism. Contemporary neo-imperialism is closely linked to the international agencies promoting development and the expansion of transnational capitalism.   JR

See also CULTURAL IMPERIALISM; POST-COLONIALISM.

Industrialisation Industrialisation refers to the processes whereby industrial activity, typically manufacturing, takes root in a particular location. In a capitalist society this can take place through indigenous processes of product development or through the location of foreign direct investment. Industrialisation in Western economies from the mid-eighteenth century onwards has been associated with the gendered division of private and public space. Scholars have suggested that the separation between WORK and HOME was engendered by processes of industrialisation. Feminists have argued that in pre-industrial societies the HOUSEHOLD was the unit of labour, ensuring that men and women worked alongside one another. With industrialisation, however, home and work became separated and each of these spaces became gendered; public space being the domain of men and private, domestic space being gendered female (Davidoff and Hall, 1987). Moreover, the introduction of the family wage in the United Kingdom during the mid-nineteenth century bolstered expectations that men were breadwinners and women were housewives (Tilly and Scott, 1989). More recently, however, this orthodoxy has been challenged by some feminists who highlight women's continual involvement in paid industrial labour, despite the rhetoric of male breadwinners (Parr, 1990). Despite its association with MASCULINITY, women have always played a key role in industrialisation, often as cheap, factory labour. Many newly industrialising countries rely on young, female labour as a key resource to attract foreign capital to export processing zones (Elson and Pearson, 1981; Pearson, 1992).

Industrialisation is spatially uneven and as new locations are opened up for economic development increased competition will lead to closures and restructuring in other locations. Scholars such as Neil Smith (1990) and David

Harvey (1982, 1989) have argued that this process of uneven development is inevitable in capitalist society. Moreover, industrialisation has been shown to be historically as well as geographically uneven, and new regimes of accumulation or Kondratiev long waves follow periods of crisis in previously established systems of industrial organisation.   JW

See also DE-INDUSTRIALISATION; ECONOMIC RESTRUCTURING; FORDISM.

Informal sector Refers to the small-scale, often low-productivity, economic activities that largely lie outside state regulation or recognition. Found in both industrial and developing economies, informal sectors demonstrate a wide variety of participants, remuneration levels, linkages into formal activities and levels of technological development (Portes *et al.*, 1989). Such differentiation suggests that informal activities (ranging from petty street trading to unrecorded services and unregulated manufacture) run along a continuum with other economic sectors. Informal activities are primarily associated with urban areas, although they share characteristics of low-capitalisation, risk-aversion, use of family labour and non-regulation with small-scale agrarian production. Just as in formal WORK, gender SEGREGATION operates in informal sector activities, resulting in wage differentials by gender. Women's lack of access to credit, and lesser mobility compared with men between formal and informal sector activities (resulting in fewer skills and sources of capital), have been identified as limiting female opportunities in the informal sector, particularly in the South. The domestic location of many female informal workers (domestic service, laundry, outwork, etc.) reflects ideologies of FEMININITY and DOMESTICITY as well as discrimination against women in gaining access to higher-productivity informal work or formal employment. Indeed the male bias of the ideal-typical model of the informal sector conceals the degree of gender differentiation within the sector (Scott, 1991). Deregulation associated with economic neo-liberalism is further blurring boundaries between informal and formal activities, just as feminisation of the labour market occurs. Ideologies of domesticated femininity are being reworked as production, and to a lesser extent services, are restructured (Wilson, 1993).   SAR

Inscription See BODY

Institute of British Geographers See WOMEN AND GEOGRAPHY STUDY GROUP (WGSG)

International division of labour See DIVISION OF LABOUR: DOMESTIC, INTERNATIONAL, GENDER, SPATIAL

International Women's Day/International Year of the Woman Started in 1908 by US suffragettes and revived by SECOND WAVE FEMINISTS, International Women's Day is celebrated on 8 March. On this day women campaign for universal rights and SUFFRAGE and celebrate women's international

solidarity. The International Year of the Woman (1975) preceded the UN Decade for Women (1976–85).   NL
See also RIGHTS/EQUAL RIGHTS/EQUAL OPPORTUNITIES.

Interpretive communities A term that relates to Stanley Fish's work on reader-orientated or reader-response literary theory and criticism (Fish, 1980). Fish writes that different reading strategies produce different meanings about texts and that such reading strategies are socially constructed in a collective more than in an individual way. For Fish, the term interpretive communities denotes groups of 'informed readers' who are able to make sense of texts because they are competent both in linguistic and in literary terms. Moreover, the reading strategies of interpretive communities both reflect and reproduce certain assumptions that bind together a group of 'informed readers'. Other commentators have questioned the HEGEMONIC status of interpretive communities and have pointed to strategies of RESISTANCE and subversion, such as 'reading against the grain'. An important area of feminist literary theory has been an examination of the ways in which men and women read as well as produce TEXTS. For example, Margaret Beetham has shown how different interpretive communities of middle class female readers have been both identified and constantly reasserted by women's magazines (Beetham, 1996). Not only are such publications addressed to particular readers but the strategies employed by an interpretive community of readers produce certain textual meanings.   AMB

Iron John See MEN'S MOVEMENT; MYTHOPOETIC

Iron Lady See THATCHERISM

Irony A rhetorical strategy whereby meaning is conveyed in terms of its opposite or a referent that is simultaneously different but similar. The use of irony can destabilise fixed meanings, but does so in a way that might be seen as exclusive and/or elitist, relying on a 'knowing' recognition of its deployment (Hutcheon, 1994). Kathy Ferguson has written about the ways in which gendered SUBJECTIVITIES may be represented in ironic and mobile terms (Ferguson, 1993).   AMB

J

Jouissance Untranslatable into English, *jouissance* is a French word that is employed by critical theorists to refer to a Lacanian concept of excess PLEASURE that is nonetheless never reducible simply to PLEASURE. Covering the fields of both law and sex in French, *jouissance* refers to enjoyment in the sense of enjoying rights, but also sexual, spiritual and physical enjoyment. In the LACANIAN schema, *jouissance* is evoked as the (imaginary) joy of unification that subjects continually seek but can never completely access in the Symbolic. In *God and the Jouissance of The Woman. A Love Letter,* Lacan (1982) claims that 'The WOMAN' does not exist because the Law of the Father restricts her to a position of fantasy for men seeking unity. However, in opposition to this fantasy of Woman, Lacan posits the potential *jouissance* of women as that which exceeds the PHALLIC order. Partially excluded from the realm of the Symbolic, women function as a supplement to language: women thus do not complement men in the sexual order, but represent 'a jouissance beyond the phallus' (1982: 145).

Feminist psychoanalytic critics have seized on this notion of *jouissance* to theorise a place for women that is both within but in excess of the phallic order. For instance, Julia Kristeva (1980) has theorised *jouissance* in connection with a desire for a return to the pre-oedipal realm of the Mother, which can only be articulated through poetry and art. Similarly, Hélène Cixous (1990) argues that poetic language can extract meanings that contest the phallic definition of the signifier. Defining *jouissance* as the realm of hysteria, Luce Irigaray (1977) considers the possibilities of the daughter's return to pre-oedipal closeness with the Mother. Jane Gallop (1982) links *jouissance* to metonymy: in opposition to the phallic order, women's *jouissance* is characterised by concentric circles of pleasure that refuse the closure of phallic desire. In these configurations, *jouissance* has the power to disrupt the coherence of the Symbolic and thus articulate feminist resistance. However, non-Lacanian feminists criticise this model for relegating women to the realm of the pre-Symbolic where they are silent without access to LANGUAGE or REPRESENTATION.   WS

See also FRENCH FEMINIST THEORY; FREUDIAN THEORY.

Justice Justice is a word with a deep ring. It resonates with ideas of fairness, EQUALITY and of doing things 'right'. Historically, philosophers have addressed the question of how to realise visions of the 'good life' through the design of guiding principles and arbitratory rules. The impulse, it seems, has been to transcend the messy and mundane aspects of daily life in search of UNIVERSAL and sovereign ideals. Yet, behind this lofty and seemingly placeless and timeless word, there is a long history of diverging goals, contested theories and very different procedures propelled forward for achieving a just society. One result is the peppering of conversation with various references to divine justice, legal or retributive justice, different schools of philosophical thinking (such as utilitarianism or egalitarianism) and even the twist of poetic justice. As with all disciplines, these shifting philosophical and popular interpretations of social justice have influenced the practice of geography, framing what is perceived as relevant and respectable research. Until recently, however, the number of geographers explicitly addressing ideas of social justice have been few: David Harvey's book *Social Justice and the City*, published in 1973, has been described by many commentators as pioneering. Harvey's goal was to restore CRITICAL THEORY (in this case Marxist theory) and examine the reproduction of social injustice through complex processes of urbanisation. His work brought to the fore questions of socio-spatial alienation, marginalisation and exclusion which many feminists had long been discussing. There has been a growing number of geographers exploring how to achieve a fair and equitable distribution of resources, including material goods as well as non-material things such as RIGHTS, freedoms and opportunities (Smith, D., 1994).

Alongside Marxist thinking, questions of social justice have been sharpened by a diverse range of social groups, including feminists, POST-COLONIAL writers, people battling for environmental justice and against environmental RACISM, movements of people with DISABILITIES as well as people campaigning around issues of SEXUALITY. These political-cultural currents have combined in controversial ways with strands of POSTMODERN and POST-STRUCTURALIST thinking to undermine the notion that we can or should appeal to universal ideas of social justice. The main convergence between these different currents of thinking has been in uncovering the ways in which abstract notions of social justice function ideologically, by promising to be above and apart from social life whilst simultaneously legitimising very particular and oppressive configurations of power relations. The fundamental divergence stems from the political implications of this radical critique of ideas of justice: while feminists and other SOCIAL MOVEMENTS seek to reconceptualise non-oppressive notions of justice, others argue that such a consensus is impossible in our postmodern times. Many feminist geographers have sought to overcome this 'spectre of RELATIVISM'. Drawing on the work of political theorists and activists, feminists have sought to contextualise our understanding of social injustice whilst retaining a normative commitment to address and transform the socio-spatial processes that reproduce injustice both in terms of distributive issues (i.e. who gets what) as well as the political and institutional contexts in which such distributive processes operate (e.g. what cultural assumptions underscore these distributive processes) (Fraser, 1997).   FG

K

Kinship An anthropological and sociological term describing people of the same FAMILY or relatives; kinship ties are those of blood (consanguinity) or of affinity, through legal marriage or other forms of relationship between individuals and families. Kinship links a wider group of people than immediate family or HOUSEHOLD members. Kinship ties often involve obligations and responsibilities, especially PROPERTY rights. For Engels (1985), the origins of the subordination of women in marriage lay in the legal requirements to pass property to a legitimate heir.   LM

Knowledge See SITUATED KNOWLEDGE

L

Labour: domestic, sweated, waged, green, surplus Waged labour is the term used to describe work performed for an employer: that is, the exchange of an individual's labour power for the payment of a wage. Waged labour is to be found within a vast array of spaces, including not only factories, workshops, offices, hospitals, schools, banks, shops, hotels, restaurants (and so on!) but also within the home, in the form of HOMEWORK, outwork or indeed domestic waged labour.

Domestic labour refers to waged or unwaged labour performed within the HOUSEHOLD which is necessary for the reproduction of labour power. Historically, the goods and services which were products of domestic labour 'were intended for direct use and they were consumed within the household' (Glucksmann, 1990: 226). During the 1930s, however, domestic labour in Britain underwent considerable changes within both middle-class and working-class households, as 'the reproduction of labour power came to be effected much more on the basis of factory produced goods than on domestic labour' (Glucksmann, 1990: 6).

In the late 1970s and early 1980s, theorising the role of domestic labour prompted considerable debate, as feminists sought to determine whether women's involvement in household labour was functional to CAPITAL or whether it arose from patriarchal relations between men and women. In retrospect, as Dale and Foster (1986: 55) have argued, 'the debate seems rather sterile', for it was impossible directly to link housework with profit. 'There can be no "iron law" that without the sexual division of labour in the home, capitalists would always make lower profits – witness the position of black mine workers, denied their families, in South Africa' (Dale and Foster, 1986: 55).

The term sweated labour is used to describe a system of subcontracting which first emerged in Britain during the nineteenth century. Within the London clothing trade, for example, a wholesale clothier would sell cut pieces of cloth to the lowest bidding sweater, or middleman. In turn, the sweater (always a man) would engage family members and other casual labourers to fashion clothing within a home-based workshop. The system encouraged sweaters continually to seek the cheapest possible labour, sharply driving down the earnings of female homeworkers (Phizacklea, 1990). By the turn of the century, it was deemed necessary to protect (particularly female) workers in sweated trades, and wages councils were introduced by the British Government in 1909 to set minimum wages in the clothing, agricultural and catering sectors. Sweated labour is not, of course, purely an historical phenomenon. It can also refer to the organisation of particular forms of work in the contemporary period which is arduous, poorly paid and hazardous to health. Indeed, the downward pressure on wages which is characteristic of labour-intensive contract service industries would seem to indicate a return to precisely those forms of work organisation which characterised the late nineteenth century, dominated by sweated capitalism and informal and fragmented work (see Reimer, 1998).

The notion of green labour has been used to describe a labour force (often within a particular geographical area) which has had little or no experience of waged employment or of a particular form of employment. Beynon (1984), for example, suggested that in opening the Halewood plant in Liverpool in the 1960s the Ford Motor Company sought to recruit men who were both new to factory work and unfamiliar with industrial action. Additionally, manufacturing employers were seen to target local reserves of 'green' female labour in northern and western regions of the United Kingdom during the 1960s and 1970s (Massey, 1984) or (at a global scale) within industrialising nations (Fröbel *et al.*, 1980). There is a danger, however, of homogenising the potentially diverse experiences of different groups of women through the caricature of green labour and thereby failing adequately to address the processes which lead to OCCUPATIONAL SEGREGATION by gender (Hanson and Pratt, 1995: 15; see also Pearson, 1986).

Surplus labour is defined primarily through an opposition to necessary labour. A distinction is made between labour which is essential for subsistence and that which is surplus to basic needs for survival; that which is 'above and beyond the costs of reproducing the worker' (Gibson-Graham, 1996: 53). Within the MARXIST tradition, the concept of surplus labour is important in defining class location. A distinction is made between social groups who are producers of surplus labour and those who appropriate it (via capitalist EXPLOITATION). Rather than defining class by social grouping, however, Gibson-Graham (1996) has recently argued that the focus should be upon the social *process* of producing and appropriating surplus labour and

of surplus labour distribution. She also suggests that we should examine a wide array of arenas in which surplus labour is produced, appropriated and distributed, including the household, the workplace and the community. Within the heterosexual household in industrial societies, for example, 'a woman produces surplus labour in the form of use values that considerably exceed what she would produce if she were living by herself' (1996: 66). SR
See also DIVISION OF LABOUR: DOMESTIC, INTERNATIONAL, GENDER, SPATIAL; WORK/WORK FORCE/EMPLOYMENT.

Labour market The 'institutional means by which the purchase and sale of labour power are arranged' (Cooke, 1983). Some of the most interesting feminist work on the labour market has emphasised the very local practices which shape microgeographies of employment within urban areas (Hanson and Pratt, 1995). SR

Lacanian theory The French psychoanalyst Jacques Lacan (1901–81) generated a version of psychoanalytic theory that remains both influential and controversial. He advocated a 'return to Freud' that focused on the functions of LANGUAGE, most especially in relation to the UNCONSCIOUS, which Lacan described as being structured like a language. Lacanian theory is often understood as a recasting of Freudian concepts in a linguistic (rather than a biological) register, made possible by structural linguistics. This shift is accompanied by continuity in the centrality accorded to VISION, to the OEDIPAL phase and to CASTRATION, as well as to the unconscious and sexuality (see FREUDIAN THEORY).

Whereas the term Freudian theory refers mainly to ideas with which Freud worked towards the end of his life, and which provided a point of departure for OBJECT RELATIONS THEORY, Lacanian theory picks up especially on Freud's early writings on the unconscious. Like Freud's work, Lacan's spanned several decades, during which he revised some of his ideas considerably. In addition, most of those in English-speaking parts of the world who engage with his work rely on translations from the original French. While these circumstances generate difficulties in defining and delineating Lacanian theory, the task is further complicated by the fact that many of Lacan's writings convey their meanings rhetorically, meanings that include a notion of unconscious dimensions of SUBJECTIVITY bound up with language through misunderstanding, misapprehension and misconstrual. In short, Lacan used words (and other symbols) to disturb, confuse and unsettle.

Like other psychoanalytic theorists, Lacan understood the human infant to be born 'prematurely' in that a baby has no sense of its existence as an entity separate from its mother's body. For Lacan, the infant experiences itself as fragmented, as a 'body-in-bits-and-pieces', but also as indistinguishable from its mother. Around the age of six months the infant enters what Lacan called the mirror stage. At this point the child begins to recognise itself in its image, whether in a mirror or in its mother's eyes. In this act of recognition the child takes up a position of bodily coherence and distinguishes itself from the world about it. The mirror stage positions the child in relation to its mother. It separates what Lacan calls the register of the 'real', which is unrepresentable, from the register of the imaginary, characterised by visual representation.

Three interrelated aspects of this moment are crucial. First, acknowledgement of absence, loss or LACK is a precondition for the child's capacity to recognise itself as a separate, bounded being. Prior to this the infant exists in a state of omnipotent unity with its mother, thereby experiencing itself as the source of all that its mother provides. Only in relinquishing this sense of completeness and plenitude can the child be born into the social world of human individuals. Second, the child recognises itself in an image. Lacan argued that this always and fundamentally entails misrecognition and alienation. On the one hand the child 'sees' a bounded, unified image but comes to it experiencing itself differently (as a body-in-bits-and-pieces); on the other hand it identifies with an image outside itself. This, according to Lacan, makes NARCISSISM (love of one's own self-image) an essential part of human existence: our sense of ourselves depends upon a paradoxical and problematic incorporation of self-image. Third, the idea of the GAZE is integral to this formulation. For Lacan the infant's captivation with (and by) its image depends upon more than seeing. It is also orientated towards lack, and driven by DESIRE, which, following Freud, Lacan construed as inherently sexual.

Within object relations theory the child's development within the mother–infant dyad is fundamental to its capacity to participate in the social world, but for Lacan the dyadic relation locks the infant within a pre-social, if individuated, mode of existence. In other words, while the mirror stage separates the infant from its mother, left on their own, the infant remains bound within an imaginary sense of exclusive togetherness. Following Freud, Lacan argued that the introduction of a third term, which precipitates the oedipal phase, is crucial to the child's 'birth' as a social being. In Freudian theory, this third term is personified by the father. So too in Lacanian theory, but Lacan conceptualised the father symbolically rather than biologically. In other words, the father is figured in language such that the child's entry into the social entails entry into the world of symbolic forms of communication (or the SYMBOLIC ORDER), in contrast to the apparently unmediated symbiosis between mother and infant. The symbolic order into which the child enters pre-exists him or her and also constitutes the child's emergent subjectivity. In this sense we are all subjected to language. Moreover, speaking consolidates our alienation, in that it necessarily separates us from what we want: it marks a gulf between ourselves and a now external world (compare with Winnicott's notion of transitional space).

According to Lacan, the symbolic order is phallocentric, or governed by the Law of the Father. Lacan conceptualised the PHALLUS as both an imaginary object and as a signifier. As an imaginary object, 'phallus' is the name Lacan gave to the object of the mother's desire. In so doing he suggested that the oedipal triangle can be understood in terms of mother, child and the object of the mother's desire, which the child wishes to be but cannot be. In this triangle, the mother desires something neither she nor the child possesses. Drawing on Freud's reading of sexual difference Lacan represented this as the father's penis. But as a signifier the phallus is not the penis; rather it is what has already been lost in REPRESSING incestuous desires (i.e. in getting through the oedipal complex). What this implies is that we are all castrated by virtue of taking up a place in the symbolic order. At the same time, Lacan insisted that phallocentrism is inescapable so that woman 'does not exist'; rather, femininity is produced only within an economy of phallocentric desire.

In describing the unconscious as structured like a language, Lacan suggested that the contents of the unconscious (repressed desires and renounced pre-oedipal wishes) can be understood in terms of signifiers, the meanings of which operate through chains of signification. But, following Freud, the discourse of the unconscious cannot be articulated consciously: it can be discerned only as a kind of interference distorting, undercutting or overlaying conscious uses of language. Lacan's conceptualisation is also infused with a BINARY distinction between masculine and feminine in which the unconscious is figured as feminine. Lacan deployed the term *jouissance* to invoke unrepresentable qualities of the unconscious, and it has been taken up by some post-Lacanian theorists as a concept through which to open up spaces for forms of femininity beyond the phallocentric.

As with other versions of psychoanalysis, Lacanian theory has provoked widely varying responses among feminists. Lacan was hostile to feminism and openly ridiculed feminist politics. In this context it is hardly surprising that many feminists have shunned Lacanian theory, dismissing it as deeply antithetical to feminism. But, as in the case of Freudian theory, other feminists have found it useful in disclosing a great deal about how the oppression of women works and persists. These perspectives are replicated within feminist geography, where those taking up Lacanian ideas have generally drawn strongly on feminist readings (see for example Blum and Nast, 1996; Bondi, 1997; Rose, 1995b).

Two of the principle appeals of the Lacanian account are its emphasis on language as the pivot between the social and the psychical, and (in contrast to object relations theory) its potential for subverting binary conceptions of gender. While the former provides a bulwark against the biologism that haunts much of the Freudian legacy, the latter has been taken up in the name of anti-essentialist understandings of gender and sexuality (see for example Adams and Cowie, 1990; Butler, 1990; Cornell, 1991; Gallop, 1982, 1985; Grosz, 1990a; Henriques *et al.*, 1984; Rose, 1987). Moreover, both themes have been developed by post-Lacanian theorists informed by feminist perspectives, including Hélène Cixous, Luce Irigaray and Julia Kristeva (see for example Grosz, 1989). Feminist readings of Lacanian and post-Lacanian theory have probably been most influential in the field of cultural studies, including especially literary studies and film studies (see especially Mulvey, 1989). This has provided an important route by which Lacanian ideas have entered feminist geography (see for example Rose, 1995b).   LB

Lack See CASTRATION ANXIETY/CASTRATION COMPLEX; DESIRE; see also FREUDIAN THEORY; LACANIAN THEORY

Laddism The social expression and development of adolescent male identity has attracted the interest of a wide variety of researchers (Kehily and Nayak, 1997; Nayak and Kehily, 1996). The association of young men with antisocial behaviour, particularly VIOLENCE, CRIME and social prejudices, appears to be a principal factor behind this concern. Paul Willis's study *Learning to Labour* (1977) provides a seminal example of such work. Willis's interviews with rebellious male school pupils in England, self-defined as 'lads', suggests that their SEXISM, RACISM and hostility to authority acts to secure their compliance to low-wage, low-responsibility positions when they leave school. Ironically, the 1990s saw the alignment of reactionary expressions of young men's identities with consumerism; laddism and shopping together shaping the content of British men's magazines founded in this period, such as *Loaded* and *Maxim*. AB

Land Women live and WORK in and on the land but rarely conquer it and are less likely to own it. In many agricultural areas farms are often passed from father to son, illustrated by the strong tradition in Ireland of 'keeping the [male] name on the land' (Beale, 1987). Unusually in Europe, a 1974 law passed in Norway gave equal rights of inheritance to women (Haugen, 1994). This has a longer history in the Commonwealth Caribbean (Momsen, 1993).

Very different people–land relationships exist in hunter-gatherer societies. Australian Aboriginal concepts of land ownership include 'owners' who trace their responsibilities through the paternal line, and 'guardians' who have maternal links. Both genders are necessary to perform ceremonies and care for the land (Young, 1992).

Land may be linked metaphorically with women's bodies, through its associations with NATURE and HOME. Often there are violent overtones, for example in the concept of 'virgin lands' awaiting penetration and conquest by male explorers and 'rape' by exploitative farming methods. Colonial mapping often involves masculine renaming in the language of the conqueror (Berg and Kearns, 1996).

Feminist LANDSCAPE art offers one medium by which masculine appropriation of the land can be challenged. Catherine Nash (1994) analyses Kathy Prendergast's exhibition *Land* which consisted of a canvas tent on which a conventional map had been painted. This metaphor 'frees conceptions of identity and landscape from repressive fixity and solidity' (1994: 242).

Women can also reclaim land to create new ways of living and relating to the 'natural' environment. Gill Valentine (1997a) describes the establishment of 'lesbian lands' in North America. From the 1970s groups of LESBIAN women practised spatial strategies of separation, choosing rural locations far from 'man-made' CITIES.   BW
See also ECOFEMINISM.

Landscape A central concept in human geography. It can refer to: a portion of the earth's surface; a way of seeing, knowing and controlling that area; and the cultural practices involved in producing the composition of that area. Geographers examine the tensions between the material and symbolic meanings and qualities of landscape (Olwig, 1996), and historically have treated landscape as a visual form of spatial knowledge. Very few women scholars are represented in the landscape tradition of human geography, and feminists have criticised the totalising view of this approach.

The etymology of the word landscape dates back to medieval England when it referred to land controlled by a lord; in medieval Germany it was a legal term defining the collective ownership of an area. By the early seventeenth century, landscape meant the representation of scenery in painting as well as the design of SPACE. With the rise of linear perspective, landscape became associated with a way of seeing that was informed by the RATIONAL SCIENCE of geometry, and legitimated the emerging ideology of capitalism and the ruling, male-dominated, bourgeois class (Cosgrove, 1984; Mikesell, 1968; Rose, 1993b).

In geography, landscape was used in the late nineteenth century by German scholars who defined the discipline (*Landschaftsgeographie*) as the classification of natural regional landscapes according to morphological forms. In the USA, Carl Sauer rejected the environmental determinism of his day and defined the concept in 1925 as an area made up of a distinct arrangement of physical and cultural forms: 'the cultural landscape is fashioned from a natural landscape by a culture group. CULTURE is the agent, the natural area is the medium, the cultural landscape is the result' (1925: 343; quoted in Leighly, 1967). The role of the geographer was scientifically to document the historical processes by which a natural landscape became transformed into a cultural one. Underlying this genetic approach was the assumption that 'NATURE' and 'culture' were separate realms – a problem that Sauer later recognised.

In addition to Sauer and his students, the English historian W.G. Hoskins and J.B. Jackson (founder of *Landscape Magazine*) were important contributors to landscape studies in the 1950s. Jackson was interested in exploring American popular culture through vernacular landscapes such as the highway strip and the shopping mall. Geographers also examined everyday landscapes as a clue to culture, as illustrated in Donald Meinig's (1979) edited volume on *The Interpretation of Ordinary Landscapes*, David Lowenthal's (1985) work on national landscapes and memory, Fred Kniffen's (1936) discussion of regional vernacular architecture, and Wilbur Zelinsky's (1973) descriptions of American culture regions. In the 1970s and 1980s, works emphasised the visible forms and spatial distributions of landscapes and explored how landscapes socially acquired meaning through time. Meinig (1976), for example, argued that depending on the needs of the user/see-er, landscape could be envisioned as nature, habitat, artifact, system, problem, wealth, ideology, history, place and aesthetic.

By the 1980s and 1990s, geographers began to examine critically the POWER relations involved in producing ordinary landscapes. Scholars criticise the emphasis on vision as a means to understand landscape in human geography (Holdsworth, 1997) as well as the processes by which 'natural' landscapes historically have been observed, surveyed, mapped and feminised (Rose, 1993b). A number of approaches to landscape studies are informed by CRITICAL SOCIAL THEORY. Denis Cosgrove and Stephen Daniels use a cultural materialist approach to define landscape as 'a pictorial way of representing, structuring or symbolising surroundings' (1988: 1). Audrey Kobayashi (1989) proposes a landscape interpretive approach based on existentialism; James and Nancy Duncan (1988) argue for a POST-STRUCTURALIST approach to reading the landscape as a TEXT. Don Mitchell (1996) advocates a neo-Marxist approach to reveal the histories of labourers whose stories have remained hidden from traditional studies of corporate landscapes.

Feminists explore the exclusionary processes by which groups create, produce and represent landscapes to legitimise gendered ideologies. Some authors examine literature and imagery produced by or of women (Kolodny, 1975, 1984; Norwood and Monk, 1987), whereas other scholars examine how cityscapes are spatially and symbolically gendered and sexed (Domosh, 1996; Hayden, 1981; Lopez Estrada, 1998; Wilson, 1991). Recent work on official, national landscapes explores how public institutions, statuary, performance, and myth reify and embody gendered notions of CITIZENSHIP and PRIVATE/PUBLIC spheres in the built environment (Dowler, 1998; Johnson, 1995; Warner, 1985). Feminists also advocate landscape-based citizen activism by arguing that marginalised groups can empower themselves by transforming taken-for-granted material landscapes to make their perspectives and voices materially visible (Hayden, 1995). KT

See also BUILT ENVIRONMENT; CITY; NATURE.

Language In psychoanalytic usage, the term generally refers to French philosopher Jacques LACAN's influential theories in which he rewrites FREUD's oedipus complex to consider how subjects are constructed in and through language. Taking advantage of insights from Sausserian linguistics, Lacan (1977b) delineates three registers: the Imaginary (which coincides with the pre-oedipal and includes the mirror stage), the Symbolic (the resolution of the oedipus complex) and the Real (that which is always inaccessible). In the realm of the Imaginary, early in the child's development, the child makes no distinctions between SUBJECTS and objects, the SELF and the external world, or its own body and that of the mother. Eventually entering into the mirror stage, the child begins developing its ego by discovering a unified image of itself reflected back through the mirror. The child thus misrecognises itself by identifying with a pleasing totality that the child does not experience in its own body. In Sausserian terms, the child perceives no gap between the signifier (its self before the mirror) and the signified (its image). When the father enters the scene, as in the oedipal scenario, he disrupts the mother–child dyad with the threat of incest. Representing what Lacan calls The Law of the Father, the FATHER's presence inserts the child into the wider familial and social network of language – the realm of the Symbolic – in which the child's role is predetermined. Whereas the child's experience of the Imaginary is marked by plenitude and (false) unity, its place within the Symbolic is defined by sexual difference and exclusion. By entering into the Symbolic, the subject is inevitably split into conscious and unconscious selves, and a gap is inserted between the signifier and the signified. Retaining the desire for unification and direct access to the mother's body, the subject must make do with substitute objects (moving from signifier to signifier) in an attempt to recover the (false) sense of unity experienced in the Imaginary. In this Symbolic realm, the phallus functions as 'the privileged signifier' or sign of what divides subjects from themselves, and inserts them into the Symbolic. The Real refers to that which is neither Symbolic nor Imaginary; it is an *a priori* state that lies beyond language as the signified or meaning and is thus never fully accessible.

Feminist theorists generally criticise Lacan for his PHALLOCENTRISM – the primacy of the phallic signifier in his theory. Many psychoanalytic feminists, including Luce Irigaray (1985a), Hélène Cixous (1990) and Julia Kristeva (1980) have complicated Lacan's scenario by arguing that female subjects occupy a different position in relation to language. Because Lacan's theory of language also includes a theory of how subjects locate themselves spatially in relation to both the mother's body and subsequent Others, feminist geographers, such as Gillian Rose (1996a), use Irigaray's intervention into Lacanian theory to reconsider the relationship between real and imaginary SPACE. Broader feminist critiques of Lacan include the charge that he constructs a totalising theory that cannot account for historical, material or geographic specificities. By constructing a UNIVERSAL model of language acquisition based on the Western nuclear FAMILY, Lacan reifies Western patriarchal rule. As Anne McClintock suggests, Lacan's theory can be seen as 'a rearguard attempt to rescue the potency of the Law of the Father as an abstraction, at the very moment when it is disappearing as a political force' because of feminist and non-Western challenges to a global order based on Western PATRIARCHAL authority (1994: 198).   WS

See also NARRATIVE; STRUCTURALISM; TEXT/TEXTUAL/TEXTUALITY/INTERTEXTUALITY.

Law Laws are norms, practices or rules which have become formalised and which attract sanctions when TRANSGRESSED. Laws must be made through processes which are deemed to be legitimate if they are to be recognised as strictures to which obedience is due. The role of codifying norms into a legislative framework, of enacting and enforcing them, is usually reserved for those who occupy positions of authority within a given community. Within Western legal scholarship the law constructs itself as an autonomous, objective/neutral, rational and self-sustaining system immune to the vagaries of social and political life, creating, notionally at least, the idea of 'a distinct legal domain delineated by boundaries which segregate legal discourse from political rhetoric, poetry, or late night conversations with friends' (Blomley, 1994: 10). Within this discourse, exercises such as law-making and legal interpretation are understood not as subjective processes but as normative ones, informed and structured by a 'deep' logic and rationality and a set of underlying principles which are universally shared (see Unger, 1983).

Challenges to this conception of the normativity of law have come from the Marxist and feminist legal scholars and from the field of critical legal studies (CLS). Although CLS scholars pursue different projects they are united by a desire to 'DECONSTRUCT' or examine the alleged neutrality, RATIONALITY and UNIVERSALISM of the concepts and principles on which the law rests. Their aim is to reveal the constructed and contingent nature of the law and the POWER relations which underpin it. No longer autonomous or objective, the law is recast, within this formulation, as a social and political construct designed to reproduce and sustain particular power relations.

Feminist legal scholars, in particular, have drawn attention to the role of law as an instrument for perpetuating PATRIARCHAL relations of domination. They argue that the law effects this in several ways: by constructing a normative concept of 'the FAMILY' which is inherently patriarchal, by helping to define and maintain a division between PUBLIC and PRIVATE space, by affording men and women comparatively more or less rights within their respective 'natural' domains, by creating legal rights which afford men particular power to regulate both the family and sexual relations and by constructing for women certain gender-specific roles that reinforce and perpetuate sexual stereotypes. As Bartlett and Kennedy (1991: 9) suggest

> the hierarchical opposition of MASCULINE and FEMININE within such BINARY pairs as public/private, objective/subjective and form/substance rest upon certain unstated notions about gender roles, sexuality, and private intentions that create and reinforce norms that favour male characteristics and value over female ones.

In this regard the law could be said to efface its own specificity. It denies the fact that decisions are made in ways that sustain male power, that, as Polan (1982: 33) argues, 'the whole structure of law – its hierarchical organisation, its adversarial format and its underlying basis in favour of rationality is essentially patriarchal'.

An agenda of legal reform particularly in areas such as ABORTION rights, sexual assault, family law, social security and employment rights now forms part of an ongoing commitment by feminist legal scholars and practitioners to effect social change through a restructuring of the canon of law.   BP

Lesbian Used in a particular sense to refer to women's same-sex sexual preferences and activities, the term lesbian does not have the same history as the term HOMOSEXUAL with its origin in nineteenth-century debates about the scientisation and medicalisation of SEX, SEXUALITY and DESIRE. In Britain, at least, lesbian sexual practices have never been legally defined and controlled. However, in the establishment of COMPULSORY HETEROSEXUALITY as the hegemonic norm, lesbians have become marginalised and discriminated against in a wide range of areas, from housing policy, workplace discrimination and inheritance. The term LESBIAN CONTINUUM was coined by Adrienne Rich to refer to a wide range of attachments between women which are denied through the dominance of HETEROSEXUAL practices and representations.   LM

See also DYKE; GAY AND LESBIAN COMMUNITIES.

Lesbian communities See GAY AND LESBIAN COMMUNITIES

Lesbian continuum A term widely associated with the writings of Adrienne Rich (1980); it refers to the range of friendships and social relations between women that shade into lesbian activities. In this way lesbianism is not solely restricted to sexual activities between women, nor is it seen as 'perverse' or entirely separate from the ways in which women establish close friendships and supportive activities.   LM

See also COMPULSORY HETEROSEXUALITY.

Liberalism/liberal feminism Liberalism is based on the belief that all individuals are equal in the eyes of the law and so should have equal access to the rights of citizenship. It developed in Europe and North America through the seventeenth and eighteenth centuries and was a powerful influence on campaigns to EMANCIPATE slaves and for women's SUFFRAGE. Liberalism was based on the rejection of tradition and the 'natural' or divine order, rejecting the right of hereditary monarchs, for example, as well as the PATRIARCHAL AUTHORITY of individual men and the state. In 1869 John Stuart Mill, a key liberal political theorist, published *The Subjection of Women*, challenging women's 'natural' SUBORDINATION.

The liberal belief in individual EQUALITY lay behind the powerful claims of women to suffrage, that is, the right to vote, which was the main demand of what has now become recognised as the first wave of the WOMEN'S MOVEMENT. In the USA and the United Kingdom suffrage campaigns were a key part of feminist politics in the early years of the twentieth century. Before that the main claim was for the right for married women to own property in their names. In the second half of the twentieth century, liberal feminists organised around claims for equality with men in other arenas, most notably in the workplace.

As liberal theory denies sexual DIFFERENCE, based as it is on an ideal of disembodied individuals, it has been the subject of a vigorous critique by feminist theorists, especially SOCIALIST FEMINISTS, who have documented the structural organisation of capitalist societies that constructs female subordination. Liberalism is based on an imaginary world of individual atoms, all making RATIONAL decisions. Further, liberalism depends on a distinction

151

between the PUBLIC or POLITICAL arena and that of the PRIVATE sphere or personal life. Liberal theorists believe in minimal state intervention and insist on the right of individuals to retain as much control as possible over their private lives. Mill, for example, turned a blind eye to the separation of the domestic lives of many women from the public lives of most men in the nineteenth century, believing that women had 'chosen' the domestic sphere.

From the late 1960s onwards, radical and socialist feminists mounted a powerful argument that 'THE PERSONAL IS POLITICAL'. So-called private issues about sexuality and reproduction, for example, were not the subject of individual choice but rather the focus of state control through, for example, state regulations about who may or may not have access to CONTRACEPTION, who may or may not have control over children. And it became to be accepted too that sexual preferences and behaviour, seemingly the most private of private issues, were also socially constructed and regulated.

In recent years, the focus of feminist theory and politics has moved away from a focus on equality towards an emphasis on difference. The whole notion of equality, it is argued, is based on an ideal of an individual who is in fact a bourgeois man. Thus claims for equal rights in the workplace fail to take into account women's particular needs during their working lives for, for example, maternity leave, and fail to recognise that at present when women still perform most of the domestic tasks they cannot participate equally in workplaces which deny personal responsibilities. As the feminist political theorist Ann Phillips has documented, the equality–difference tension runs right through the long history of feminist struggles and raises dilemmas for women:

> Advocates of ... equality have argued – with considerable force – that once feminists admit the mildest degree of sexual difference, they open up a gap through which the currents of reaction will flow. Once let slip that pre-menstrual tension interferes with concentration, that pregnancy can be exhausting, that MOTHERHOOD is absorbing, and you are off down the slope to separate spheres. It was with good reason that prominent suffragists argued against emphasizing women's maternal role: the whole point of the movement was to get women out of their stereotyped domesticity, to assert their claims in the public sphere (1987: 19). LM

See also DEMOCRACY; ENLIGHTENMENT/ENLIGHTENMENT THEORY; RATIONALITY; RIGHTS/EQUAL RIGHTS/EQUAL OPPORTUNITIES; SUFFRAGE/SUFFRAGE CAMPAIGNS.

Libido A term used to refer to drive or energy, often, but not solely, that associated with masculine sexual desire. LM

Life course/life stage/life cycle The sequence of phases through which an individual passes (but can be applied to a unit such as a family or business) from birth to death and is analysed in terms of events such as marriage or childbirth. The duration at or between various stages and the number of people in these stages are used as an analytical indicator of trends such as fertility and female labour-force participation. Feminists are critical of the ideological use of the concept in legitimating oppressive policies.   **PM**

Life histories A qualitative methodology which attempts to capture the whole story of a person's life and/or the details of particular events within it. The term is often used interchangeably with 'oral history', 'biographical interview' or 'personal NARRATIVE', but unlike these methodologies, life histories seldom rely solely on the memories of those involved. Documents, such as letters, diaries and newspaper reports, are sought to expand on the oral information provided, as well as photographs which add 'life' to the life history (Plummer, 1983; Russell, 1989; Thomas and Znaniecki, 1984).

Life histories can play an important role in recovering 'lost geographies' and airing the VOICES of 'others' in academic texts (Miles and Crush, 1993). Until recently, however, their part in human geography has been mainly implicit (Eyles and Perri, 1993). The current interest in this technique can be attributed to feminist attempts to depart from conventional, standardised, masculine interview styles (Miles and Crush, 1993). Feminist geographers favour a life-history approach, particularly in obtaining knowledge about women in developing countries and economically and socially deprived women in 'the West'. By capturing the specificities and richly textured experiences of individuals' lives, life histories avoid UNIVERSALISM and challenge ANDROCENTRISM (Keegan, 1988; Stubbs, 1984).

Few geographers have subjected life histories to serious methodological scrutiny. Notable exceptions are Miles and Crush (1993) who recognise that life histories cannot be treated as unproblematic 'sources' for the construction of 'real' events and experiences. Like other ETHNOGRAPHIC methods, life histories can give a distorted presentation and may not be representative of broader patterns. A more central problem is the 'insider–outsider syndrome'. Some writers have challenged the authority of WHITE middle-class Westerners to represent the 'other' (hooks, 1992; Spivak, 1988). The challenge, therefore, for feminist geographers is to assess critically their POSITIONALITY when adopting a life-history approach.   **RK**
See also AUTOBIOGRAPHY; ETHNOGRAPHY; FEMINIST METHODOLOGY.

Literary criticism Euro–American feminist literary criticism has traditionally focused on the politics of PATRIARCHY as it is represented in LANGUAGE, especially in literary works such as novels. Widely understood in Western feminism to have begun with Kate Millet's *Sexual Politics* (1971), the goals of feminist literary critics were to uncover the misogyny of the (male) literary establishment; learn to recognise and resist the SEXISM of literary works; create a literary history of women writers, often including the works of 'unknowns' in non-conventional genres; and theorise about women's creativity and ways of writing and reading texts (Gilbert and Gubar, 1979; Showalter, 1977).

'DIFFERENCE' emerged in the 1980s as an overarching theme of many literary critics, and their focus turned to the heterogeneous and historically and spatially contingent subjectivities of BLACK, LESBIAN, Hispanic, Native American, Caribbean, 'THIRD WORLD' and working-class women's literature (Moraga and Anzaldúa, 1981; Smith *et al.*, 1982). Criticism dating from the mid-1980s also shared with geography complex theorising about social life drawn from wide-ranging Marxist, POST-STRUCTURALIST, PSYCHOANALYTIC and POST-COLONIAL CRITICAL THEORIES. Marxist feminist criticism began by analysing how literature historically represented categories of GENDER and CLASS, and many Marxists now employ the DECONSTRUCTIVE methods of post-structuralism to problematise the category of gender itself or the unitary SUBJECTIVITY of authors (Moi, 1985; Showalter, 1985). Many literary critics draw on French feminists such as Julia Kristeva (1991) to deconstruct the PHALLOCENTRIC language and theories of gender identity formation postulated by FREUD and LACAN. Post-colonial literary criticism, perhaps more than any other, has been integrated into feminist geography, especially in works that theorise authorial and SUBALTERN subjectivity in the context of colonialism and imperialism (Spivak, 1985, 1988a, b and c). KM

Local/locality Complex words that are used in different contexts to refer to a place or a setting for particular social relations. The words local and locality have attracted considerable controversy within human geography, particularly with regard to debates about ECONOMIC RESTRUCTURING and GLOBALISATION. During the mid-1980s a number of research projects explored the impact of economic restructuring on particular places and these became known as the localities projects.

As Massey (1991a) points out, the localities research projects were designed to explore the geographical dimensions of economic restructuring in the United Kingdom. Structural changes in the national economy were having different effects in different parts of the country, and local people made different responses to those developments. As Massey explains:

> It became important to know just how differently national and international changes were impacting on different parts of the country. Something that might be called 'restructuring' was clearly going on, but its implications both for everyday life and for the mode and potential of political organising were clearly highly differentiated and we needed to know how. It was in this context that the localities projects in the United Kingdom were first imagined and proposed. It was research with an immediate, even urgent, relevance beyond academe (1991a: 269–70).

As a result, a series of detailed research projects were undertaken, exploring economic, social, political and cultural developments in different localities (see Allen *et al.*, 1998; Cooke, 1989; Murgatroyd *et al.*, 1985). Much of this empirical work highlighted the way in which economic restructuring was reshaping GENDER roles in different parts of the country (see, in particular, McDowell and Massey, 1984; Murgatroyd *et al.*, 1985). As heavy industry, mining and manufacturing employment continued to decline, increasing numbers of women were found in service jobs while male, blue-collar workers were often left in long-term unemployment (see DE-INDUSTRIALISATION).

Criticisms of this research raised a number of important theoretical questions about the status of the local, or localities, in geographical thought. Particular criticisms included N. Smith's (1987) suggestion that the focus on the local and the particular was detrimental to theorising ongoing changes in capitalist development, Jackson's (1991) view that the economic was prioritised over cultural change in this research and Duncan and Savage's (1989) arguments about the danger of ascribing political agency to locality. More generally, the research was also used as a mechanism to demonstrate the value of critical realism in research and to theorise the local as part of wider networks of social relations 'stretched out' across space (see Allen *et al.*, 1998).

In the 1990s, with the so-called CULTURAL TURN in human geography, the term locality has tended to be used less, and the subject of debate has become the 'local'. This concept is more ambiguous than locality and is applied in all aspects of human geography. The 1990s debate around the notion of globalisation has often been framed in terms of the relationship between the global and the local scales. Many writers have argued that the process of economic and sociocultural globalisation have made the local much more significant (cf. Harvey, 1989, 1996; Lash and Urry, 1994). For example, the local has become a focus of study for those interested in 'place-competition' in increasingly globally integrated capitalist activity (Gold and Ward, 1994; Massey, 1995a), for those examining the rise of new social movements (e.g., women's environmental movements; see Jacobs, 1996b; Shiva, 1989; Shiva and Mies, 1993) or for those examining the impact of global cultural and information flows (Castells, 1996a and b, 1997). Feminist writers have been involved in all of these debates: considering, for example, how women's life EXPERIENCE has been influenced by the changing social conditions of globalised cities (Lyons, 1996; McDowell, 1997a and b; Rose, 1989) and in terms of the local context of empowering political projects (e.g. Massey, 1991).

However, the conception of local still retains many of the epistemological problems outlined in the locality debate: in particular, the question of delimiting the local and defining in what sense processes are local rather than regional, national and global.   JW, AJ

Local state Processes of uneven development inevitably force the STATE to organise control of its territory through some form of local autonomy (Taylor, 1993). Hence, many state functions, such as the provision of HOUSING, education, transportation and land-use planning, are organised at a local level. The local state is commonly elided with municipal government although it also includes local judiciaries and quangos. The concept of the local state is somewhat ambiguous, however, because it does not refer to sovereignty but rather a set of relations that encapsulates the formal POLITICS of a LOCALITY. The concept of the local state was introduced by Cynthia Cockburn (1977) in her study of the POLITICAL practices of a London borough. She viewed it in terms of its functions of social REPRODUCTION, but others challenged the notion of the local state simply as an agent of the national state (Saunders, 1984). More recently, Duncan and Goodwin (1988) accentuated the contradictory role played by the local state both as part of the state apparatus and as a challenge to it, representing the interests of the locality. Feminists' coverage of the local state varies tremendously between countries but overall they have paid much less attention to it than to the central state (but see Chouinard, 1996; Evans and Wekerle, 1997; Fincher, 1991; Halford, 1988). LP

Logocentrism See PHALLOCENTRISM; PHALLOGOCENTRISM

M

Macho/machismo A Spanish term related to maleness and/or MASCULINITY which has now permeated many other languages; frequently associated with male POWER, through the demonstration of physical strength generally, sexual power in particular, and the protection/ABUSE of the less powerful (especially women) and of values, traditions, institutions and virtues (as defined by dominant cultures) including, but not limited to, women's virginity, male HETEROSEXUALITY and family honor; argued by Ana Castilo (1994) to have roots in multiple transcontinental PATRIARCHIES, colonialisms (especially Spanish COLONIALISM in the Americas and Arab conquests of Spain), religious traditions (especially Catholicism and Islam) and pre-Catholic and pre-Islamic patriarchal religious and cultural practices (especially North African and European).

Machismo has been argued by some, including Gloria Anzaldúa (1987) to be 'false' in that its roots in the disempowerment of some men and all women by other men lead to behaviours motivated by caring and loss; others, including Castilo (1994: 67), reject this rationalisation as 'dangerous'. LK

Male The concept of male is central to many different explanations of human difference and social development. Within the natural sciences male is usually interpreted as a natural category, biologically configured through the possession of one or all of the following: a male set of chromosomes; male sex organs; male behaviour. The implications of these characteristics for the social organisation of humans have been disputed within natural science, but the perspective that has prevailed is that men's social dominance is, at least in part, a natural state of affairs. This chain of association has been challenged by many feminists who have related it to male dominance within society and SCIENCE itself (Haraway, 1989).

Although an enormous variety of perspectives of what male means exist within feminist work, the dominant tendency within mid-to-late-twentieth-century feminism is to offer explanations for maleness that do not rely on natural CATEGORIES or processes. Within the 1960s and 1970s psychological approaches gained influence. Thus the notion of an internalised male sex role was developed by feminist writers as a necessary complement to the critique of female sex roles. Such roles were posited as social norms that act to constrict the authentic expression of the self. This approach was subjected to a variety of critiques in the 1980s and 1990s (Connell, 1995; Pleck, 1981). Weaknesses that were identified included: vagueness over the meaning of 'role'; the implication of discrete, homogenous and ultimately biologically rooted male and female norms; and the marginalisation of issues of POWER and social structure. Newer approaches emphasised diversity, transformation and RESISTANCE within maleness. A related focus on historical and ethnographic methodologies emerged, approaches more suitable to tracing the varied formations of maleness across time and within particular places. AB See also BIOLOGY; BODY; FEMALE; GENDER; MASCULINITY/MASCULINITIES/ MASCULINISM; SEX/SEXUALITY.

Maps See CARTOGRAPHY

Margins A locational term relating to an area or region, part inside, part outside the dominant area or region; also used metaphorically to define the position of an individual on the margins, part inside and part outside the group. The concept has been applied to many geographical areas and social groups who trangress established norms (Sibley, 1995a). These include women, people of colour and subjugated peoples of the South. It has also been associated with migrants in which the disorientating effects of marginality are stressed by virtue of the mistrust, often the persecution, they encounter from members of the societies they enter. Whilst the process of creating social and geographical margins in both symbolic and material terms is deep-rooted in many cultures, it has been associated recently with an increasingly interconnected GLOBAL economy and the practices of Western IMPERIALISM whereby the world was mapped around the imperial powers and, in the process of drawing boundaries, marginalised non-dominant others and regions were excluded. Within Western political theory the notion of democratic CITIZENSHIP has been interpreted similarly as inclusive of some whilst marginalising or excluding others. As questions of ETHICS and the politics of DIFFERENCE are increasingly implicated in contemporary political and social debates, feminist interventions have sought to subvert the more bounded and proprietary space of the master subject and have questioned those forms of theory which render GENDER invisible and which have invariably assumed that the individual or citizen is implicitly Western, WHITE and MALE (Tronto, 1993).

Within this more critical context the concept of margins has also been employed in a positive sense, referring to the displacement of ANDROCENTRIC, ETHNOCENTRIC 'grand narratives' from the dominant centre to the margins, as the voices of multiple OTHERS have challenged the 'universal' claims of Western theorists. Stuart Hall (1991) argues that the margins have become a powerful space as new subjects, genders, ETHNICITIES and regions, hitherto excluded from major forms of cultural representation as decentred or SUBALTERN, have emerged and acquired the means to speak for themselves. With reference to Western intellectual decolonisation, the ability of Western scholars to write on behalf of those on the margins has been questioned (hooks, 1990c; Spivak, 1990b). Evidence for this has been drawn, for example, from the manner in which THIRD WORLD feminisms are incorporated within some Western feminist discourses (Mohanty, 1991a).

Some recent critiques have questioned the DUALISM implicit in the concept of margins itself. The tendency to interpret and locate particular institutions, values and practices inside the centre or Occident and others outside in the margins, the periphery or the ORIENT, has been challenged not only as exclusive and excluding but also for its tendency to essentialise the inner content and meaning of each CATEGORY. Thus it is argued that the concept of margins relies on the notion of an ESSENTIAL, AUTHENTIC core that requires the elimination of all that is considered foreign or not true to the original, authentic self. In order to move away from this exclusivity, emphasis is increasingly placed upon the porous quality of division and the fluidity of boundaries as new forms of space are proposed based on notions of 'between' and 'around' which are both supportive and enabling in contrast to notions of 'distance and separation' (see HYBRIDITY). Recent reassessments of imperialism, for example, focus on the interactions between centre and margins and the complex power relations between metropolitan and colonial cultures. In a contemporary context this emphasis on interactions is strengthened by the flows of capital and labour associated with late CAPITALISM which disrupt the significance of state boundaries and promote the movement of large numbers of people from the margins to the centre.   MB

See also EMPIRE.

Marriage Institutionalisation of one-to-one heterosexual relationships through the marriage contract. Historical research emphasises the legal and economic constraints of marriage for women especially in terms of PROPERTY and employment rights (Pateman, 1989a). Socialist feminists point to the (oppressive) conditions of marriage including MOTHERING (Rowbotham, 1989), HOUSEWORK (Barrett and McIntosh, 1982) and DOMESTIC VIOLENCE (Dobash and Dobash, 1992).   HC

See also FAMILY; KINSHIP.

Marxist geography This came into being alongside the NEW LEFT. In the early years of the 1970s, radical geographers sought to apply estabished techniques to new problems, making geographical research socially useful. It was not until the 'break through to Marxism' that this group of scholars established a new set of theoretical concepts to apply to the study of CAPITALISM. The publication of David Harvey's *Social Justice in the City* (1973) marked the first use of Marxist concepts in geography and this book generated enormous debate and further development. Marxist geographers have since concentrated their research work on the political–economic dynamics of capitalism and urbanisation (see Harvey, 1982, 1989; Massey, 1995a), IMPERIALISM and development (Blaut, 1987; Watts, 1983) and the production of NATURE (Fitzsimmons, 1989; Smith, 1990).

Feminist geographers have always had a slightly ambivalent relationship to Marxist theory and scholarship. Marxism has tended to unite the struggle for women's equality with working-class politics and organisation, negating the value of gender politics as an independent cleavage of struggle (Barrett, 1988; Sargent, 1981). In the early years of the twentieth century women such as Clara Zetkin, Rosa Luxemburg and Alexandra Kollontai (see Kollontai, 1984) tried to integrate the struggle for women's liberation with the struggle for SOCIALISM, remaining within the Marxist tradition. During the 1960s and 1970s 'SECOND WAVE FEMINISTS' were also attracted to Marxism as a revolutionary platform to eliminate women's OPPRESSION. However, by the 1980s, the honeymoon was largely over, as Hartmann puts it:

> The marriage of marxism and feminism has been like the marriage of husband and wife depicted in English common law: marxism and feminism are one, and that one is marxism. Recent attempts to integrate marxism and feminism are unsatisfactory to us as feminists because they subsume the feminist struggle into the 'larger' struggle against capital. To continue the simile further, either we need a healthier marriage or we need a divorce (1981: 1).

Geographers had a taste of this argument about class and gender politics during the late 1980s (Foord and Gregson, 1986; McDowell, 1986) and more recently, in reaction to David Harvey's (1989) *The Condition of Postmodernity*. Massey (1991c), Deutsche (1991) and Rose (1993a) argued that Harvey neglected the struggle for women's liberation in his prioritisation of class politics and the battle for socialism. **JW**

See also PATRIARCHY; SOCIALIST FEMINISM.

Masculinity/masculinities/masculinism Until relatively recently, the meaning and enactment of male identity was not a central concern of feminist debate. A focus on women's roles, subordination and resistances dominated discussion, a focus that implied but rarely explicitly engaged ideologies of masculinity. However, in the 1980s and 1990s a move towards analysing normative and dominant identities was witnessed across a number of academic fields. Thus, for example, within race and ethnic studies the issue of WHITENESS came to the fore and within research on SEXUALITIES the theme of HETEROSEXUALITY. Similarly, within gender studies, the topic of men and masculinities gained considerable ground, particularly amongst feminist male scholars (Brod, 1987; Brod and Kaufman, 1994; Collier, 1998; Connell, 1995; Mac an Ghaill, 1994; Morgan, 1992).

Work in this area has been characterised by historical, ETHNOGRAPHIC and sociological approaches that attempt to trace the contingent nature of masculinity, highlighting its changeable and contradictory character. One of the most influential writers on masculinity is Robert Connell. Connell suggests that ideologies of masculinity first arose amongst the Western 'gentry'. He notes that 'All societies have cultural accounts of gender, but not all have the concept of "masculinity"' (1995: 67). Thus Connell connects the notion that men have qualitatively different characters to women – that men and women each have their own separate and discrete spheres of social experience and POWER – to the formation of Western bourgeois society. Connell also offers four definitions of masculinity that may be summarised as follows:

- *essentialist* definitions of masculinity attempt to identify an unchanging and static core to maleness, often of a biological nature;
- *positivist* definitions of masculinity are based on empirical, fact-finding, attempts to record objectively 'what men do';
- *normative* definitions of masculinity posit a social ideal of maleness, a norm that is disseminated and enforced through the media and other socio-cultural forms;
- *semiotic* definitions of masculinity focus upon the realm of social symbolism. Semiotic approaches tend to view maleness as constructed in opposition to femaleness, that is, as not-female.

Connell's own approach may be added to this list as a fifth definition:

- *Connell's approach* focuses on masculinity as a social process and social practice. In other words, he is interested in the various ways masculinity is formed and reformed within the context of changing social and gender relations.

These various definitions indicate the complexity of the field but also its mirroring of the approaches taken to the subject of FEMININITIES. The distinctive character of studies of masculinity lies less in their theoretical originality than in their ability to reconfigure, and otherwise shed light on, ideologies and practices that have traditionally been considered unrelated to matters of gender. Within the late 1980s and 1990s this process became increasingly debated within academic geography, a subject dominated by men but where 'gender issues' have conventionally been delimited to a subfield whose remit was restricted to a familiar repertoire of matters understood to concern women. Gillian Rose's *Feminism and Geography* (1993a) took the male dominance of the geographical tradition as its central theme. Drawing examples from across a wide range of geographical study, from LANDSCAPE studies to TIME geography, Rose argued that academic geography is socially and ideologically structured by various forms of male dominance, a set of processes she cohered under the term 'masculinism':

> Geography is masculinist. ... Masculinist work claims to be exhaustive and it therefore thinks that no-one else can add to its knowledge. ... Masculinism can be seen at work not only in the choice of topics made by geographers, not only in their

conceptual apparatus, not only in their epistemological claim to exhaustive knowledge, but also in seminars, in conferences, in common rooms, in job interviews (1993a: 4).

It is important to note that Rose's definition suggests that masculinism works within geography not only as an intellectual pursuit but also as a social and pedagogic setting. In other words, it is not only geographical knowledge that is subject to masculinist bias but also the discipline's career structure and the arrangement and conduct of geographical teaching and learning.

The attempt to expose and analyse the way geographical studies contain a 'taken-for-granted' masculinism has encouraged and structured feminist critiques of both contemporary geographical literature and the history of the discipline. An example of the former are the critical readings of David Harvey's *The Condition of Postmodernity* (1989) by Deutsche (1991) and Massey (1991c; for Harvey's response, see Harvey, 1992). The re-examination of the history of geography in the light of the critique of masculinism offers some of the most interesting examples of feminist work in the discipline. Such work includes critiques of those surveys of the development of modern geography that omit reference to its domination by men. Thus, for example, Rose argues that Livingstone's acclaimed history *The Geographical Tradition* (1992) offers a

paternal tradition [that] can be used as a kind of legitimation process, in which would-be great men cite men already-established-as-great in order to assert their own maturity: what might be described as the 'dutiful son' model of academic masculinity (1995a: 414).

The role of ideologies of masculinity within earlier models of geographical endeavour has also come under scrutiny. The relationship of masculinity to 'exploration', 'overseas adventure' and military skill is of particular importance within disciplines, such as geography, with strong connections, both historical and contemporary, to these endeavours. The theme of geography as 'manly adventure' is central to Phillips's *Mapping Men and Empire* (1997; see also Phillips, 1995). Phillips takes Victorian works of fiction as his empirical focus (see also Dawson, 1994). However, as Rose has shown, ostensibly purely academic geographical work is also amenable to this form of enquiry.

Work on masculinities within geography has been dominated by feminist cultural geography. For many, Jackson's (1991) influential paper 'The cultural politics of masculinity' introduced them to the topic. Jackson's work has opened up a number of empirical lines of enquiry, including the mutable relationship between masculinity, LABOUR division and geographical location. The association of research into masculinities with CULTURAL GEOGRAPHY is enabling a number of useful and provocative lines of research to be developed within that subfield. However, it would be ironic if what started as a critique of the limitation of feminist analyses to 'women's issues', became itself shackled within intra-disciplinary boundaries. The subject of masculin-

161

ity could, and many would say should, be on the agenda of every branch of geography, cultural and economic, human and physical.   AB

Masquerade In a well-known article which was originally published in 1929 but reprinted in 1986, psychoanalyst Joan Riviere argued that in a patriarchal society FEMININITY or what she termed 'womanliness' is a mask which is worn by women in order to masquerade as a type of woman acceptable to men. This masquerade is a PERFORMANCE to please men and means that women have to disguise their 'true' feelings. In an interesting analysis of a woman academic Riviere shows how, after a competent oral presentation, her subject presents a mask of traditional feminine ineptitude and flirtatiousness so as to deflect the anxious responses of her male audience, faced with evidence of female professional ability and rational argument: characteristics purportedly restricted to men. Riviere suggests that the masquerade is a psychological defence against unconscious achievement of MASCULINITY, a way to deflect fear of criticism for the possession of the PHALLUS.

In more recent work, Judith Butler (1990, 1993) criticises PSYCHOANALYTIC interpretations from a POSTMODERN perspective and suggests that the term masquerade should become seen as a conscious PARODY of femininity, a mimicry of conventional notions of masculinity and femininity in order to destabilise what she terms the regulatory fiction of COMPULSORY HETERO-SEXUALITY. She analyses drag as a key parodic political act, suggesting that all gender identities are performances. The idea of gender as a performance is influential in recent work by feminist geographers and has been developed in the context of social workplace relations as well as in the growing body of work about GENDER and SEXUALITY (see for example the papers in Bell and Valentine's (1995) collection).   LM

See also SELF; SOCIAL CONSTRUCTION OF REALITY; SUBJECT/SUBJECTIVITY.

Masternarrative See NARRATIVE

Materialism Theory that everything that really exists is material in nature, that is, it occupies SPACE and exists for some time. Materialist theories disregard concepts and ideas and insist upon the importance of material foundations for social action.

Historical materialism stresses the succession of modes of production in human history as the basis for CLASS struggle. The dialectics of this struggle (between workers and their material conditions) provides the basis for history. Harvey (1982) has posited a historico-geographical materialism which places GEOGRAPHY next to history at the centre of MARXIST analysis to insist upon the importance of productions of – and resistance to – space, PLACE and NATURE, within the historical dialectic.

Materialist arguments often suppose an opposition of materialism to idealism. This BINARY has been deconstructed by cultural theorists, however, who argue that it is impossible to separate the two terms and that concepts of discourse and text, for example, are inherently and necessarily material.

Feminist materialists have argued for the CORPOREALISATION of feminist theory, taking the BODY as the prime site for feminist POLITICS and theory. JS See also BODY; CORPOREALITY; MARXIST GEOGRAPHY; SOCIALIST FEMINISM; STRUCTURALISM.

Medicalisation Medicalisation involves the processes whereby biomedical categorisations are inscribed, for example, on a particular social group or upon an individual BODY or mind. Invariably medicalisation involves the capturing of a condition or state-of-being by medical science in a vocabulary that speaks of 'illness', 'dysfunction' and 'pathology'. Once a state-of-being is pathologised by medical knowledge it becomes amenable to medical treatments and understandings. Often it is the case that social deviancy is subject to processes of medicalisation, and hence there is medical coding of elements of human behaviour, appearance, movement and speech constituted as socially normal and other elements considered as abnormal. Social abnormalities are often defined as medical problems by increasingly sophisticated ranges of clinical interventions. Different geographical and temporal settings give rise to different responses to social deviancy, but it is increasingly the case within WESTERN capitalist societies that deviancy is medicalised. Good examples of the difference that understandings and languages of medicalisation can make are evident in the case of 'madness' (different mental states). Foucauldian accounts of madness have critiqued practices of psychiatry as a particular framework of understanding which pathologises socially abnormal speech, movement and behaviour into varieties of 'mental illness'. Geographical understandings of the processes of medicalisation emphasise the PLACES and SPACES through which such medicalisation is engendered. In particular, geographers have paid attention to the asylum, the health clinic and the hospital as institutional spaces which carry powerful meanings in terms of medical inscriptions. In recent feminist geographical literature, attention to processes of medicalisation and their inscriptions upon the body have become more prominent (Cream, 1995a; Longhurst, 1994), at the same time as different conditions, sites and spaces which are amenable to such processes have widened.   HP

Memory A subject that has become the focus of considerable debate in recent years in which the traditional view that memory is simply about recall, reflecting the past as it happened, has been widely disputed. Instead, memory is understood as a social process involving the organisation of the past in relation to the present and as an active force bound up with the constitution of consciousness and senses of IDENTITY and history. Memory is not fixed but dynamic, entailing forgetting and negating as well as remembering, and has histories and geographies of its own. Halbwachs (1980) first began to explore how social groups constructed memories during the 1920s (on social memory, see Burke, 1997; Fentress and Wickham, 1992). His discussions of 'collective memory' emphasised the significance of a spatial frame, and recently geographers and others have studied memory, space and LANDSCAPE in relation to a range of settings, media and technologies of transmission. Recognition of memory's mutability – the fact that it is 'always a re-membering, a temporary putting together of disparate elements' and an 'unstable configuration', which 'can be attained by different paths to reveal diverse stories' (Chambers, 1997: 237) – along with an acute sense of its politics have particularly come to the fore in challenges to received histories by social groups marginalised along lines of GENDER, SEXUALITY, CLASS and 'RACE'. Here feminist 're-visions' of the past, to use Adrienne Rich's term, also involve rereadings and reappropriations of the meanings of spaces and places, as shown in such examples as struggles over public memorials and MONUMENTS and projects developed alongside critical studies of 'the CITY of collective memory' (Boyer, 1994) to promote and give form to memories associated with those underrepresented in urban landscapes (Hayden, 1995; see also Ganguly, 1992; Samuel, 1994).

DP

Men's movement, mythopoetic The late twentieth century has witnessed the development of a number of 'men's movements' (Messner, 1997). All have emerged, to a lesser or greater extent, in response to FEMINISM. However, the different movements tend to have different interpretations of feminism. Thus, for example, whilst those within the 'men's rights' movement, which campaigns against discrimination against men, characteristically see feminism as socially destabilising and an unwelcome threat; adherents of the 'antisexist' men's movement view feminism as offering liberation both to men and women. A somewhat more complex set of interpretations is presented by the mythopoetic men's movement, which contains both anti-feminist and pro-feminist men.

The mythopoetic movement began in North America in the early 1980s as an attempt to find new forms of masculine IDENTITY. It proposed a critique of the male role types and identities available within Western CAPITALISM, suggesting that these forms were alienated from the real essence of MALENESS. This essence was associated with 'primitive' and 'more natural' societies, historical periods and LANDSCAPES. Thus the mythopoetic movement has drawn extensively on gender symbolisms associated with 'indigenous cultures' and primal, wilderness landscapes in order to create an alternative subculture that is capable of celebrating the 'male spirit'. The strong environmental and geographical themes within this subculture have been studied (Bonnett, 1996) and it has been suggested that the movement is engaged in a creative process of adopting and adapting COLONIAL notions of non-Western and pre-modern CULTURE.

The central text within the mythopoetic movement is Robert Bly's *Iron John: A Book About Men* (1990). Bly attacks the feminisation of Western men, pouring particular scorn on the mother-dependent 'soft male'. Drawing on the work of the psychoanalyst Jung, Bly proposes the use of mythologi-cal archetypes which men may explore and adopt in order to locate a psychosocial pathway back to AUTHENTIC masculinity. The title of Moore and Gillette's *King, Warrior, Magician, Lover: Rediscovering the Archetypes of the Mature Masculine* (1990) encapsulates this particular project.   AB

Mestizo/Mestiza Mestizos are people of European and Native American racial ancestry. Mestizo is used both for men and to refer to the group, whereas Mestiza refers only to women. Within the racial hierarchies of Central and South America, Mestizos are seen as superior to indigenous people and Africans, but inferior to those of pure Spanish lineage. There are numerous terms that categorise Mestizos more finely, depending on the proportion and source of Spanish lineage.

The origins of *mestizaje* date back to the conquest of Mexico and Central America by Spaniards in the sixteenth century. Intermarriage and the sexual exploitation of native women by Spaniards, together with the decimation of the Indian population by smallpox and other European diseases, created the new HYBRID race which now dominates Central and South America. Indeed the Mestizo race has its origins in the unequal relations between Spanish men and indigenous women (Anzaldúa, 1987; Blea, 1992; Moraga, 1986).

A central figure in the legendary history of Mexican Mestizos is Malintzin, Hernán Cortés's concubine and translator. In Mexican legend she is commonly called *La Malinche* (the traitor) or *La Llorona* (the weeping one). Originally sold by her parents, she was one of 20 women given to Cortés by Moctezuma. In legend she is held responsible for betraying her people to Cortés and then drowning her own son rather than taking him back to Spain with Cortés. Latina feminist scholars argue that the legacy of *La Malinche* has been blaming women for the betrayal of their race and the subsequent 'bastardisation' of the indigenous peoples of Mexico; a legacy that has perpet-uated male privilege and the portrayal of women as sexual traitors.

La raza mestzia (the Mestizo race) also enters into contemporary feminist discourse as a symbol and example of racial and cultural HYBRIDITY and toler-ance for ambiguity. Gloria Anzaldúa writes of the consciousness of the Mestiza, an IDENTITY forged from the marriage of oppressor and oppressed. Embracing both identities and both histories, Anzaldúa argues, creates a new consciousness and epistemology that can move us individually and collectively beyond the DUALITIES of subject–object, man–woman, and WHITE–BLACK that oppress and dehumanise us all.   DM

Metanarrative See NARRATIVE

Metaphor A figure of speech that refers to bringing together seemingly unrelated concepts or subjects in order to draw out a comparison or new way of thinking about them or the world. Metaphors figure prominently in writings about the history of geography, as in Buttimer's (1984) search for geography's root metaphors (as map, mechanism, organism and arena). Currently, a widespread appeal to spatial metaphors of positionality, location, BOUNDARIES, BORDERS and MARGINS appears throughout critical literary, feminist and postmodern writing, but geographers challenge the absolute, objective and inert concept of SPACE on which these often rely (Smith and Katz, 1993; Pile and Thrift, 1995). They argue for spatial metaphors that extend earlier calls for conceptualising space as both medium and outcome of social relations (after Lefebvre, 1991 and others). Feminist geographers draw attention to the problematical gendering of spatial metaphors and argue that a recognition of corporeality and a fluid concept of difference can effectively challenge claims to UNIVERSAL KNOWLEDGE and truth made by writers who fail to 'ground' themselves (e.g. Pratt, G., 1992).

One of the most enduring metaphors in human geography is 'LANDSCAPE as text', a trope that emerged out of early-twentieth-century empiricist cultural geography but is now widely discussed in POST-STRUCTURAL critiques of economic, political and cultural signifying practices. Geographers highlight the extent to which written and spoken words, maps, paintings, cultures and other 'TEXTS' are not only reflective of social and spatial processes and realities but also are constitutive of them (Barnes and Duncan, 1992). Sexual metaphors of landscape and place abound in PHALLOCENTRIC and masculinist writing and culture. Anglophone feminist geographers have drawn widely on critiques of the language of 'conquest and penetration of virgin lands' during Euro-American empire-building (Kolodny, 1975, 1984), as well as the metaphor of land or NATURE itself as woman (Merchant, 1980).   KM

Migrant/migration Migration is the semi-permanent or permanent movement of individuals or groups from their usual place of residence. A migrant is a person involved in the move. Migration, together with fertility and mortality, is regarded as one of the primary influences on population size and composition. Migration is analysed in various ways, including: the reasons for the move, the distance people have moved and the duration of the move. Reasons for migration are predominantly economic but also social, physical, cultural and political. In terms of distance, distinctions are made between internal (domestic) and international migration, rural to urban and urban to rural migration, inter-regional migration and intra-urban or intra-rural migration. In terms of the duration of the move, Chant and Radcliffe (1992: 10) distinguish between permanent and temporary migration. Permanent migration occurs when people remain in the place they migrated to. Temporary migration is more complex. Chant and Radcliffe (1992) identify four different types of temporary migration; namely, seasonal, oscillating, relay and circular/return migration.

Theoretical approaches to migration have in the main neglected gender despite the fact that migration is not gender-neutral (Chant and Radcliffe, 1992: 19). The experiences of those left behind are also often neglected. A gendered analysis suggests that women have common experiences when left behind of autonomy mixed with vulnerability; men's destinations are more varied and of a greater distance, and, although many men do remit to their families, women tend to maintain stronger social and economic links with FAMILY in origin areas (see Chant, 1992: 198). PM

Military/militarism Militarism is a system which a society adopts to sustain peace and/or prepare for war (Elshtain and Tobias, 1990). Feminists have long argued that militarism plays a major role in shaping GENDER identities and gender relations, which then reinforces male privilege within a society (Enloe, 1983; Elshtain and Tobias, 1990; Sharoni, 1994). Jacklyn Cock contends that in order to understand war we need to understand how the military shapes gender identities within a society. She explains, 'understanding the military involves examining gender relations – in particular the way MASCULINITY and FEMININITY are defined. The military mobilizes gender identities and in periods of war this process is sharpened' (Cock, 1993: viii).

Militarism intensifies the predisposition to perceive men as action orientated – the heroes of the battlefield – and women as supportive to the male soldier – the custodians of the homefront. These gender designations do not denote the actions of women and men in a time of war but function instead to recreate and secure women's position as non-combatants and that of men as soldiers (Dowler, 1998). Rosemary Ridd has investigated the relationship between the 'power' of men and the 'powerlessness' of women and contends that in wartime gendered images become antithetical and exaggerated to the point that the representations of women become ever more symbolic in form, rendering them as the mothers of a NATION (Ridd, 1987: 12).

Militarism can create a demand for women to acquire new skills, thereby granting them more access to the PUBLIC arena. However, feminist scholars contend that women's claim to the public arena, whether as individuals or as a national symbol, is usually temporary. Cynthia Enloe argues that militarism, 'on the one hand, can transform women's role and sense of self-worth, while on the other hand, sustain the social order that in the past has ensured the reproduction and nurturing of the next generation' (Enloe, 1983: 166).

Although feminist scholars agree that militarism reinforces the gendering of public and private spaces, they also maintain that a more fluid understanding of the interdependent nature of these spaces will result in transcending these spatial designations. In other words, when deconstructing traditional gender roles it is critical not simply to subvert or privilege either the public or the private arena. Instead, as Nancy Duncan argues, 'the destabilizing of this boundary is a countervailing force working to open up not only private space but to reopen public space to public debate and contestation' (Duncan, 1996b: 127). Therefore an understanding of the interdependency of these two spheres can invert patriarchal power structures within war zones. LD

Mimesis A mode of representation that is both referential and (conventionally) realistic. Rorty (1979) and other POSTMODERN philosophers and feminists have been critical of the modernist concept that LANGUAGE is merely a reflection (mimetic representation) of reality, as opposed to a constitutive aspect of it. The closely related term 'mimicry' refers to an imitative speech or action. Feminist geographers have drawn on POST-COLONIAL theories to argue that colonised peoples confront and resist colonial intervention through the tropes of mimicry and PARODY in particular. Bhabha (1984a, 1984b), for instance, argues that effective subversion of COLONIAL AUTHORITY and knowledge takes place through the very appropriation of its signs.　　　　　　　　　　　　　　　　　　　KM

Mobility The term mobility can be used in relation to people's physical movement and also, as in the term social mobility, in relation to changes in social status.

Physical mobility: in geography the general term mobility can be used in relation to the long-term movements of migration and the shorter-term, circulatory movements involved in everyday life such as travelling to and from work, shopping, visiting friends and going on holiday which are addressed by studies of TRAVEL behaviour.

In relation to MIGRATION, feminist scholars have criticised many migration studies for ignoring the gendered power relations within households and in the labour market which can affect not only who migrates but also why, where and whether migration occurs and have shown that these issues are highly significant to migration (Boyle and Halfacree, 1998; Radcliffe, 1991; Chant, 1992).

In relation to shorter-term movement, feminists have criticised many studies of travel behaviour for ignoring the gender of travellers, the significance of life-course stage and issues of gendered power in accounting for travel patterns (Rosenbloom, 1993). They have pointed out that women's mobility is often highly constrained not only by lack of money or access to motorised transport but also by the time and space constraints imposed by the demands of housework and caring for children or elderly family members (Pickup, 1985). Furthermore, they have established that because of these domestic demands women's travel patterns are frequently more complex than those of men, involving both more multipurpose trips and a greater variety of locations (Rosenbloom and Raux, 1985). Relatively few travel-behaviour studies explore the experience of travelling, but those that do have emphasised the stresses of travelling with children and the importance of issues of safety from sexual attack by men.

Social mobility: This term refers to changes in social status either of an individual during their life-time or of children relative to their parents' social status. Many studies of social mobility have been criticised by feminists because they have measured status solely in terms of the male 'head of household'. Thus older empirical studies of social mobility often ignored the status of daughters and mothers and concentrated on sons' social status relative to their fathers' status.　　　　　　　　　　　　　　　　　　　SB

See also CLASS/CLASS RELATIONS.

Modern/modernism/modernity/high modernity The term 'modern' has long been used in Western culture to mark a break with the past and to designate the 'newness' of the present. Its meanings have varied according to place and time but the qualitative claim of novelty came to be stressed during the ENLIGHTENMENT and afterwards, when modern became connected with ideas of 'progress' and an open future and with looking forward in a manner counterposed not only with the ancient but also with tradition in general (Osborne, 1995, Chapter 1). While 'modernity' is typically traced back to the seventeenth century, as a quality or experience of the latter notion of the modern, it is identified particularly with developments from the late eighteenth century, with industrialism and associated economic, social and cultural changes, and with a cultural self-consciousness that gave new meaning to innovation, progress, and history itself. The direction of development was meant to be guided by the application of REASON, RATIONALITY and elements of the 'Enlightenment project' so as to deliver the individual from the bonds of tradition; but the social tumult unleashed by processes of change was also emphasised by critics such as Marx, who connected modernity with the dynamics of CAPITALISM and with a world in which 'all that is solid melts into air' (see Berman, 1983). 'Modernism' is often depicted in dialectical relation to such changes, as a mode of REPRESENTATION tied to movements in the arts from the late nineteenth century, linked to crises of modernity and increasingly apparent experiences of 'the ephemeral, the fugitive, the contingent' (Baudelaire, 1964: 13; see Gregory, 1993; Harvey, 1989).

The assignment of favourable connotations to 'modern' during the nineteenth and especially twentieth centuries means that it is 'a prestigious word, a talisman, an open sesame', notes Lefebvre (1995: 185), yet 'if we ask what it means, no answer is forthcoming; indeed, it is a word which we are not even permitted to question'. Assumptions about the inherent progressiveness of modernity have in fact been closely scrutinised and contested, not least in relation to the ways in which its constitution as a social order has been dependent upon forms of VIOLENCE, including through connections with COLONIALISM and IMPERIALISM. Indeed, much classical social theory has centred around aspects of modernity's 'dark side', and Marx's work has inspired numerous critiques, among them Lefebvre's own reading of modernity's presentation of newness in terms of repetition and the reproduction of relations of production (Lefebvre, 1976). More recently Giddens (1990, 1991) has identified modernity broadly with four institutional clusters of capitalism, industrialism, SURVEILLANCE and MILITARY power and has argued that the present represents a condition of radicalised or 'high modernity', characterised by inherent reflexivity and high-consequence risk. Many of the classical social theories have in turn been questioned, however, about their neglect of the gendered processes of change and of the gendering of the spaces often associated with modernism. Feminist theorists have been influential in developing critiques at an EPISTEMOLOGICAL level of modernity's implication in forms of Enlightenment rationality and BINARY modes of thought, leading into notions of POSTMODERNISM (for example, Nicholson, 1990); but, rather than dismiss modernity and modernism altogether, many have also problematised and worked critically with these terms, seeking to revise their meanings and histories in relation to FEMINISM and experiences of women.

Feminist critics have therefore argued for the need to historicise binary or dualistic categories that underpin many theories of modernity, such as PUBLIC–PRIVATE, individual–society, family–economy, and to rethink categories of social theory in a way that recognises the restructuring of gender relations as a fundamental characteristic of modernity (Marshall, 1994; Wolff, 1985). Some, such as Barbara Marshall, have further asserted that feminism provides a means by which the 'project of modernity' might be challenged and reformulated while defending its EMANCIPATORY aims. But as Marshall makes clear, the process of 'engendering modernity' has also revealed significance of gender within discourses of the field even if the concept has often been 'unthought'. Increasing interest has thus focused on the relationship between representations of modernity and notions of FEMININITY and MASCULINITY and on the positionings of gender within cultural readings of the modern (Buci-Glucksmann, 1994; Felski, 1995; Huyssen, 1986). While noting that most contemporary theories of modernity are male-centred, Felski emphasises the field's complexities and ambiguities and attends to different stories opened up around conjunctions of femininity and modernity. She is particularly interested in the ways in which women drew upon, contested or reworked dominant representations of gender and modernity, and as such her account connects with feminist re-visionings of modernism (see also von Ankum, 1997).

Modernism in the arts is usually located between around 1880 and 1940, emerging particularly in cities in Europe and North America and involving diverse currents that challenged realism and romanticism. It may be identified with certain common features including aesthetic self-consciousness; simultaneity, juxtaposition and montage; paradox, ambiguity and uncertainty; and the demise of an integrated individual subject (Lunn, 1985, Chapter 2). Much feminist research has sought to rediscover women writers and artists excluded from historical accounts, to uncover communities and networks of women modernists and to consider the distinctiveness of women's modernism (Benstock, 1987; Gilbert and Gubar, 1988, 1989; Scott, 1990). Challenges to modernism as an 'official' cultural institution by feminist art practices since the 1970s have also been influential (Pollock, 1987). This process of recovery and contestation has involved attempts not only to rewrite histories of modernism, but also to rethink modernism as a DISCURSIVE and historical field, and in some cases to '*appropriate* the term and – by teasing out its aporias, its gender blindnesses, its *constructedness* – put it to feminist use' (Elliott and Wallace, 1994: 16; see also Wolff, 1990). Research has highlighted issues of gender politics and has explored spaces of modernism in relation to GENDER and SEXUALITY and their mutual inflections with other axes of power around CLASS and 'RACE' (Pollock, 1988). Similar rereadings are focusing on modern architecture and PLANNING, which saw the emergence of a 'high modernism' in the 1950s and

1960s, and, collectively, these studies are part of an increasing current of interest in critically addressing the plurality and heterogeneity of the modern. It is a project that resonates with developments in POST-COLONIAL theory and discussion of 'counter-cultures of modernity' (Gilroy, 1993) and one in which the tracing out of different spaces of modernities promises to be of considerable importance. As Felski (1995: 212) argues, 'The history of the modern is thus not yet over; in a very real sense, it has yet to be written.'   DP

Modes of production/modes of reproduction The ways in which production and reproduction are organised in society. This broad concept incorporates social relations, technologies, political structures and cultural processes, and the term is used to describe the historical development of production and reproduction. Each mode of production or reproduction will tend to have a different geographical structure. In CAPITALIST society, the spaces of production have been gendered male and those of reproduction (at HOME, with the FAMILY) are widely assumed to be FEMALE.   JW
See also INDUSTRIALISATION.

Monument/monuments Components of the LANDSCAPE, often relatively small-scale, intentionally conveying symbolic messages to collectivities of people both locally and nationally. Feminist geographers have been particularly interested in those which were part of the NATION-building process in the nineteenth century. These took the form of sculpture located in PUBLIC places, which could reach a wide spectrum of the population. Their meanings have begun to be decoded but little has yet been written on how these are popularised, consumed or challenged at the popular level (Johnson, 1995).

The imagery of national monuments reflects HEGEMONIC structures and is thus strongly MASCULINE. Men are figured or named as individuals, as heroes and leaders, or collective actors. By contrast, women are portrayed allegorically, not as individual people or even as real women (Warner, 1985). The readings of such representions are complex and contradictory. For example, the Statue of Liberty in New York ironically uses the image of a woman to represent freedom. A particularly contentious monument is the statue of Anna Livia Plurabelle in Dublin, which represents both womanhood and the City of Dublin (Smyth, 1991).

Women are usually absent from war memorials, though the shame of military defeat may be deflected by representations of weeping women (McClintock, 1993). Only after lengthy debates was a memorial constructed in Washington explicitly honouring women soldiers and nurses who died in Vietnam (Monk, 1992).

The massive scale of commercial monuments, such as Crystal Palace and the Eiffel Tower, also celebrates the intertwining of CAPITALISM and masculinism, though many other meanings become attached to them over time (Shurmer-Smith and Hannam, 1994).   BW
See also NATION/NATION STATE/NATIONALISM.

Morals/moral reasoning Paradoxically, women are assumed both to have weaker powers of moral reasoning and to be morally superior to men. The first assumption is clear in FREUD's belief that women are morally inferior to men because they have a weaker superego. However, women's moral superiority lies behind the DOMESTIC IDEOLOGY that became important in Victorian England. The nineteenth-century writer and painter John Ruskin believed that women are naturally good and must therefore be protected from corruption by the hurly-burly of indutrial capitalism. Ruskin placed woman on a moral pedestal: a position which actual women found hard to occupy. However, there is an important strand of moral superiority in RADICAL FEMINISM's definition of feminine virtues – women are seen as more moral, peaceful and caring than men. The Harvard psychologist Carol Gilligan (1982) developed this argument in her book *In a Different Voice* in which she distinguished two types of moral reasoning. She argued that the first type – based on an ethic of justice and rights – was a typically masculinist formulation, whereas the second type – based on an ethics of care or responsibility for others – was typically found among women, whom, she believes, are more likely to make particularistic and contextual moral judgments than are men. Feminist philosopher Seyla Benhabib (1987) has developed these ideas in her explorations of abstract and concrete versions of JUSTICE. The emphasis on a particularistic or contextual version of moral reasoning has strong resonances with more recent work by POSTMODERN theorists (see for example Benhabib, 1995; Squires, 1993). LM

Mother Earth See EARTH AS MOTHER

Motherhood/mothering/maternity Mothering is the term used to denote the various ways in which women mother, the day-to-day organisation of children's lives and the care women provide for them. Implicit within it are the notions that mothers are responsible for good child development and that there is a key emotional bond between mother and child. Motherhood encompasses the meanings which being a mother have for women and the ideologies which give shape to these. Thus, within the feminist literature, motherhood is acknowledged as a crucial, defining aspect of women's lives and identities, yet something which carries many (problematic) normative prescriptions. Thus, the 'good mother' is seen to have children when falling herself within a particular age-range, when in a stable (heterosexual and preferably married) partnership and in acceptable social circumstances. Moreover, she is expected to put the needs of the child before those of herself and to enable successful (defined as non-deviant) child development. In all this the pervasiveness and assumed preferability of maternal care in the home remains central, although there have been a number of studies in geography which have looked at different ways of mothering, for example those which employ substitute labour in the home and those which involve day-care services. In contrast to these readings are feminist psychoanalytic accounts of motherhood, notably those of Chodorow, Dinnerstein and the French feminist theorists, Irigaray and Kristeva. It remains to be seen how the development of new reproductive technologies might transform the meanings attributed to and inscribed in mothering and motherhood. Maternity is a (rarely used) term of slippery usage, conveying both pregnancy and mothering/motherhood. In a wider usage the terms 'to mother' or 'mothering' are associated with the daily care, love and attention that a woman provides for children within the space of the home and so the term also includes women who are the mothers of adopted or fostered children. However, as Stack (1974) pointed out in her studies of African-American families, the assumption that joint residence is a necessary requirement for mothering is an ethnocentric one which ignores the wide range of class-based, ethnic and cultural differences in patterns of mothering and CHILDCARE. It is frequently asserted that SECOND WAVE FEMINISTS ignored motherhood as an interesting question for research and analysis until they themselves had children. However, mothering has been a particularly important emphasis for radical feminists who celebrate women's ability to give birth as a crucial characteristic of femaleness. This position is sometimes criticised as ESSENTIALIST as the 'natural' features of motherhood are celebrated (although social variations in mothering practices are recognised). Indeed, women who celebrate their fertility and fecundity and their construction of self through motherhood are sometimes referred to as 'earth mothers'. In a reversal of this image, the earth and NATURE are often portrayed as female or as a mother (Merchant, 1980; Seager, 1993).

The geographical literature about motherhood has, in the main, focused on issues of childcare, both by mothers (Tivers, 1984) and by paid DOMESTIC WORKERS, employed both inside and outside the home (England, 1997; Gregson and Lowe, 1994; Pratt, 1997).

Throughout the twentieth century the restrictions on biological motherhood have been redefined by innovations and reproductive techniques such as surrogacy, *in vitro* fertilisation and embryo transplantation for postmenopausal women, leading to an era that the feminist anthropologist Marilyn Strathern (1993) has termed *After Nature*. Given the increasing attention being paid to the body within contemporary geographical writing, and to notions of leaky, messy bodies which disrupt boundaries, it would seem only a matter of time before geographers' attention is turned to the pregnant body.   NG
See also EARTH AS MOTHER.

Motherland See NATION/NATION-STATE/NATIONALISM

Multicultural/multiculturalism The terms 'multicultural' and 'multiculturalism' are complex, reflecting the inherent contradictions within cultural processes and the contested conceptalisation of CULTURE. The term multicultural is a descriptive term referring, for example, to the demographic diversity of a population. This cultural diversity must be recognised as socially constructed within particular contexts. Multiculturalism refers to a policy which endorses the principle of cultural diversity and supports the right for different cultural or ethnic groups to retain a distinctive cultural identity rather than assimilating to a dominant society's cultural 'mainstream'. Multiculturalism thus seeks to promote cultural diversity and equal rights.

Multicultural policies are most evident in the spheres of education and the arts. Multiculturalism, expressed in policies such as the Swann Report (HMSO, 1985), is based on the premise that intolerance of minority groups by the majority is a problem of individual attitudes of prejudice. The sympathetic teaching of 'other cultures' is intended to dispel ignorance and therefore build tolerance (Rattansi, 1992).

Multiculturalism has been criticised by those who espouse an assimilationist approach to ETHNIC and cultural diversity. The institution of multicultural policies has been seen as a challenge to the autonomy of educational institutions – reflected in the so-called 'cultural wars' and accusations of 'political correctness', particularly within universities in the USA (Bloom, 1987; Graff, 1992). Multiculturalism has also been criticised for its weak liberalism and for a celebration of cultural diversity which is largely symbolic, characterised as 'saris, samosas and steel bands' (Troyna, 1990: 404). Multiculturalism is seen to apply only to minority cultures and not to challenge the majority culture or to seek to secure equal rights. This opposition has been expressed through an alternative policy of 'anti-racism' which concentrates less on cultural diversity and more on identifying and challenging the mechanisms and institutions by which racialised groups are discriminated against and disadvantaged (although this policy too has been strongly criticised, see Rattansi, 1992).

Multicultural policies have been important vehicles for securing the RIGHTS of minority groups to important government resources. Yet multiculturalist policies can also be seen to work against the interests of some members of minority groups, particularly women. Multiculturalism relies upon an unproblematised and coherent notion of culture. The categorisation of minority communities through their cultural identity, often by focusing on religion, assumes that minority groups are internally unified with no CLASS or GENDER differences or conflicts (Anthias and Yuval-Davis, 1992; Yuval-Davis, 1992). This means that women's demands for EQUALITY or freedom may be defined as inappropriate because they are seen to be outside assumed 'cultural traditions'. Multicultural policies may also act to reinforce PATRIARCHAL leaderships as the 'AUTHENTIC' representatives of minority communities.

While geographers have focused primarily on analysing the geographical effects of multicultural policies (e.g. Kobayashi, 1993), recently some geographers have sought to deploy the notion of 'multicultural imaginaries' (Kahn, 1995). While denying the ontological reality of 'culture', such work seeks to explore the ways in which discourses of multiculturalism are constructed, deployed and used within particular contexts, such as, for example, the circuits of culinary culture (Cook *et al.*, in press). Such perspectives suggest the possibilities of reworking multiculturalism not necessarily as an ideology or discourse but as a perspective for rewriting or reworking notions of cultural identity which emphasise webs of connections across transnational boundaries (Hall, 1996).   CD

Myth/mythology The reproduction of male power is seen by many feminists as being enabled both by economic and by ideological processes. Within the latter arena, creation myths concerning how the world was formed and what values its constituent social and natural forms embody have been understood as particularly influential. Thus, for example, the privileging of men, and symbols of masculine power, within Christianity has been subject to a diverse range of feminist analyses, both from non-Christians and from Christians (Daly, 1968, 1973).

Recognising the importance of mythology in the reproduction of social IDEOLOGIES some writers have sought to develop counter-myths that celebrate and privilege women. This form of feminist literature is, at least in part, enabled by the assertion that, whilst male IDENTITY within CAPITALIST societies tends towards alienation and rationality, FEMALE identity can be characterised, both in the past and in the present, as more natural, TRANS-GRESSIVE and responsive. This distinction draws on a variety of sources, some of which are of a reactionary and anti-feminist nature. However, it has found fertile ground within those forms of feminism that seek to align women's movements with the protection of NATURE and 'MOTHER EARTH'. These projects have tended creatively to rework a variety of existing mythologies, as well as to seek to invent new ones. Reworked myth has drawn on notions of the GODDESS, WITCHCRAFT and magic (Daly, 1987; Starhawk, 1982). Where original mythological structures have been developed they have tended to draw on PSYCHOANALYTIC WORK, particularly the archetypal schema of Jung (Pratt, 1982).

The creative adoption and adaptation of mythology within feminism has influenced the emerging MEN'S MOVEMENTS of the 1980s and 1990s. The mythopoetic men's movement, in particular, can be seen to have reworked existing feminist, female-centred, articulations of mythology in order to develop new, male-centred, ones. In what amounts to both a challenge and a tribute to mythopoetic feminism, Kipnis (1992: 163) calls on men to reject alienation and rationality: 'The Earth Father welcomes us, challenging us to become stronger and deeper as men. We are at home in nature'.

Both feminist and men's movement uses of mythology may be classified as primitivist. In other words, they posit a more NATURAL, more primeval, social realm (usually in the distant past and/or within non-Western societies), in order to develop a critique of existing, Western society. This approach to social critique may be criticised for its tendency to romantic distortion, as well as for a characteristic erasure of issues of POWER and social structure. However, whether more academic, and self-consciously politicised, forms of gender politics are themselves completely devoid of mythological content and structures is open to doubt. Themes (or are they legends?) of redemptive power and utopian transformation may be found within many liberal, socialist and anarchist feminist traditions. It might be suggested, then, that mythology is not merely an anachronism that can be simply dispatched; that it is something that all those involved in gender politics are, at some level, engaged in reworking. AB

See also MEN'S MOVEMENT, MYTHOPOETIC.

Mythopoetic men's movement See MEN'S MOVEMENT, MYTHOPOETIC

N

Narcissism A term derived from the Greek figure Narcissus who fell in love with his own image reflected in a pool of water. In PSYCHOANALYTIC theory, narcissism refers to the extreme love of SELF, and for Freud the 'pathology' of homosexuality. JS

See also LACANIAN THEORY; FREUDIAN THEORY.

Narrative An account of a series of events. Geographers have drawn on literary and artistic narrative sources, such as novels, diaries, essays, travelogues, oral histories, films, paintings and dance, to extrapolate spatial relationships, representations of landscapes and meaning of place (e.g. Barnes and Duncan, 1992). Feminist geographers have drawn especially on personal narratives to reconstruct sites of OPPRESSION and to call attention to them. Personal narratives allow women and others who have been denied access to public venues of disclosure to give their own statements and become active 'agents' in the production of knowledge (hooks, 1989), even if the unified SUBJECTIVITY of authorship has been questioned in DECONSTRUCTIVE methods. Feminist geographers rely extensively on personal narratives to understand the intersections between GENDER, SPACE and PLACE, for instance in examining whether the gender of the author, researcher, 'others' discussed in the narrative and reader makes any difference in how the work gets produced and received. Narratives about women's experiences of WORK, HOME, FAMILY, TRAVEL and MIGRATION spatially and temporally locate authors and others within their cultural and material circumstances. These have been important in locating institutional, economic and cultural sites of oppression and in interpreting how gendered identities and unequal POWER relations shape distinctive places and gendered experiences through space (Jones *et al.*, 1997; Kay, 1991; Valentine, 1993b).

Narrative can also be more broadly understood as a method for describing historical and geographical phenomena, for instance in the chronicling of events unfolding as an historical geography of a given place, or in writing the history of geography as an academic discipline. Often embedded in this method are EPISTEMOLOGICAL foundations or 'metanarratives' which directly or indirectly claim UNIVERSAL truths that exclude the experiences of most of the world's population. Challenges to these master or metanarratives are fundamental to POSTMODERN philosophy (Lyotard, 1984), and a number of feminist critics have argued that the grand narratives of Western civilisation (e.g. of 'progress') speak only of the EXPERIENCES of white, male, heterosexual protagonists in Europe and America (Nicholson, 1990). In fact FEMINISM itself has been understood as an attempt to recover that which metanarratives have excluded (Jardine, 1985). Postmodern and feminist theorists agree that there is no single historical narrative truth to be uncovered, and that to write as if there were is to make false claims to neutrality, objectivity and legitimacy. Donna Haraway (1988) put the brakes on the endless relativism that these philosophies implied, however, by arguing for the existence of 'SITUATED KNOWLEDGES' or legitimate partial perspectives. Feminist critics have been skeptical of not only the content of geography's historical narratives but also the rhetoric and structure of them (Bondi, 1990a; Rose, 1993a; WGSG, 1984). Feminist geographers working within cultural or post-colonial studies also discuss ways that historical NARRATIVES function to create inclusionary and exclusionary concepts of NATION and CITIZENSHIP, and address the historical narrativity of such phenomena as public monuments and spectacles that shape, reproduce and contest dominant understandings about place and belongingness (e.g. Anderson and Gale, 1992). **KM**
See also TEXT/TEXTUAL/TEXTUALITY/INTERTEXTUALITY.

Narrative closure In literary criticism closure refers to the moment in the NARRATIVE or drama when resolution of conflict occurs. While some feminist critics advocate rewriting traditional closures that have been scripted by PATRIARCHY, others regard closure itself as a MASCULINIST trope that falsely presents the world as ordered, disciplined and contained within a linear and dualistic logic. Many resist narrative closure and opt for a more open-ended textuality that values ambivalence, complexity and multiple positionalities (Gregory, 1989; Rose, 1993a). Some post-colonial critics also find the absence of closure as characteristic of COLONIAL DISCOURSE (Ashcroft *et al.*, 1995). **KM**

Nation/nation-state/nationalism A nation, according to the widely accepted definition by Benedict Anderson (1991) is an 'IMAGINED COMMUNITY' of people who are bound together by a common territory and culture. Unity is an overriding characteristic of the nation and is constantly being recreated through PERFORMANCE and REPRESENTATION, for example by ceremonial events, media images and monuments. Although unity is apparently unmarked by GENDER, in reality it is implicitly MASCULINE. This is epitomised by Anderson's description of the nation as a 'fraternity' and his illustration of its profound importance in people's lives by their willingness to die in battle to defend it (Walter, 1995).

This emphasis on unity submerges diversity, which is often naturalised in ways which disguise the different interests and positions of constituent groups. Different genders, CLASSES, ETHNICITIES and generations do not relate to, or belong in, the nation in the same ways. Gender differences in particular are harnessed to buttress the project of unity. Anne McClintock (1993) argues that the nation is imagined temporally as having a golden past, representing mythical shared origins, and a progressive future, its destiny. The past is gendered feminine, signifying tradition and NOSTALGIA, whilst the future is gendered masculine, depicted as forward-thrusting and MODERN. The nation also has specifically gendered SPATIALITIES, for example through the use of domestic imagery to represent national space and of women's BODIES to represent the boundaries of the nation (Radcliffe, 1996).

Women are thus excluded from full participation in the present and future concerns of the nation by their relegation to the symbolic realm. Men on the other hand have a full and active part to play in material life. Female imagery is often used to depict the nation, for example in the terminology of 'MOTHER-LAND' and the choice of female emblems such as Marianne, Hibernia and even Britannia, although the nation may also be referred to as the fatherland (see Nast, 1998b). The trope of the FAMILY serves to naturalise the social hierarchy of the nation, apparently sharing common interests but in reality under the control of the father.

The symbolic gendering of the nation has real effects on women's lives, removing their freedom of choice. For example the 1937 Irish Constitution enshrined women's role in the home as a keystone of the independent Irish nation (Beale, 1987). National policies on reproduction may directly reflect anxieties about the future of the nation, as in child support benefits in France introduced after population losses in World War II and the outlawing of abortion in Croatia in 1992 (Sharp, 1996).

However, not all women within the nation are constructed in the same way. Class is often used to differentiate between them, so that in eugenicist discourses upper-class, educated women are encouraged to have more children whilst poor, disabled and ethnic minority women are prevented. In Ecuador, for example, white-creole elite women from the large cities were represented positively in the media whilst indigenous rural women were not the basis of national pride, and BLACK women were excluded altogether (Radcliffe, 1996).

Nation-states are rarely contiguous with nations. They are territorial units defined by political boundaries. These boundaries may cut across peoples who imagine themselves as members of the same national community, or include more than one nation. Members of nation-states are bound together by their shared CITIZENSHIP, which is nevertheless experienced unequally by gender as well as class, race, SEXUALITY and location.

Immigration policies are a key area where gender differences in state belonging are registered, for example where women are admitted as 'dependents' (Klug, 1989). But the processes of both state-building and nation-building, through the creation of citizens, are also highly gendered. In state education policies, for example, hegemonic gendered as well as raced and classed messages are instilled. School geography texts play a role in this process (Madrell, 1998). State policies and practices mediate women's experiences in wide-ranging and complex ways, not always disempowering, but the privileging of particular classes, ages, ethnicities, sexualities and abilities is usually implicit (Fincher, 1993).

The nation-state is constituted by a body of institutions which have powers of coercion and control. VIOLENCE is a key constituent of nation-building, through territorial expansion, defence of national territory and suppression of internal dissent. There are thus strong military and policing elements involved. Women and men are positioned differentially and participate differently in policing and MILITARY activities. It is mostly men who are selected to fight and be killed whilst women continue to sustain all other facets of social life (Enloe, 1983).

Nationalism refers both to individuals' sense of a national identity and to social movements based on the political ideology of rights to a separate territory for a nation. The two are intertwined as people's identities are constituted through social contests.

Although Anderson (1983) argues that nationalism required the arrival of print, in order to disseminate ideas outside the immediate locality, McClintock (1993) puts more weight on the importance of spectacle for organising collective unity. She argues the elite are most influenced by print and that popular nationalism takes shape through the visible ritual display of objects such as flags, uniforms and anthems. These are most evident in masculine events such as team sports and military tattoos. The ways in which women consume, refuse and negotiate male rituals of national spectacle are as yet little understood.

National identities are frequently defined overtly or implicitly on grounds of RACE, usually asserting the genealogical rights of 'insiders' (Miles, 1993). However, counternarratives from autochthonous groups within national boundaries are emerging more strongly. In Aotearoa/New Zealand, for example, Maori demands for equal access to health care are being acknowledged (Dyck and Kearns, 1995).

Feminists have been placed very differently in relation to nationalism according to their COLONIAL and POST-COLONIAL positioning. Women in hegemonic FIRST WORLD collectivities have experienced their invisibility in the patriarchal national project as the major issue, or have sought to challenge inclusion in national projects such as wars and armament programmes. However, women in national liberation struggles have faced the dilemma of a more complex set of subjugations, which has placed them in alliances with men whilst simultaneously pursuing feminist aims

(Jayawardena, 1986). An example of the different strategic choices made by feminists is women's roles in Irish independence struggles of the early twentieth century (Ward, 1989).   BW

Native Refers to one of the original inhabitants of a country whose rights are the subject of increasing political and intellectual debate. In Western thought the term has also been associated with marginalised groups of non-European origin where the native was allegedly subhuman or uncivilised (Young, R., 1990). Used in this pejorative sense it has been linked with the ideology of cosmopolitanism and its gender, class, race and locational dimensions where the cosmopolitan (travellers) are contrasted to the rest who are merely local (natives) (Gregory, 1994).   MB

Natural Used to distinguish products, processes or tendencies originating outside of human cultures, from those constructed or manufactured within them. Whereas 'nature', in the couple NATURE–CULTURE, is usually less highly valued (cf. savage–civilised), in the couple natural–artificial 'natural' is the preferred term. Thus herbal products, although processed, are described as 'natural' in contrast to synthesised vitamins; organic farming is praised for its use of 'natural' processes, and the tendency for parents to prefer their biological children over others is seen as 'natural' and therefore right and proper. Like 'HUMAN NATURE', the term is used to legitimise certain forms of social organisation. However, human nature is generally understood as fixed and as explaining why human beings inevitably revert to anti-social or selfish behaviour. In contrast, 'natural' has a positive resonance, and the context generally suggests that humans are all too capable of prolonged 'unnatural' behaviour, which is decadent rather than altruistic.

Feminists have long denied the naturalness of the sexual division of LABOUR and of the exclusion of women from sites of privilege and power. FEMININITY and MASCULINITY have been deconstructed and reconceptualised as normative constructions, initially by women's studies and then more widely within social science. Only in some versions of RADICAL FEMINISM (e.g. Daly, 1978: 337) does 'natural' have positive connotations in reaction to PATRIARCHAL culture. The 'feminine woman' is 'man-made'; a mutant 'painted bird' who double-crosses and attacks natural, wild females who have dared to strip away their false male-imposed identities (see Tong, 1989: 107).

In gay politics the idea of naturalness has been put at the service of HOMOPHOBIA: natural–unnatural is also coded as natural–perverse. There have been two main responses from GAYS:
- to reclaim the language of naturalness for themselves, claiming HOMOSEXUALITY or bisexuality as valid, universal variations of sexual orientation;
- to see the category 'homosexual' as a modern social construction, and, following Weeks (1977, 1985) and Foucault (1978), to reject the idea that some sexual behaviour is more AUTHENTIC than others because rooted in some extra-cultural realm which can ground our sexual morality.

SOCIAL CONSTRUCTIONISTS point out that to see HETEROSEXUALITY as natural implies it is given, while other sexualities require a (distorted) process of PRODUCTION. Psychoanalysis, of course, has always seen all forms of sexuality as developmentally produced (Sayers, 1986). Constructionists believe that all sexualities are constructed *within discourse* – and so is the notion of naturality itself (Richardson, 1996). QUEER THEORY rejects not only normative heterosexuality but also the naturalising of gay, lesbian and BISEXUAL identities, and the allocation of SEXUALITY to such fixed categories in the first place. Although Foucault's later work refers to the natural BODY as the site of cultural inscriptions (Sawicki, 1991: 67), strong social constructionists also deny the naturalness of bodies (Butler, 1993, 1995). For them, the 'natural' is invented within culture and then used to impose certain conventions. Such a thoroughgoing denial of the natural arguably has drawbacks. It means we cannot distinguish between things which are entirely artefacts of culture and those which, like bodies, have extra-discursive features given meaning, worked on and transformed by cultural practices. CN

Nature A deeply ambiguous and resonant term, the many contested meanings of which have political implications (Harvey, 1996: 118). Everyday meanings in Western cultures include *wilderness*, that is, parts of the earth free of human presence and apparently of its effects. There is arguably no longer any wilderness in that full sense, and much of what is enjoyed as such is conserved and managed. The term is also used to refer to the *countryside*, to rural as opposed to urban localities. Here nature means not only what is green and alive, but even by extension ways of life governed by the rhythms of the seasons and by biological necessity. Past or rural cultures can be included in nature when it is understood as the origin, as the state of affairs preceding human action and furnishing its raw material. Usually, things which do not owe their existence to deliberate human activity are seen as part of nature, in contrast to human manufactured products and the associated waste. Nature has always been personified, usually as a female figure or MOTHER, and the idea of Nature as an order with moral implications is still current in such phrases as 'That's against nature'. All these common meanings depend on some form of opposition of nature to humanity, so that animals, plants and other life forms are part of nature, and so are non-sentient features of the planet, whereas humans and human culture are wholly or partially non-natural. In *What is Nature?* Kate Soper argues that the 'OTHERNESS' of nature is fundamental to this highly unstable concept. Changes in the idea of nature, she argues, reflect changes in dominant ideas of what it is to be human. Partially or completely socially excluded groups are always seen by dominant groups as 'closer to nature' – for example 'barbarians', women, 'idiots', black people (Soper, 1995: 74). Here too humans and human cultures can be included in nature, because they are seen as not fully human or are seen as 'uncivilised', so not genuinely acculturated.

Theorists in several disciplines have attempted to trace the shifting meanings of nature across several centuries of European history (e.g. Leiss,

1974). Lovejoy describes the medieval cosmology of the 'Great Chain of Being' which persisted into the eighteenth century. All beings were hierarchically organised, from maggots to God, with humans somewhere in the middle (1964). Harvey argues that it is too simplistic to blame ecological destruction on the ENLIGHTENMENT project of domination of nature. Eighteenth-century and nineteenth-century thought was not homogenous, expressing the sense of alienation from nature and questioning the aim of human self-realisation and its implications. He concludes that the political economy of hegemonic CAPITALISM entails 'a triumphalist attitude to nature' (1996: 131).

Ecology developed from the late nineteenth century as the SCIENCE of relations between organisms and their environments – including each other. In the 1970s the idea of nature as an interconnected web made sense of rising concern about environmental degradation, and ecological ideas of nature began to be widely disseminated. 'When we try to pick out anything by itself, we find that it is bound by a thousand invisible cords ... to everything in the universe' (Muir, quoted in Colborn *et al.*, 1997: 168). Deep ecologists argue that humans are part of this interdependent totality of nature. To save nature we must come to understand it as part of ourselves, as an extension of our bodies. We are part of nature, who have destructively misunderstood ourselves as conquerers at worst, stewards at best, but in either case as 'other' (Sessions, 1995).

ECOFEMINISTS also put forward a RELATIONAL concept of nature, but criticise deep ecology for failing to recognise that the split between man and woman must be recognised and healed in order to heal the split between humanity and nature. Merchant, in her influential book *The Death of Nature* (1980), traces the transition, from the sixteenth century on, from nature conceived as an omnipresent mother and living feminine principle, to nature as still FEMALE but now a terrain to be subjugated. She attributes this shift to the 'disenchantment' of the world which resulted from Enlightenment thought, to the rise of Western rationalist science and its legitimation of developing capitalist economic systems. Its end result has been 'the death of nature', the patriarchal 'reductionist' view of nature as 'a system of dead, inert particles moved by external, rather than inherent forces' (1980: 182). Vandana Shiva contrasts the Cartesian view of nature as ENVIRONMENT and resource, to Indian cosmology in which nature is the primordial energy which produces all things: 'ONTOLOGICALLY, there is no divide between man and nature, or between man and woman, because life in all its forms arises from the FEMININE principle' (1989: 40). Ecofeminists see the Western rejection of this feminine principle as being at the root of ecological crisis.

Since the meanings of nature are culturally variable, some writers have focused on the social construction of nature instead of engaging in the debate over the contested perspectives sketched above (e.g. Hannigan, 1995; Yearly, 1992). Soper, Benton and other realists emphasise the importance of distinguishing between nature the concept and the reality of nature. Nature is one of the concepts through which humans think about their being and its meaning by distinguishing its scope, limits and effects. But the structures of the physical world, the causal processes which result in observable natural features, animals, LANDSCAPES and so on, are not reducible to concepts. How we think about nature affects nature, but they are not one and the same thing.

CN

New Age The millenarian and utopian connotations of this term are suggestive of the forms of feminism to which it is often applied, such as 'mythopoetic' feminism and, to a lesser extent, ECOFEMINISM. New Age philosophies and social practice tend to be characterised by the search for new forms of spiritually grounded and self-consciously socially 'alternative' community and experience. Such currents tend to draw on primitivist imagery in order to articulate their distinctness from 'mainstream' society. Although often regarded as a distinctly contemporary and Western phenomena it is should be noted that marginal socio-spiritual communities and fashions have existed for many centuries within both Western and non-Western societies.   AB
See also MYTH/MYTHOLOGY.

New cultural geography Also known as the 'cultural turn', this refers to a meeting of social and cultural geography based upon a shared interest in social, cultural, literary and PSYCHOANALYTIC theories. Whereas Berkeley School cultural geography was primarily interested in describing the patterns of cultural influence on the landscape, new cultural geographers have focused more on the contested nature of CULTURE, studying not only how culture works through the LANDSCAPE, but also less tangible, and certainly less visual, concerns of cultural identity as an aspect of social relations (for a review see Cosgrove and Jackson, 1987). Rather than being regarded as a totality, culture is seen as a signifying system: a process through which power relations are enacted and meanings contested. Drawing on the work of cultural theorists, especially Edward Said, new cultural geography recognises a politics of REPRESENTATION and attempts to open up a space for those whose meanings are represented (and marginalised) by hegemonic figures. Feminist concerns are integral to new cultural geography which understands GENDER relations to be involved in the production of culture, as well as documenting the signification of gender norms in cultural systems.

There has been a reaction to the cultural turn which variously argues that the Berkeley School tradition has been misrepresented and that 'there's no such thing as culture' (Mitchell, 1995; Price and Lewis, 1993). Some feminists fear that the continued focus upon landscape of some new cultural geography perpetuates the tradition of geographers' privileging of the visual, a form of knowledge that has been seen as inherently masculinist (see Rose, 1993a). Given the range of theoretical bases and empirical sources for new cultural geography, it might be more appropriate to speak of 'cultural turns' in the plural rather than understanding there to be a coherent project or movement.   JS

New Left A politics born from disillusionment with Stalinism during the 1950s. Following the Hungarian Revolution in 1956, the Prague Spring in 1968 and popular protest in France during 1968, socialists began to rethink their commitment to the Communist Party and to lay the seeds of a 'new left'. *The May Day Manifesto*, originally published in 1967 (Williams, 1968) and the *New Left Review* (originally called the *Universities and Left Review*) were key publications in laying the foundations of a New Left agenda. Without tight party organisation, the New Left emerged as a network of different SOCIAL MOVEMENTS which sought to challenge particular oppressions and imperialisms. The flourishing anti-war movement and civil rights campaigns of the late 1960s and early 1970s allowed the New Left to take shape through activity at an international scale. In contrast to previous adherence to the working class as the key agents of socialist transformation, the New Left embraced new social actors as central to this New Left agenda. The women's movement was key to these developments, overturning many long-held assumptions about GENDER relations (see Rowbotham, 1997; Rowbotham *et al.*, 1979). FEMINISM played a key part in New Left thinking and this period is generally associated with 'SECOND WAVE FEMINISM'.

Within geography, the launch of the journal *Antipode* in 1969 made explicit connections to the New Left agenda. In his article in the first issue of the journal, Peet (1969: 4–5) argued that New Left geographers can make three contributions; first, helping to design a more equitable society and create 'a whole new geography truly based upon the precepts of EQUALITY and JUSTICE'; second, taking part in radical change 'employing all the techniques at our disposal for the purposes of shattering and then rebuilding the structure of conventional opinion'; and, third, organising within academic geography to democratise the discipline. In some senses feminist geography emerged as part of this agenda, although it was also part of the critique of the New Left as MASCULINIST in its focus and in its personal politics.　　JW

New man A term from the 1990s to refer to a feminised version of caring manhood, not afraid to reveal emotions and feelings or to assist with childcare and domestic labour; perhaps more a media construction or a figment of the imagination than an empirical phenomenon.　　　　　　　　　　LM
See also LADDISM; MEN'S MOVEMENT, MYTHOPOETIC.

New Right A politics which rose to prominence in the 1970s/early 1980s with the electoral triumph of Margaret Thatcher and the Conservative Party in the United Kingdom and Ronald Reagan and the Republican Party in the USA. At this time the right wing of the political spectrum was rejuvenated, generating new ideas and supporting new think tanks (such as the Adam Smith Institute in the United Kingdom). Thatcherism became associated with a new political platform designed to roll back the state, deregulate the labour market and prioritise the power of the market. Moreover, this economic LIBERALISM was combined with extreme social conservatism and support for the nation, FAMILY and Christian values (see Levitas, 1985). As McDowell (1991b) suggests, the British Conservative Party laid great ideological store by the family, as Thatcher explained:

> The family is the building block of society. It is a nursery, a school, a hospital, a leisure centre, a place of rest. It encompasses the whole of society. It fashions our beliefs. It is also the preparation for the rest of our life. And women run it (quoted in McDowell, 1991b: 413).

It is ironic then, that the ascendency of the New Right witnessed record levels of births to unmarried mothers, divorce and family breakdown in the USA and the United Kingdom (see McDowell, 1991b). Economic liberalism, welfare cuts and raised expectations undermined the very social institutions upon which the New Right based its agenda.   JW

New social movements See SOCIAL MOVEMENTS

Nimble fingers/manual dexterity The term nimble fingers was introduced into geography through the work of Diane Elson and Ruth Pearson (1981a, 1984) in their research on the FEMINISATION OF THE WORK FORCE in the new international division of labour. They analysed how specific 'natural' attributes of women were emphasised when recruiting for jobs in 'footloose' manufacturing industries. In particular, they focused on world market textile factories and new jobs in electronics industries requiring manual dexterity in ASSEMBLY LINE production. In these jobs nimble fingers and 'docility' are often seen by multinational employers as 'natural' advantages in a female work force. Such essentialised stereotypes fuel the search for international cheap labour and influence the location of new industries in 'THIRD WORLD' countries.

'Nimble fingers' research, however, illustrates that wage differentials in world market factories are often more indicative of 'male bias' (Elson, 1995) than levels of 'natural' skill. Elson and Pearson argue that nimble fingers are socialised skills rather than 'natural' ones – girls learn to have nimble fingers by copying tasks taught by their mothers in domestic spheres. Consequently, human capital theory arguments supporting pay differentials between genders (whereby 'learnt' male skills are paid more and women's 'natural' skills are paid less) are not valid. Some authors (see Lim, 1990) have criticised the ways in which Western feminists have conceptualised 'nimble fingered' women in world factories as passive victims. Other research has shown how the gender stereotyping of tasks and downgrading of women's skills occur in a variety of ways across different paid work environments (Laurie *et al.*, 1997).   NL
See also LABOUR: DOMESTIC, SWEATED, WAGED, GREEN, SURPLUS.

Nomad/nomadism/nomadic space From the writings of Deleuze and Guattari, especially their two-volume work on capitalism and schizophrenia (1983, 1987), nomadism refers to the idea of the rootless subject that is freely able to traverse global space. For Deleuze and Guattari, the state apparatus is part of a cultural drive towards immobility and fixity. Nomadism is a fluid positionality which blurs BOUNDARIES and subverts stable definitions. The mobility of the nomad can refer either to actual movement across SPACE or to a metaphorical state of being. The figure of the nomad resists settled patterns of thought and as such has been held up as the decentred or fragmented subject of POSTMODERNISM and POST-STRUCTURALISM. Deleuze and Guattari draw their theorisation of the nomad from PSYCHOANALYSIS. For them, paranoic DESIRE represents a territorialising, repressive channelling of desire into institutions (such as the STATE, FAMILY and school) whereas schizophrenic desire subverts these fixed and bounded identities by deterritorialising: transcending borders and resisting any attempts to contain or discipline. These desires also have spatial expressions: territorialisation produces 'striated spaces' of control and limitation, while deterritorialisation produces 'smooth spaces' of movement.

This understanding of the mobile subject has been criticised by Kaplan (1987) who argues that such mobility is only available to privileged white males (see also Braidotti, 1994b). Certain women (and minorities) do not have access to the technologies of mobility, and are often very much situated in place. Feminists have also criticised the romanticisation of HOMELESSNESS and psychological disorders in the ideas of nomadism, arguing that the metaphorical use of these terms denies the pain of their physical reality (see Parr and Philo, 1995). JS
See also DETERRITORIALISATION; POWER GEOMETRY; TIME–SPACE COMPRESSION.

Normalisation Normalisation refers to the production of homogenous forms of selfhood and SUBJECTIVITY which is a characteristic aspect of the workings of modern disciplinary POWER. It is evident in our preoccupation with 'socialisation' into appropriate roles and IDENTITIES and in the tendency of all institutions to sediment a set of practices, values and beliefs around a given object. Contemporary relations of DOMINANCE are typically sustained not through force or by forbidding certain behaviours but by telling people what they can and must be and hence through the regulation and surveillance of the most intimate and minute elements of daily life.

For Foucault, an early example of 'disciplinary' power was Bentham's Panopticon: a prison in which all the individual cells were to be directly observable by one guard in a central tower. Instead of relying on displays of physical force or violence, power operated through SURVEILLANCE which was continuous, anonymous and automatic. The prisoner, in a position of feeling himself to be constantly under observation, was obliged to participate in procedures to train, 'correct' and induce states of docility.

The Panopticon, though never actually built, may be seen as a model for the variety of training techniques which Foucault calls 'disciplines'. They reflect a wider emphasis on rational procedures as the most effective way of producing docile subjects who do not need to be controlled 'from the top down'. This is not to say that the processes of normalisation only produce homogeneity. As power becomes more anonymous and more functional, those on whom it is exercised tend to be more strongly individualised. Normalisation thus functions perfectly within a system of formal EQUALITY, in which it is possible to measure gaps by providing a measure of differentiation. Disciplines such as medicine, psychiatry and the social sciences, along with the mass media, investigate and police the crucial dividing line between the 'normal' and the 'abnormal'. By comparing individuals with each other, and by a continuous assessment of each individual, discipline exercises a normalising judgement. The system offers rewards as well as punishments, since all behaviour can be assessed in terms of good and bad works. When it is operating most effectively, 'individuals' actively seek to construct themselves as 'normal'. In an increasingly image-dominated culture, homogenised images function as models against which the self continually measures, judges, 'disciplines' and 'corrects' itself.

A number of feminists have used Foucault's conceptualisation of a 'disciplinary' society to study the production of 'docile' female BODIES through practices such as dieting, exercise and cosmetic surgery (Bordo, 1993: 28). In some versions it is argued that these practices are imposed directly by men or by the 'male norms' on which PATRIARCHAL society is based. But more often than not, women are willing participants in these practices and may even experience them as EMPOWERING. The surveillance is anonymous and functions by comparative measures that have the 'norm' as the reference point. Power relations are never seamless but are always spawning new forms of culture and subjectivity, new opportunities for transformation. Individuals neither create nor control these institutions and practices in which they find themselves EMBEDDED AND IMPLICATED. RP

Nostalgia A yearning for a past time or a regretful MEMORY about an earlier period or place. Used in the late seventeenth-century to describe a condition of extreme homesickness, it also became identified by doctors in the nineteenth century with a retreat from the present and attachment to the past, often attributed to disruptions of the MODERN era and associated with idealised images of femininity (Davis, 1979; Felski, 1995, Chapter 2). Current use frequently carries sentimental or reactionary connotations, but the term's histories and connections with perceptions of change suggest that its meanings and politics are more ambiguous and complex than broad dismissals imply (see also Wilson, 1997). DP
See also HERITAGE; NATION/NATION-STATE/NATIONALISM.

O

Objectivity The assumption that knowledge is produced by individuals who have the ability to transcend their historical, political and social worlds and to remain wholly detached from the object being studied. Objectivity is central to POSITIVISM. Philosophically, it relies on the Cartesian notion of the SELF as a conscious, unitary and rational being; this assumption influenced the development of the humanities more broadly (Lloyd, 1984).

Feminists are amongst the fiercest of critics of the Cartesian definition of objectivity. First, they oppose the implicit notion of an autonomous self because it is an ideal developed by people – men – who have not been in a position of having primary responsibility for others. Objectivity is thus highly gendered: it is defined in opposition to Woman and perpetuates a masculinist view of the world (Griffiths and Whitford, 1988; Haraway, 1988; Nicolson, 1990). Second, feminists argue that theory grows from EXPERIENCE: researchers cannot and should not divorce themselves from the social relations in which knowledge is produced. So, women's SUBJECTIVITIES and experiences of everyday life are central to knowledge production and become the site of redefinition of patriarchal meanings and values and of RESISTANCE to them (Smith, D., 1987; Stanley, 1990).

These critiques have enabled feminist geographers to counter the dominant and envisage other kinds of possibilities. Examples include rereading the landscape tradition in geography (Rose, 1993b); retelling the history of geography (Domosh, 1991; McDowell, 1992a and b) and a vigorous debate about the role of subjectivity in the research process (McDowell, 1992c). HC

Object relations theory One of two main bodies of post-Freudian psychoanalytic theory, the other being LACANIAN THEORY. It may be more accurate to speak of an object relations tradition rather than object relations theory, in that many psychoanalytic writers have adopted its basic premises but in diverse and sometimes conflicting ways.

Whereas FREUDIAN THEORY emphasises the influence of drives or instincts in psychological development, and in so doing attaches psychical life more closely to biological existence than to human relationships, object relations theory reverses this emphasis, hence the term 'relations'. The term 'object' is carried over directly from Freudian theory: Freud assumed that drives are directed towards targets or 'objects', for example that HETEROSEXUAL desire is directed towards a person of the opposite sex. The 'object' of the drive is incorporated within the subject's inner world (or psychical reality) in the form of representations, images and fantasies. Object relations theory offers a perspective on human subjectivity and psychological development in terms of this idea of inner worlds peopled by representations, images and fantasies of 'real' others.

A move away from a theory based solely on drives to one in which object relations come to the fore can be traced within Freud's own writings but gathered pace among a second generation of psychoanalysts working in Britain during the interwar years. It was closely linked to the development of psychoanalytic work with young children (especially by Melanie Klein and D.W. Winnicott) and with a shift in focus from OEDIPAL, or three-person, relationships, to the earlier, pre-oedipal relationship between infant and primary caregiver (generally assumed to be the mother).

According to object relations theorists, a very young infant has no sense of its own bodily BOUNDARIES, hence Winnicott's (1964: 88) famous phrase 'There is no such thing as a baby'. Thus, the earliest weeks and months of life entail a process of coming to terms with the infant's emergence from its mother's body. In elaborating this process, object relations theory offers a psychoanalytic theory of embodiment. Winnicott introduced the notion of 'transitional objects' to convey something of the infant's psychical experience of learning the distinction between itself and the world beyond, or 'me' and 'not me'.

Object relations theory has many important ramifications, both theoretically and practically, about which feminists have had a great deal to say. By focusing on the mother–infant dyad, object relations theory privileges the 'maternal object' within psychological development. This has been hailed by some feminists as an important and welcome break from Freud's focus on the father, conceptualised as central to the oedipus complex. But it also has disadvantages. At a practical level, by emphasising the infant's absolute dependence on its mother, together with the need for continuity of care, object relations theory gave succour to dominant and oppressive views of MOTHERHOOD. And theoretically it takes for granted an existing GENDER order.

Several feminist writers have taken up the emphasis on the maternal within object relations theory but have supplemented it with a more critical account of gender roles. Most notably, Nancy Chodorow (1978) has explored the formation of gender identities in the context of parenting by one sex, namely women. She argues that a preoccupation with separation from the mother (*en route* to adult autonomy) in psychoanalytic thinking reflects key issues in the development of MASCULINITY, at least in Western cultural contexts. But in these contexts FEMININITY is characterised more by connectedness, and she argues that through mothering this gender difference tends to reproduce itself. Developing this broad approach, feminist writers, including feminist geographers, have drawn on object relations theory to examine gender-inflected dualisms, for example between OBJECTIVITY and SUBJECTIVITY, and between WORK and HOME.

While feminist engagements with object relations theory have proven productive in a number of fields, critiques have also emerged. In particular, this perspective does not offer fundamental challenges to ESSENTIALIST accounts of gender difference: masculinity and femininity are understood as culturally produced rather than biologically given, but the approach offers little scope for thinking about gender outside a BINARY framework. In addition object relations theory tends to downplay the effects of the UNCONSCIOUS. For these reasons some feminists have found Lacanian theory to hold more potential, precipitating debates within feminist theory between proponents of the two perspectives.   LB

Oedipus/oedipal The Greek myth of Oedipus has an important place within FREUDIAN THEORY. In the myth, the baby Oedipus is left to die because of an oracle warning that he would murder his father and marry his mother, but he survives and is brought up by adoptive parents. As a young man he learns of the oracle and, not knowing that he had been adopted, he leaves his adoptive parents' home hoping to avert his fate. He quarrels with a man he meets on the road and kills him. He proceeds to marry the man's widow. Unwittingly he has murdered his biological father and married his biological mother. When he discovers what he has done he blinds himself. Freud took this myth to be symbolic of key elements in human psychosexual development: it flags issues of repressed incestuous and murderous desires ('oedipal' desires), and the accompanying unconscious guilt, in the context of the taboo on incest. That Oedipus blinded himself also flags the significance of vision within psychoanalytic accounts of sexual difference. In Freudian theory the oedipal triangle is understood as necessarily HETEROSEXUAL (the son unconsciously wishes to kill his father in order to marry his mother) so that the term 'oedipal' is now closely linked to a normative account of 'family' life. See FREUDIAN THEORY for the broader context of these ideas.   LB

Ontology Ontology asks what exists in the world. Any form of understanding is based upon assumptions about what constitutes the world and what it is like. This ontological foundation allows the establishment of theories of the world (EPISTEMOLOGIES).   JS
See also SITUATED KNOWLEDGE.

Oppression Used to denote the coercive maintenance of injustice for particular social groups defined by their class position, race, ethnicity and/or gender. Feminists have prioritised empirical and theoretical analyses of the ways in which women have been suppressed and repressed. Different feminist approaches have emphasised specific arenas of oppression, for example women's economic positions, biological reproduction, women's sexualities, psychological constructions of identities and representations of femininity as well as constructions of knowledge about women (Tong, 1992).

Feminist geographers' use of the concept of oppression has been as a descriptor of women's unequal condition. Rather than theorising oppression's geographies much of the work undertaken by feminist geographers has documented, often inadvertently, locational and temporal differences in the practices of women's 'oppression' (Duncan, 1996b; Katz and Monk, 1993; WGSG, 1997).

Early feminist use of this term derives largely from Marxist analyses in which human history is seen as a series of antagonistic struggles between social classes over the division of labour within a specific mode of production. Oppression, for Marxists, is manifested in the *economic* exploitation of one CLASS by another (serf by lord; worker by capitalist) and, although Marxists recognise that the particular conditions of oppression change over time and with shifts in the mode of production, oppression remains based in *economic relations*. For Engels (1845) women's oppression only emerged with the production of surplus within society and its association with male property ownership and patrilineal inheritance. Men exploited women and devalued women's labour through the imposition of monogamous marriage which was used to ensure transfer of a father's private property to his children. This economic basis for gender relations implied, for Engels, a 'class' division between 'propertied men' and 'propertyless women'. Women's oppression would cease only with the dissolution of private property and of the elimination of women's economic dependence on men.

Marxist-feminists attempted to refine and broaden Engels's analysis of women's oppression. Some challenged Engels' assumption that women's oppression would end with women's entry into waged labour by documenting the oppressive trivialising and undervaluing of women's work in waged workplaces (Beechey, 1987). Others focused on re-evaluating domestic work as a site of economic *production* and therefore of economic oppression (Dalla Costa and James, 1972; Delphy, 1984). Hartman (1981) went further to suggest that women's oppression rests not only in waged capitalist labour relations but also within HETEROSEXUAL PATRIARCHAL relations in which the *material* benefits from women's waged and unwaged labour accrue directly and personally to men. While recognising that the particular manifestations of women's economic oppression vary by social group, ethnicity and culture these analyses define all women as an 'economic class' and their oppression as materially based.

Radical feminists' analyses of women's oppression prioritise non-economic sites of suppression – BIOLOGY, psychology and SEXUALITY. Emphasis is placed on identifying *men*, rather than 'society', as the perpetrators of oppression and on male domination of women as a blueprint for all other oppressions (of class, ethnicity, race). In a critique of Marxist analysis, Firestone (1971) placed the roots of women's oppression in biological difference. For her, women's oppression lay in a *sex class* system underpinned by men and women's biological roles in reproduction. The control over reproduction, including particular practices and cultural meanings, was retained by men, giving rise to women's lack of determination over childbirth, sexuality and therefore women's identity. Overcoming oppression required a 'biological revolution' in which women could separate themselves from the restrictions of biologically determined reproduction through the control and application of reproductive technology. The aim was a society based on androgyny in which genitalia had 'no cultural meaning'.

However, Rich (1986a) saw this as a rejection of women's nature. While agreeing that women's oppression is often maintained through male control of the *institutions* of biological reproduction she argues against a technological fix and advocates women determining their own 'natural' rhythms and priorities in reproduction, sexuality and identity by rejecting the institutionalised forms of gendering/mothering/reproduction.

A pervasive theme has been the identification of language and KNOWLEDGE as a means of maintaining women's oppression through constructing women as 'different' (Gilligan, 1993; Mitchell, 1984). Here the legitimation of women's oppression through the social and intellectual framing of women's psychology and morality is challenged. The 'OTHER' in masculinist analysis maintained women as 'second' and as the necessary antithesis to masculine centredness (de Beauvoir, 1974). In post-structuralist feminist analysis, understanding 'Otherness' has been used as a means of redefining central feminist questions such as the nature of oppression (Barrett and Phillips, 1992). The condition of 'Otherness' enables individual women to affirm DIFFERENCE between women and to criticise norms, values and practices that the dominant culture (oppressive patriarchy and oppressive orthodox feminisms) seek to impose on everyone. Here 'Otherness', despite its associations with oppression as imposed inferiority, is turned on its head, becoming a way of being and of claiming different knowledges, not only of the practices of oppression but also of the positive identities which challenge their legitimacy (Evans, 1995; Tong, 1992). JF

Oral history See AUTOBIOGRAPHY

Orientalism A Western construction of the East, imposed from outside and derived from an epistemology that divides the world into two asymmetrical parts, the West and the East, Europe and the Orient. Orientalism has become a broad term with numerous meanings. It is the academic study of a geographical area, the Orient, the boundaries of which have shifted, sometimes restricted to the Middle East, to the Arab and Islamic worlds, at other times including territories east of Suez, the Far East or most of Asia.

It is a style and AESTHETIC which sees the distinction between East and West as one of the fundamental divisions of the world where orientalism is the inverse of all things European. It is a DISCOURSE, a Western mode of dominating, restructuring and having authority over the Orient. Used in this way it is associated with Edward Said's influential text, *Orientalism* (1978), which established and encouraged wide reconsiderations of empire through post-colonial studies. Said argues that since the late eighteenth century Western writers have constructed an image of the Orient centred around the distinctiveness of the 'Oriental mind' as opposed to the 'Occidental mind' and that this has been achieved through the manipulative amassing of information by a politically and economically superior West. Said contends that, whilst this imagery has been a significant part of intellectual thought in the West since the ENLIGHTENMENT, it corresponds to no empirical reality and reduces to insignificance the varieties of language, culture, social forms and political structures in the 'Orient'.

Central to orientalism as discourse is what Edward Said calls an 'IMAGI-NATIVE GEOGRAPHY', a set of contradictory ideas and attitudes about the East which combine an admiration for Arab and Islamic cultures with a set of derogatory stereotypes about 'Orientals'. Orientalism works ideologically through a series of oppositions between 'us' and 'them', the familiar and the exotic, which are produced and reinforced by seemingly impartial academic disciplines and knowledge. The highly gendered nature of orientalism has been discusses by Kabbani (1986). She argues that in European narratives the lascivious sensuality and inherent violence of the 'Oriental' male combines with the double demeaning of the 'Oriental' woman as both woman and Oriental. Whilst these imaginative representations were significant in medieval thought and continue to be voiced at the present time, they were also based on material foundations in the history of European imperialism. During the nineteenth century, they found deliberate expression in imperial confrontation and served to justify Europe's civilising mission. Kabbani (1986) emphasises that although not all representations were pernicious, the fiction of the erotic East framed both European political and popular culture. Through TRAVEL literature, painting and photography, it offered a prototype of the SEXUAL in a repressive age. Notions of Western civilisation as repressive and restrictive have been widely associated with Orientalist traditions, with RACISM and with the use of sexual and gendered metaphors to characterise the feminised colonies (Stoler, 1995).

Notwithstanding the significance of these analyses, numerous critiques have been offered of their tendencies to communicate a vision of both the Orient and the Occident as unified and monolithic and of a failure to suggest alternatives to orientalist DISCOURSE (MacKenzie, 1995; Smith, N., 1994). Melman (1992) questions the exclusion of Western women as actors from Occidental interpretations of orientalism and of their representation as passive victims in imperial discourses. By reference to the writings of women TRAVELLERS to, and residents in, the Middle East during the eighteenth and nineteenth centuries, she highlights alternative visions of the Orient which emerged outside formal networks of power and which challenged the BINARY OPPOSITIONS implicit in East–West. Mills (1991) also offers a gendered analysis of Orientalist critiques. An attempt to give form and character to decolonised knowledge is suggested in the work of SUBALTERN studies. Central to this is the recovery of the lost historical voices of the marginalised which contest elite versions of the past. The difficulties involved in producing these counter-histories are highlighted by bell hooks (1990b). Gayatri Chakravorty Spivak (1990b) similarly warns against interpreting POST-COLONIAL study as merely retrieving a subaltern history, a history of the excluded, the VOICELESS, who were previously at best only the object of colonial knowledge and fantasy. She points to persistent absences even in radical historiography, not least the failure of revisionist historians to address

193

the histories of native subaltern women. She also argues that as an object of investigation in Western scholarship, MARGINALITY may be reconstructed as an object of control – hence the danger of a 'new orientalism'.   MB

Other/otherness Refers broadly to the opposite or opposed element in a BINARY opposition such as SELF–Other, East–West or masculine–feminine. More specifically it can be related to another person or persons existing distinct from oneself, different in identity, in kind or quality. Associated with the dichotomy between the autonomous rational UNIVERSAL and the particularist non-rational other, within Western political theory it has been argued that ideals such as formal EQUALITY and universal RATIONALITY and impartiality create a logic of identity which denies and represses difference. Those marked by differences derived from their sex, skin colour, age, sexuality, physical capabilities or other variations from the posited 'norm' do not qualify for full participation in liberal democracy (Tronto, 1993). Whilst many Renaissance, classical and modern constructions of ALTERITY can be identified, during the nineteenth century the European construction of the other as a figure of 'non-reason' represented a powerful moment in the self-constitution of the European subject as the sovereign figure of reason and normality (Pratt, M.L., 1992). It was central to the ideology and practices of European imperialism. Recent analyses have addressed the complex processes of othering which accompanied the interplay between gender, 'race' and class differences in various imperial contexts (for example, McClintock, 1995; Ware, 1992).

Links between forms of POWER and the production of human bodies as human subjects have extended the concept of otherness to the female BODY through an emphasis on its relations with nature. During the late nineteenth and early twentieth centuries within a range of scientific fields, including biomedicine, the view that sex and reproduction were more fundamental to woman's than to man's nature resulted in the emergence of gynaecology as a separate branch of medicine devoted to the diagnosis and treatment of the female body (Oudshoorn, 1996). In response to 'the woman question', Mosedale (1978) points to the role of a range of sciences in uncritically lending their theories and professional authority to woman's traditional roles and to the conceptualisation of woman as an ontologically distinct category as Other. A feature of these discursive and institutional processes of othering has been, it is suggested, the establishment of gendered subject–object relations in which men possess the subject position and women are considered as objects of scientific enquiry (Fox Keller and Longino, 1996).

Whilst early generations of feminists sought access for themselves and other Others to the male-dominated sphere, more recently writers have stressed the need to move away from dichotomies and to theorise the notion of multiple DIFFERENCE. They have suggested that the EPISTEMOLOGICAL frameworks of the Western human, natural and social sciences need to be transformed to accommodate an inclusion of women not as a special case deviating from the norm but as one of many different groups in an open and heterogeneous universe (Duncan, 1996a). Whilst recent acknowledgements of diversity have challenged the gendered subject–object relations, Mohanty (1991b) criticises the production of the 'Third World woman' as a dominant REPRESENTATION and an implicit victim in some recent Western feminist DISCOURSE. Similarly Oudshoorn (1996) suggests that within the reproductive sciences the othering of people of colour in countries of the South persists where population control is deemed to be a precondition for the solution of environmental problems.   MB

See also EMPIRE; ORIENTALISM; WEST, THE/WESTERNISATION.

P

Parody A term closely related to that of PERFORMANCE, developed by Judith Butler to refer to an exaggerated and theatrical display of HETEROSEXUAL-ITY, often through drag, to destabalise naturalistic assumptions about gender identity.   LM

See also COMPULSORY HETEROSEXUALITY.

Partial truth Early SECOND WAVE FEMINISTS sought to challenge what they saw as the incomplete and distorted knowledge of the world produced by masculinist science and to replace it with a 'true' understanding of the world that drew on women's knowledge. However, this endeavour was soon understood to be highly problematic in the light of arguments that knowledge is always mediated by an individual's social position, experience and culture at a particular time and place. Once this idea is accepted it is clear that knowledge is necessarily partial. Proponents of STANDPOINT THEORY suggest, however, that the knowledge generated by women as an oppressed group can start to move us towards a less partial and more 'truthful' understanding of the social world (Hartsock, 1983; O'Brien, 1981). Other feminist commentators argue that all knowledge is SITUATED KNOWLEDGE, necessarily and always partial, and that the notion of a unitary knowledge generated by women cannot be sustained (Flax, 1987; Haraway, 1991). This position poses problems for feminist politics in that it undermines the rationale for feminist political action based on a commonality of interests amongst women (Bondi, 1990b).   SB

See also DIFFERENCE.

Participant observation As the term suggests, a participant observer not only observes the behaviour of the group that she or he is studying but also participates, as much as possible, in the daily lives of the community members. SUBJECTIVITY rather than objectivity characterises the ideal relationship between a feminist researcher and her or his subjects. Consequently, many texts and articles discussing FEMINIST METHODOLOGY have privileged participant observation as the preferred method (McDowell, 1993). The strength of participant observation is derived from an understanding of the everyday EXPERIENCE of the respondent, by means of putting people at ease and by spending sufficient time with a small group of people. Participant observation is therefore ideologically compatible with a feminist research methodology. One commonly cited problem associated with participant observation is a loss of detachment, which comes from being intimate with a group of people over an extended period of time. Such a loss of detachment is inevitable, in that one becomes involved, either positively or negatively, with the group. Ironically, one of the most significant problems stems from the strength of the method: its in-depth quality. Owing to the time spent developing depth and detail, the sample size is very small. The question that arises then is to what extent can one generalise from the small group that one has studied in depth? For this reason, feminists who advocate quantitative methods feel they can be helpful in adding breadth to this type of study. Most feminists agree that it is best to beware of generalising from such a sample, recognising that what one gains in depth, one gives up in breadth. Participant observation should not be regarded as a technique that can be effectively employed in isolation from other research procedures.   LD

See also AUTOBIOGRAPHY; ETHNOGRAPHY; ORAL HISTORY.

Paternalism See FATHER/FATHERHOOD; PATRIARCHY

Patriarchy In its most general sense, the term patriarchy refers to the law of the father, the social control that men as fathers hold over their wives and daughters, but in its more specific usage within feminist scholarship patriarchy refers to the system in which men as a group are assumed to be superior to women as a group and so to have authority over them. Whereas some theorists regard patriarchy as a universal system, which is analytically distinct from CAPITALISM or other MODES OF PRODUCTION, others suggest that it takes specific forms at different times and is not a separate system but is part of the mode of production, so that a specifically capitalist patriarchy or a patriarchal capitalism might be distinguished (McDowell, 1986). These different positions tend to be referred to as the dual systems model and the single systems model. In a debate in *Antipode*, the geographers Jo Foord and Nicky Gregson (1986) argued that patriarchy is constituted by four interrelated but analytically distinct sets of relations between women and men – biological reproduction, HETEROSEXUALITY, marriage and the nuclear FAMILY. Adopting a REALIST position, they argued that while the latter two factors are historically contingent the former two factors – biological reproduction and heterosexuality – are universally necessary. Their argument generated a heated debate about the necessity or specificity of gender relations (Gier and Walton, 1987; Johnson, 1987) in which the ABSTRACTION of both dual and single systems models was criticised. Indeed, there seems to have been a general turn among feminists from abstraction towards studies of the specificity or particularity of gender relations in different circumstances. This move is evident in the work of the feminist sociologist Sylvia Walby who has had a significant impact on the theorisation of gender relations.

In her book *Theorising Patriarchy* (1990) Walby suggests that patriarchal relations in advanced industrial societies are constructed and maintained by six sets of analytically separable structures in which men dominate and exploit women. The six structures she identifies are household production (in which men appropriate the value of women's unpaid domestic labour), patriarchal relations in waged WORK (in which women are segregated into particular occupations and are less well paid), patriarchal relations in the STATE (in which men dominate the institutions as well as produce gender-biased legislation), male VIOLENCE, patriarchal relations in SEXUALITY (male control of women's bodies) and patriarchal relations in cultural institutions (male domination both of the production and the form of different media).

In advanced industrial societies, therefore, there are numerous ways in which patriarchal superiority and control is constructed and enforced: through the legal system for instance, which falls into Walby's third category. Throughout the nineteenth century in Britain and well into the twentieth, women's legal status was as dependents, and their lives and property were in the hands of their father or their husband. In Britain, until the Married Women's Property Act was passed in 1882, for example, on marriage all a women's worldly goods passed into the hands of her husband. In the United Kingdom women could not vote until after World War I, they could not have a legal abortion until 1967 and could not gain access to mortgage finance without a male guarantor until the 1970s.

These sets of relations, Walby suggests, are connected in different ways in particular circumstances but together constitute what she terms a GENDER REGIME. Walby distinguishes two main regimes: a domestic regime, or private patriarchy, and a public regime, or public patriarchy. This is how she describes the two regimes:

> The domestic gender regime is based upon household production as the main structure and site of women's work activity and the exploitation of her labour and sexuality and upon the exclusion of women from the public. The public gender regime is based, not on excluding women from the public, but on the segregation and subordination of women within the structures of paid employment and the state, as well as within culture, sexuality and violence. The household does not cease to be a relevant structure in the public form, but it is no longer the chief one. In the domestic form the beneficiaries are primarily the individual husbands and fathers of the women in the household, while in the public form there is more collective appropriation. In the domestic form the principal patriarchal strategy is exclusionary, excluding women from the public arena; in the public it is segregationist and subordinating. In both forms all six structures are relevant, but they have a different relationship to each other. In order to understand any particular instance of gender regime it is always necessary to understand the mutual structuring of class and ethnic relations with gender (Walby, 1997: 6).

Although these regimes may be analytically distinct, Walby recognises that they coexist and that women are differently involved in each structure:

> Different forms of gender regime coexist as a result of the diversity in gender relations consequent upon age, class, ethnicity and region. ... [In Britain] older women will be more likely than younger women to be involved in a more domestic gender regime. Women whose own occupations place them in higher socio-economic groups are more likely to be in a more public form. Women of Pakistani and Bangladeshi descent are more likely to be in a domestic form and BLACK Caribbean women more likely to be in a more public form than white women' (6).

Although Walby's empirical work is primarily in Britain and the European community and her theoretical model seems to reflect this, her arguments have some resemblance to an earlier distinction between two broadly different forms of patriarchy by Deniz Kandiyoti (1988). In her work, Kandiyoti was anxious not only to differentiate patriarchal structures on a broad geographical basis (distinguishing African from Asian patriarchy, for example) but also to explore the reasons why women in the majority accept rather than rebel against patriarchal structures. Her work allows women's agency to be recognised; women may be subordinate but are not necessarily subservient.   LM

Pay equity See COMPARABLE WORTH

Peace movements The relationship of gender to violence and war has often been a principal concern of feminists. Although the association of men with aggression, and of women with peace, has been seen by some critics as a natural distinction, the predominant tendency within the twentieth century has been to focus on the way social structures and IDEOLOGIES sanction and encourage male violence (Woolf, 1938). Since MILITARISM, and the practice of war, were both central to the development of modern geography it is, perhaps, somewhat surprising that feminist geographers have, to date, shown relatively little interest in the geography of, and geographers' participation within, anti-war resistance.

It is within the sphere of social activism that the closest relationship exists between feminism and peace movements. Campaigning groups that assert women's ability to act as peace-makers have emerged in many modern conflicts. Geographical research on the development of these groups in particular places, their appropriation and use of space and their use and development of ideologies of environmental concern is at an early stage. The formation and representation of women's peace camps has, however, provided a focus for debate (Cresswell, 1994).   AB
See also GREENHAM COMMON.

Penis One part of male genitalia, the penis has come to epitomise maleness. In FREUDIAN THEORY the penis is the marker of sexual difference. Its symbolic significance is developed within LACANIAN THEORY through the notion of the PHALLUS.   LB

Penis envy The idea of penis envy is central to the account of FEMININITY elaborated within FREUDIAN THEORY. According to Freud it is a response to the CASTRATION COMPLEX experienced by little girls when they come to recognise sexual difference. Many feminists have repudiated the idea, but others have argued that it has value as a metaphor for women's feelings about gender inequality. LB

Performance POST-STRUCTURAL critiques of IDENTITY have argued that GENDER is constituted not in any ESSENTIALIST characteristics or through a founding act 'but rather a regulated pattern or repetition' (Butler, 1990: 145). For Judith Butler the performance of identity establishes the fiction of a SUBJECT who is performing. Butler argues that HETEROSEXUALITY is only established as the norm through a performance of heterosexual acts which give this identity the image of AUTHENTICITY.

This dynamic understanding of identity allows a possibility for POLITICS and for transformation, for if there is always a compulsion to repeat, 'repetition never fully accomplishes identity' (1990: 24). Furthermore, identity is only secure when performed 'correctly', and, as Butler's work has demonstrated, this offers a great potential for subversion. However, a question arises as to who is directing the performance if there is no subject: the notion of agency is diffuse.

Importantly for geographers, the concept of performance requires an understanding of SPACE: it is important *where* subjectivity is performed as this influences the outcome and meaning of the performance and so the meaning and effect of subject identity (a point noted by Goffman in his work on the 'presentation of self in everyday life' in 1969). Geographers have developed these ideas about performance in studies of, for example, GAY and LESBIAN identities (Bell and Valentine, 1995) and 'doing gender' in the workplace (McDowell, 1997b). JS
See also COMPULSORY HETEROSEXUALITY; MASQUERADE.

Personal is political, the Articulated through consciousness-raising groups and networks of friends, the premise that women's personal and intimate experiences were integral to, and indeed politically determined by, social life was a constant rallying call for second wave feminists. The conviction that the private is POLITICAL and the political is personal was viewed as a necessary belief for the process of self-empowerment (see MacKinnon, 1987). Its underlying assumptions though have revealed an emphasis on unproblematised notions of EXPERIENCE and of women that post-imperial and post-structuralist feminists have subjected to increasing scrutiny. Mohanty (1992) claims the idea that women share similar experiences (of OPPRESSION) leading to shared interests and political unity (constructed as feminists linked together in a global sisterhood) is based on an essentialised understanding of woman which erases material and ideological POWER differences between and within groups of women, especially between First and THIRD WORLD women.

It is not the experience of being female, but the meanings attached to the foundational categories of GENDER, RACE, CLASS and age at various historical moments and geographical places that is of significance. In other words, it is not experience that determines POLITICS but politically structured relations that shape SUBJECTS' experiences. Critical analysis demands then that the (often fragmentary and even contradictory) discursive experience of the self be spatialised and historicised. As Joan Scott (1992) emphasises, the meanings attached to experience are not transparent. Feminists need to ask what representational categories mean and how they operate. How do the bodily IDENTITIES they produce emerge and transform over time and space and become politically mobilised? The personal can then be revealed in relation to its imbrication with the social-political as well as in its spatial and historical variability. It is in this vein that many feminists have argued for an expansive definition of politics that encompasses issues classically viewed as personal. For feminist geographers 'the personal is political' has manifested itself in two major ways: as a methodological practice that draws on women's accounts of their personal experiences of active engagement in the politics of place construction (Nagar, 1997; Pulido, 1997) and as a strategy for explaining, and sometimes also claiming, their distance from the masculinist, able-bodied and heterosexist projects of Geography (for examples, see Chouinard and Grant, 1995; Domosh, 1997). LP

Personal narratives See AUTOBIOGRAPHY

Phallogocentric A portmanteau term achieved by combining the feminist term PHALLOCENTRISM and the term logocentrism. The combination term phallogocentric was devised by Derrida in his critique of LACAN. Lacan was guilty of this, according to Derrida, as he unproblematically associated the letter in Edgar Allen Poe's *The Purloined Letter* with the phallus. Derrida argued that such a direct one-to-one relationship is impossible. Feminists often use the term interchangably with phallocentricism. LM

Phallus/phallocentric/phallocentrism Terms, particularly associated with psychoanalytic thought and cultural criticism, referring to conscious as well as to unconscious symbolic representations of male POWER and privilege and of particular (sexist) conceptualisations of MASCULINITY; usually, but not always, manifested in penis-like representations (as the physical embodiment of such power and privilege, along with a myriad of other cultural values and meanings associated with this); in feminist thought in particular, it tends to refer to *oppressive* meanings and values associated with male power and privilege.

Phalluses may be represented directly as objects or indirectly as abstractions or suggestions (e.g. in literature, 'he was a pillar of society'); they may be flaccid, erect, or at some stage in between, and they may symbolise power, penetration, potential violation or (less commonly) weakness and impotence (which may or may not be associated with their level of erectness). Except in some feminist REPRESENTATIONS, they are almost always represented as a standard or norm against which absence, loss or incompleteness is measured.

When a narrative, argument or creative expression involves the explicit, implicit, conscious or unconscious deployment of masculinist norms as a standard against which all else must be measured, it is sometimes said to be 'phallocentric'. However, some theorists and analysts (including many feminists) prefer to reserve use of the term for those situations in which there is a clear reference to, or depiction of, a specific phallic symbol.   LK
See also FREUDIAN THEORY; LACANIAN THEORY; PENIS, PENIS ENVY.

Pill See BIRTH CONTROL

Place There is a tendency to see place as a bounded piece of SPACE or TERRITORY with certain identities and characteristics inherent to it which differentiate it from surrounding areas (see REGIONAL GEOGRAPHY). Humanistic geographers have suggested that different places provoke a particular 'sense of place' (sets of feelings, emotions and attachments to the place) on the part of residents (past or present) (Relph, 1976). Sense of place may be articulated in art, literature and histories or may become part of individual or group memory.

Geographers have recently become critical of this exclusive definition of place which implies a border between those who belong and those who do not. They argue that place is politically reactionary. FEMINIST GEOGRAPHERS have highlighted the different relationships to place that men and women will hold (McDowell and Massey, 1984): their differential access to different spaces and their differing relationships to POWER will influence how place is perceived. As Massey puts it, 'a woman's sense of place in a mining village – the spaces through which she moves, the meeting places, the connections outside – are [*sic*] different from a man's' (1991b: 28). Clearly, sense of place will be further complicated by issues of 'RACE', SEXUALITY, CLASS, AGE and so on.

There have been challenges to understanding place as necessarily reactionary, however, most notably by Doreen Massey in her conceptualisation of a global or progressive sense of place (Massey (1991b); see also the structurationist account in Agnew (1987), and Pred's (1984) conceptualisation of place as a contingent process). Massey regards place as a process rather than as a stable and clearly defined entity. Place is unique, she argues, for its particular combination of intersections of global processes. This 'extroverted sense of place' can be imagined to form around networks of relations and connections rather than being enacted by BOUNDARIES and through exclusions.   JS
See also GEOGRAPHY.

Placelessness The fear of a loss of distinctiveness with the rise of (modern) global CULTURE. This assumes that global(ising) processes are necessarily and always homogenising. The term entered the literature through Relph's (1976) book *Place and Placelessness* which discussed the creation of places (e.g.

shopping malls and theme parks) which were 'inauthentic', in the sense that they related only tangentially to the social history of that PLACE (as articulated through labour history, rituals, religion and so on).

However, more recent geographical analyses have indicated that global processes are differentiating: places continue to exhibit difference as a result of the articulation of place-bound specificity with GLOBAL processes (Appadurai, 1990).   JS

See also AUTHENTIC; HERITAGE; NOSTALGIA.

Planning A term used mainly to refer to the practice of town and country planning and to the development of the BUILT and natural ENVIRONMENT. Town and country planning can be separated into two main activities – the control of development and the preparation of strategic policies and plans for land use and service provision. Planning is a statutory function of local authorities in the United Kingdom but is also undertaken by a range of organisations and consultants, particularly in the context of major development proposals.

During the late 1980s and 1990s planning was criticised for neglecting the specific needs of women and for reinforcing the MASCULINE nature of the built environment. Planning, it was suggested, paid too little attention to issues such as the design and location of residential and retail spaces, the provision of CHILDCARE and public transport and the production of safe environments, and, although these are not solely the concerns of women, their neglect has a greater effect on the lives of women than on the lives of men. The male focus of planning practice was seen to be related in part to the DOMINATION of men in the planning profession – particularly in more senior positions.

In response to the perceived neglect of women's needs, some local authority planning departments have introduced specific women's initiatives. These initiatives have attempted to address issues such as TRANSPORT, safety and childcare, focusing more directly on the identified needs of women rather than on traditional perspectives. Typically, women's initiatives in planning have sought to increase the availability of public transport, especially in residential areas, to ensure that new retail development incorporates baby changing facilities, to site bus stops in well-lit areas and to show an awareness of women's safety in the landscaping of new residential and commercial development. They have also sought to increase women's involvement in the planning process – both in terms of the planning profession and as members of the public. Some local authorities have introduced 'planning for real' exercises in which local people are asked for their views on the development and servicing of a community. Such exercises are designed, in part, as a way of increasing women's participation.

Women's initiatives have mainly been concentrated in urban, Labour-controlled local authorities – particularly those authorities with a recognised women's committee. In the early 1990s less than one quarter of all planning authorities in England and Wales had introduced specific initiatives aimed at addressing women's needs. Feminist analyses of the planning process have argued that the failure of most authorities to focus directly on the needs of women must be seen in the context of the unequal power relations between men and women and the continuing dominance of the traditional gender division of space. JL

See also BUILT ENVIRONMENT; CITY; SAFER CITIES.

Pleasure In a specific sense, this is a term in Freudian theory. According to Freud new-born infants' behaviour is dominated by the pleasure principle, which is replaced by the reality principle with increasing maturity, except in dreams where pleasure still dominates. In an everyday sense pleasure is associated with enjoyment, especially of a sensual nature. Some feminists have argued that male forms of pleasure are emphasised at the expense of explicitly female forms of pleasure. LM

See also DESIRE; SEXUALITY.

Pluralism A POLITICAL philosophy, developed by English liberals and socialists in the early twentieth century, about the nature of the distribution of POWER in society. It is also commonly taken to refer to a political system in which power is diffused among various groups and institutions. Pluralists emphasise a world of consensus in which small, closely knit groups respect each other's differences and no one group is able to dominate decision-making in the STATE or claim its values as the norm for society as a whole. The role of the state in pluralist theory, first posited by Dahl (1961) and commonly depicted as an adjudicator between competing interests, came to be the focus of critique. The notion of the state as a neutral arbiter serving a wide range of interests – although the balance of interests served altered as governments changed – first came under attack by Miliband (1969) who proffered instead a view of the state as an instrument of the ruling classes. Pluralists have continued to point to public participation and the right to organise and protest as evidence of the existence of pluralism, but critics claim it is a political ideal as opposed to a practice. LP

Political The term political refers to those DISCURSIVE PRACTICES and relations that have consequences for the exercise of POWER among people and places. All feminist political DISCOURSES arise from praxis, that is from the dialectical relation between thoughts and actions; as Nancy Fraser (1989: 6) so succinctly puts it, 'you can't get a politics straight out of epistemology'. The desire by feminists to extend the BOUNDARIES of what is counted as political (see POLITICS; PERSONAL IS POLITICAL) has led to the charge that when the political encompasses all referential domains the specificity of the political recedes, leading to a withdrawal from the political. This, in turn, has led to the counter-charge that to see everything as political presupposes the political's prior determination. Feminists are thus currently addressing how to develop a political imaginary that can go beyond the impasses of predetermined and naturalised categories and metanarratives and enable a politics of DIFFERENCE which recognises that all standpoints are partial and based on situational knowledges. This has involved addressing, among other issues, the following: the meaning for politics of a deconstruction of the SUBJECT, and hence, the categories 'man' and 'WOMAN'; the necessity of breaking down barriers between women through an interrogation of their relational connectedness; resisting the challenging of needs definition in terms of the traditional spheres of PUBLIC and PRIVATE; democratising access to material and discursive (such as interpretation and communication) resources; reading RACE, ETHNICITY, SEXUALITY and CLASS as more than mere attributes of the unified subject; and the development of a wider democratic politics (see Barrett and Phillips, 1992; Butler and Scott, 1992; Fraser, 1989; Mouffe, 1991). Although there are (obviously) different emphases among feminists there does appear to be agreement that these issues cannot be addressed through the privileging of any one version of IDENTITY POLITICS, including gender, or of any one site or terrain of struggle, including the STATE. Rather, as Mohanty declares, 'it is the current intersection of anti-racist, anti-imperialist and gay and lesbian struggles which we need to understand to map the ground for feminist political strategy and critical analysis' (1992: 87).

In the subdiscipline of political geography it is hard to understand the lack of progress beyond masculinist constructions of the political and its enduring concern with the public sphere when successful challenges to the public–private divide have been raised in cultural, economic and social geography (but see Kofman and Peake, 1990; Marston, 1990; Staeheli, 1993). Within geography the most immediate political task is the expansion of the geopolitical imagination, challenging HEGEMONIC constructions of space and of politics. Critical human geographers, with their concern with spaces of DOMINATION and OPPRESSION and the power relations of DISCRIMINATION and disadvantage, reproduced through the social production of NATURE, SPACE, PLACES and LANDSCAPES, are well placed not only to reveal these sociospatial processes but also to challenge their legitimacy (Barnes and Gregory, 1997). One such challenge is the understanding that representations of the world are political, serving particular ends. Slater (1997), for example, claims our analytical frameworks, with their enduring binary spatialisations – inside–outside, North–South, West–non-West, foreign–domestic, centres–peripheries – can no longer dwell on the mechanisms of power employed by centres (us), seeing peripheries (them) only as passive receivers.

The spatialisation of the political, witnessed through a concern with BORDERLANDS, boundaries, Thirdspaces, MARGINS, politics of location and so on, is encouraging of representations that can be liberationary. Harvey (1996) asserts that the political dialectics of space, place and environment have much to teach us about how to construct alternative futures: not only is context uncontainable but where there is power there is RESISTANCE. The terrains of struggle, moreover, are constantly changing, but it is vital to go beyond the particularities and to recognise patterns: '"Only connect" is still one of the most empowering and insightful of all political slogans' (Harvey, 1996: 431). **LP**

Political correctness The act of behaving in a non-discriminatory way. In the early 1990s it became caricatured as an accusatory practice of denouncing individuals and institutions (often those at the forefront of critical practices and coalitions) that had supposedly acted in discriminatory ways as defined by affirmative action and positive discrimination legislation. **LP**

Politics Politics is about how people exercise POWER through both material and DISCURSIVE practices. Definitions vary widely, and all are, necessarily, POLITICAL (Randall, 1987), but conflict and consensus both appear as critical elements in approaches to the study of politics. As an activity it has traditionally been linked to the STATE, political parties, TRADE UNIONS and other groups associated with the PUBLIC sphere, encompassing individual and collective, legal and illegal, formal and informal activities. Kate Millett (1971: 23) gave the first feminist definition of politics as 'power-structured relationships, arrangements whereby one group of persons is controlled by another'. Feminist political philosophers, such as Jean Elshtain (1981) and Iris Young (1987), have outlined the ways in which women's association with the PRIVATE sphere has been utilised to legitimise their exclusion from the public realm of politics. They have also sought to broaden what is seen as political activity, and in particular to transform leftist politics, through an emphasis on the articulation of relations, particularly the gendered politics of personal life (see Randall, 1987).

While feminists have consistently developed gendered accounts of political ideologies – such as SOCIALIST FEMINISM (Barrett, 1984), RADICAL FEMINISM (Daly, 1978), LESBIAN feminism (Fuss, 1991), POSTMODERN feminism (Hutcheon, 1989) and ECOFEMINISM (Seager, 1993) – many feminist analyses of politics have centred on the PUBLIC–PRIVATE DIVIDE. Accordingly, some feminists have focused on women's politics in the public sphere – in political parties (Campbell, 1987), government organisations (Watson, 1990) and trade unions (Cockburn, 1983; McDermott and Brisken, 1993). Yet others have emphasised the routinised practices of everyday life (Fraser, 1989), most recently with a recognition of the contextual nature of politics and IDENTITY (Radcliffe and Westwood, 1996), BODY politics (Grosz, 1995) and of the ways in which inequality and OPPRESSION are inscribed in space (Laws, 1997). **LP** See also PERSONAL IS POLITICAL, THE.

Polyphonic Literally many-voiced. A term derived from the work of Bakhtin (1981), in which the VOICES of those represented in texts and discourses are placed in relationship to the listeners in particular ways through a network of beliefs and POWER relations. Recent discussions of feminist methods have emphasised the need to try to include multiple voices in their work and so produce polyphonic accounts. **LM**

Population pyramid A double barchart diagram representing the sex and age structure of a population in an area. The numerical or percentage distribution by sex is represented along the horizontal axis, with age-groups represented along the vertical axis. Different characteristics of a population's demography can be identified, so that the impact of past and current events can be noted, particularly in terms of their gendered impact, such as war, famine or MIGRATION.   PM

See also DEMOGRAPHIC POLICY.

Pornography Literally, 'depictions of acts of prostitutes'. To feminists, pornography is not explicit sexual material *per se*, but sexist material especially depicting violence or coercive sexual acts, or involving degradation, subordination and/or harm. Pornography reflects ideologies about the SEXUALITY of women, children and people of colour. Pornographic images (in workplaces, the home or on public billboards) demarcate space as masculine and define HETEROSEXUAL arenas and can be experienced as a form of HARASSMENT. Pornography became a prominent feminist issue in the late twentieth century, highlighting deep divisions. In arguing that pornography reinforces the PATRIARCHAL oppression of women, radical feminists such as Dworkin (1981) reverse the idea that it distracts men from RAPE. Instead, they suggest that it legitimises sexual VIOLENCE by creating a climate in which women's bodies are objectified and in which women are viewed as enjoying rape. Other feminists have taken an anti-censorship stance. The objection of many is not just to the product but to the industry, as some women are coerced or forced by need into working in the sex trade. The making of child pornography always involves actual abuse.

Diverse views on and reactions to pornography exist among women and men. Increasingly, pornography is marketed for heterosexual women (though the imagery is far less explicit or coercive than male equivalents), and GAY and LESBIAN pornography is more widespread, including sadomasochistic material. There is a growing argument that although heterosexual men may tend to define women (and pornography) in a certain sexualised way, women need not do likewise.   RHP

See also ABUSE; VIOLENCE.

Positionality Discounted by jargon-bashers, positionality is nevertheless a useful term for describing the social and psychological *context* of historical and geographical AGENCY. It functions in this respect as a methodological reminder about the need to go beyond a humanist assumption of so-called Man's self-understanding and self-determination (also known as *self-presence*) and explore instead the diversity of people's historical-geographical subject positions as located in the midst of overlapping sexual, economic, political and cultural processes. To some extent, therefore, the GENEALOGY of positionality as a theoretical concept might be traced back to Marx and Engels's (1968: 173) argument that 'It is not the consciousness of men that determines their existence, but their social existence that determines their consciousness', as well as to Marx's still more famous *Eighteenth Brumaire* statement that 'Men make their own history, but they do not make it as they please; they do not make it under circumstances chosen by themselves, but under circumstances directly encountered, given and transmitted from the past' (1968: 96). However, in contradistinction to the exclusively male pronouns and the almost exclusive focus on *class* contexts in Marx's work, positionality has been developed by feminists wishing to explore some of the complex ways in which class relations are overdetermined, which is to say combined with and thereby altered by a sexualised and highly gendered 'politics of location' (Rich, 1986b). Its great advantage but perhaps also its biggest problem is that as a concept it is empty with regard to content. It is different from any particular position insofar as it is processually open-ended and can thus be defined on the basis of someone's position *vis-à-vis* anything else. This has much merit insofar as it allows us to convene a description of someone's positionality in relation to multiple social processes, and thus seemingly pre-empt the risk of turning one particular dynamic such as class into the only social relation of any analytical importance. However, all the theoretical talk of positionality itself can risk abstracting SUBJECTIVITY altogether into an anaemic realm of almost mathematical neatness where different vectors of social formation come together like so many lines on a graph. For geographers attentive to the density of geographical relations such an anaemic geography predicated on the instrumentalisation of geographical concept-metaphors such as location and position raises very real question marks. And it is in this context that the tradition of feminist geographical empiricism charted by Linda McDowell (1993a) serves as a useful corrective and reminder of the often painful messiness of positionality in actual oppressions and exploitative relations in everyday life. Michael Brown's (1997) still more recent attempts to chart the geographies of such relations as they impact communities living with AIDS also points to how positionality may well provide a basis for rethinking and rejoining political and CULTURAL GEOGRAPHY, thereby putting the human in all its diversity back into a more caringly human geography.   MS

Positivism A philosophy of science devised by a group of philosophers working in Vienna in the 1820s and 1830s, the most influential of whom was August Comte (Guelke, 1978).

Gregory (1978) outlines five precepts of Comte's positivism:
- *empiricism*: the scientific status of knowledge is guaranteed by direct experience of an immediate reality;
- *unity of scientific method*: sciences share a common method so that disciplines are distinguished by their object of study and not by their method;
- *development of theory*: science proceeds by formally constructing theories, testing these against empirical reality and developing laws;
- *use value*: all scientific knowledge has a technical use and can be applied to the study of society;
- *the search for truth*: scientific laws can be unified and integrated into a single system of knowledge and truth.

Positivism was the dominant philosophy guiding the redevelopment of the discipline as a SPATIAL SCIENCE in the 1950s and 1960s (Harvey, 1969). This dominance ended in the 1970s when positivism came under attack from RADICAL GEOGRAPHERS. By using structural approaches such as MARXISM and REALISM these geographers countered the empiricism of positivism and drew attention to the abstract processes governing social life. HUMANISTIC geographers attacked the objectivity of scientific methods by raising questions about human AGENCY and interpretation. Bondi and Domosh (1992) add a specifically feminist voice to these critiques. Drawing on POSTMODERNISM they show that: positivism produces partial knowledge; masculinity is embedded in its power–knowledge nexus; and SUBJECTIVITY is a central aspect of the research process. HC

Post-coloniality/post-colonialism Post-coloniality and post-colonialism refer, respectively, to the historical condition of living in a time after colonialism and to theoretical positions which draw on experiences of this condition, influenced by POST-STRUCTURALISM. Post-colonial perspectives have entailed a critique of Western feminism, along with other Western-centric forms of thought and politics. An intervention in literary studies, *The Empire Writes Back* (Ashcroft *et al.*, 1989) sought to decentre Western literature studies in English by pointing out the power and value of writing from former colonial countries. Further work in literature also sought to excavate the influence of the COLONIAL past on writing from the centre, suggesting that Western culture and knowledge has been as much shaped by the colonial past as have former colonised countries (Spivak, 1985).

A post-colonial account of colonialism, then, is concerned to trace the impacts of the process of colonisation in the place of the coloniser and also to identify the agency of the colonised in the making and transformation of colonial societies. Homi Bhabha, for example, rereads Frantz Fanon's *Black Skins White Masks* to point out the irony of his 'turning the European existentialist and psychoanalytic traditions to face the history of the Negro which they had never contemplated, to face the reality of Fanon himself' (Bhabha, 1986: xxiv). Bhabha finds even in the processes of alienation and stereotyping characteristic of colonial relations the possibility of subversion. More generally, historians of colonial times have documented the ways in which colonised men and women interacted in creative and subversive ways with colonisers, transforming and remaking the nature of colonial POWER relations (see for example, Comaroff and Comaroff, 1993). The colonisers, too, had varied and often emotionally charged relationships with the colonised – as Gail Ching-Liang Low's (1996) *White Skins Black Masks* demonstrates, especially with respect to administrators and explorers, who occasionally even turned to disguise and cultural CROSS DRESSING in their involvement with the world of the colonised.

The complexity and ambivalence of the relations between coloniser and colonised which post-colonialism has highlighted have also been pointed out by feminist writers, who have suggested that the experiences of colonialism were gendered. Coming from different positions in the West and in colonial society from men, WHITE women's perspectives and experiences of colonialism may have sometimes enabled them to adopt less distant or objectifying relationships with colonised peoples. The colonial relationship in general has been understood in gendered frames, with paternalistic colonisers feminising the colonised in and through their assertion of dominance. Post-colonialism has contributed to a critique of Western Feminism, especially highlighting its ethnocentricism. BLACK FEMINISTS in the West and Third World feminists have criticised FEMINISM for its assumptions concerning women's interests, women's AGENCY and appropriate forms of organising. Different women in different places face a variety of different forms of gender relations and contest these in a variety of different ways. Western feminists' assumptions that women in various THIRD WORLD societies were passively accepting of tradition has been strongly criticised, for example, by Chandra Mohanty (1987, 1988). Some attempt has been made to explore the idea of 'Womanism' as capturing a politics of women's interests, which may not be 'feminist' – but a post-colonial feminism would seem to suggest the importance of acknowledging the possibility of many different feminisms, especially in the wake of post-strucutralist critiques of the (feminist) subject which have made it difficult to define feminism's ambitions and subject.

Given the strong connections between geography's pasts and colonialism there has been some interest in re-imagining geography from a post-colonial perspective and pursuing a variety of geographical aspects of colonialism in a critical vein, including travelling, photography and cities (Blunt, 1994; Driver and Gilbert, 1998; Ryan, 1997). This includes not only exploring colonised geographies but also examining the impact of colonialism upon the metropolitan country, for example, in popular culture or urban design. Politically, the MIGRATION of many people from former colonies to the metropoles has meant that in a very physical sense the colonial past is present. Living after colonialism means negotiating the political identities and power relations which this past continues to imply – and always in specifically gendered ways. If in the centre a post-colonial politics of IDENTITY and difference challenges metropolitan political identity in former settler or colonial societies the consequences of post-colonialism have been explored in the complex and HYBRID cultural and political identities which characterise the post-colonial condition (Gilroy, 1993; Jacobs, 1996). Finding a politics suitable to a post-colonial moment requires an imaginary which acknowledges the mobility and unsettled nature of positions – the impossibility of recovering an AUTHENTIC pre-colonial past, and the instabilities of imperial or settler positions challenged (or, following Jacobs, 'haunted') by a repressed indigenous history. There is also a tradition of considering in a more historical sense the character of post-colonial societies and states, that is, how particular forms of political authority and practice have emerged since the independence of colonies (see for example, Mbembe, 1995).

There has been much criticism of post-colonialism and the post-structuralist approaches with which it has been associated (for example, McClintock, 1995). One of these is very pertinent to continuities in geography's long relationship with places outside the WEST. The suggestion is that the practice of post-colonial studies has replaced the 'OTHER' and 'other places' as the object of study of Western academics, simply replaying the same relations of domination which characterised colonialism; rather than renegotiating the relations of production of KNOWLEDGE, post-colonialism has once again silenced the voice of the colonised and represents the interests and concerns of a Western-based intellectual elite (Parry, 1987). There is also concern that the enthusiasm for 'post'-colonialism masks the persistence of colonial relations in the present, and that the emphasis on DISCOURSES in post-colonial studies detracts from an assessment of the more material ways in which this persistence can be observed (see Loomba, 1998, for a summary of this debate). As post-colonial studies becomes increasingly institutionalised, these criticisms are likely to become more pertinent and insistent.   JR

Post-feminism Suggests that feminism has succeeded in shifting the terrain of cultural politics, if not actually succeeded in all of its political aims. Post-feminism is sometimes interpreted as meaning that feminist politics are no longer relevant and so has contributed to the backlash (Faludi, 1992) against FEMINISM as an identity and political position.   JS
See also GIRL POWER.

Post-Fordism Like the other 'posts', this term implies a movement *beyond* Fordism:

> In contrast to the mass production, mass consumption, mass provision and modernist cultural forms of Fordism, post-Fordism is characterised in terms of a homology between 'FLEXIBLE' production techniques, differentiated and segmented CONSUMPTION patterns, a restructured WELFARE STATE and postmodernist cultural forms (Burrows *et al.*, 1991).

While the conceptualisation of *post*-Fordism can be attributed to some members of the French REGULATION SCHOOL, other regulationists have preferred to use the term *neo*-Fordism.

Over the past two decades, debates over the dimensions and nature of post-Fordist socio-economic change have generated a vast and heterogeneous literature (for useful starting points, see Amin, 1994; Macdonald, 1991). Indeed, some of the more recent regulationist-inspired accounts within geography appear to have abandoned the 'post', referring instead to transitions 'from' or 'after' Fordism (Bakshi *et al.*, 1995; Tickell and Peck, 1995). Nonetheless, feminist contributions to the post-Fordist debate have demonstrated most clearly that any attempts to understand and explain current restructuring processes (whether described as post-Fordist or not) must centrally address the gendered nature of contemporary socio-economic change (see, for example, Leslie, 1993; McDowell, 1991b).   SR

Postindustrial/postindustrial society Generally used to refer to aggregate shifts in economy, society and culture to a new realm, beyond the 'industrial' (see also Beck, 1994). While Bell's (1973) discussion emphasised the growth of skilled white-collar occupations and the formation of knowledge elites, Touraine's (1971) MARXIST account stressed emergent divisions between those with access to information and technology and those without, and highlighted the *de*skilling of marginalised workers at the bottom end of the 'postindustrial' LABOUR MARKET (see also Gorz, 1989). Given the dramatic labour-force changes that a shift to a service economy have entailed, it is remarkable that changing gender divisions of labour were very rarely deemed worthy of comment within initial geographical accounts of postindustrialism (e.g. Ley, 1980). SR

Postmodernism/postmodernity The terms postmodernism and postmodernity have been used in diverse and confusing ways. Although the two are often conflated, they can be usefully distinguished, respectively, as an intellectual movement and as an era (see Duncan, 1996c). Postmodernism, as a mode of thought is sometimes seen as one of the defining characteristics of post-modernity as an era. One may accept the existence of postmodernity without accepting the validity of postmodernism (a particularly lucid example is David Harvey, 1989). Alternatively, one may find postmodernism compelling without believing that Western society has experienced a significant depar-ture from MODERNITY and without seeing any structural or causal relation between postmodern ideas and late CAPITALISM or globalising societal struc-tures (see Harvey, 1989). In fact among the ideas most commonly associated with postmodernism are the rejection of periodisation, especially the totalis-ing or essentialising of eras, the search for origins and the idea of progress. Origins and periods constitute a MODERNIST problematic that is contrary to the spirit of postmodernism. Postmodernists also tend to reject simple BINARY distinctions including that between modern and postmodern. Postmodernism assumes instability of meaning and thus it is difficult to define postmodernism (or postmodernity) from within the DISCOURSE itself.

Confusion regarding postmodernism may be reduced by distinguishing between strong (philosophically rigorous) and weak versions of postmod-ernism and then arguing that only the stronger version is sufficiently distinc-tive as a philosophy to warrant the label. While weaker versions are more plausible to many feminists, it could be argued that they are too easily assim-ilated to modernist intellectual perspectives such as MARXISM or liberal humanism to be usefully labelled postmodern (e.g. Dear, 1986; Soja, 1987). The strongest version of postmodernism is antifoundationalism (for an excel-lent defence of postmodernism in its logically rigorous sense, see Strohmayer and Hannah, 1992). Antifoundationalism entails a rejection of the corre-spondence theory of TRUTH and gives rise to what has been called a 'crisis of REPRESENTATION' (for a critical analysis of this crisis, see Barnes and Duncan, 1992). It is the relativism of such a radically sceptical perspective that causes

211

the most vociferous reaction by feminists and others who are concerned about the political and ethical implications of the refusal to ground postmodern critique in progressive principles (see Nicholson, 1995b).

Weaker (philosophically less rigorous) versions of postmodernism are characterised by: a scepticism towards grand, universalising discourses of LIBERAL HUMANISM (metanarratives of HUMAN NATURE), a rejection of unity, coherence, elegance and simplicity as criteria for judging good social theory, the celebration of fluidity, ambiguity, marginality and DIFFERENCE in gender, sexuality and culture, the rejection of normalising ETHNOCENTRIC (especially Western), ANDROCENTRIC, linear, hierarchical and binary discourses, a questioning of ENLIGHTENMENT faith in the power of disembodied, disembedded (read privileged male) reason, and a questioning of the faith in OBJECTIVITY to arbitrate between discourses. Weaker versions also include postmodernism as a style of architecture and a rejection of the modernist tendency to separate high and popular culture or art. They are also associated with a loss of faith in the ability of SCIENCE to improve steadily the human condition. This loss of faith is often attributed to such twentieth century phenomena as the Holocaust, the atom bomb, and the threat of nuclear annihilation, all of which have severely shaken modern beliefs in social progress.

Postmodernity is usually defined as a societal condition or era within which postmodernism and POST-STRUCTURALISM are dominant cultural components. Postmodernity, usually associated with late capitalism, is often seen as a new era characterised by the commodification of information, signs, spectacles and other ephemeral and AESTHETICISED productions and global homogenisation through the spread of such Western cultural forms. It is also characterised by an unprecedented degree of TIME–SPACE COMPRESSION which has very uneven social and geographical effects. It has resulted in the strengthening of place-based identity, ethno-nationalism and the expansion of the HERITAGE industry as a reaction against the global penetration of capitalism.

Many feminist geographers are attracted to postmodernism's rejection of REASON as being gendered, historical and ethnocentric rather than disembodied and neutral, as is usually claimed. Others agree with postmodernism's refusal of ESSENTIALISM and the replacement of POSITIVIST and empiricist theories of representation and truth with a diversity of viewpoints. Feminist STANDPOINT THEORY (Harding, 1993) for example assumes that culturally and historically specific perspectives can lead to a greater understanding of such phenomena as oppression. The issue of postmodernism's anti-essentialism is a contentious one, dividing RADICAL FEMINISTS who see femininity as biological and stable from those who subscribe to social constructionist theories of fractured, fluid and negotiated gender identities and relations. Perhaps postmodernist thinking can help feminists reconsider such simplistic dichotomies allowing for a social constructionist perspective which allows a reduced role for BIOLOGY. The question of whether postmodernism is compatible with feminism is a contentious one which depends in part upon which definition of postmodernism one is referring to. The stronger version may prove politically paralysing because of its refusal to find firm ground on which to take a political stand. The weaker version may not be so clearly distinguishable from modernism, but may reflect a renewed modernist spirit of openness to difference and desire to use a recognition of difference to problematise, destabilise and deconstruct all totalising and essentialising theory.   ND

Post-socialism Post-socialism, or post-communism, generally refers to the period following the end of communism after the reforms and mass protests of the late 1980s and early 1990s in Eastern Europe and the former Soviet Union (EEFSU). Actual developments vary considerably across EEFSU but 'transition' (Smith and Pickles, 1998) generally involves a rapid retreat of the state from social, economic and welfare sectors and an implementation of free-market economic reforms. For women, as for other social groups, reforms have increased disparities. Those in well-paid jobs make full use of their new-found access to Western-style consumer goods while those in poorly paid employment or affected by the drastic decline in female employment account for the rapid feminisation of POVERTY and hardship (Bridger *et al.*, 1996).

Interpretations and reactions vary. For some, the loss of paid employment constitutes a realisation of the desire to 'escape' from the double burden of WORK and family commitments to traditional gender roles and the 'luxury' of private spaces (Tóth, 1993), although this 'escape' may also represent a rational response to the economic squeeze between spiralling CHILDCARE and social costs and low wages for many women (Bruno, 1996). That this has not provoked more protest is partly explained by the fact that under communism the 'HOME' was a sphere of relative autonomy from state and party influence, a retreat from the politicisation of everyday life, and at the same time PRIVATE spaces formed a key site in oppositional politics (Einhorn, 1993; Smith, 1998). These experiences suggest some reconsideration of the supposedly UNIVERSAL mapping of PUBLIC–PRIVATE, political–private and work–home divides and the generally positive valuing of the former of each pair in much feminist theory may be necessary. However, for many women, the inability to achieve the new ideals of CONSUMPTION and lifestyle represents a failing of transition economics (Renne, 1997). The failings of these policies are also stressed by those women who maintain a desire to combine family and work (Quack and Maier, 1994) and who struggle to find new forms of employment, often through entrepreneurial activities or retraining (Bruno, 1996).

Reactions therefore encompass 'dissent and orthodoxy' (Bridger *et al.*, 1996). Some women welcome the continuity of dominant gender divisions and the retreat of state involvement in gender equality and see transition economics as an opportunity to establish neo-traditional gender roles, to reduce family planning availability (Einhorn, 1993) and, combined with forms of ethnic NATIONALISM, may stress women's role as (pro)creators and embodiments of the NATION (Sharp, 1996). In this context rape as a weapon of ethnic conflict achieves its own terrible logic. Divisions between women about support of or resistance to such processes indicate the impossibility of asserting any one experience for women in post-socialism (Buckley, 1997; Funk and Mueller, 1993; Marsh, 1996; Renne, 1997).

Renegotiation of women's involvement in POLITICS includes varied responses to feminism. Reports of widespread 'allergy' (Tóth, 1993) to FEMINISM, particularly in reaction to the forms of 'emancipation' practised by communist governments, a general decline in social movement organisation and the subordination of gender interests to the 'cold realism' of market restructuring all account for difficulties reported amongst feminist movements in EEFSU. However, reactions also suggest difficulties and divisions between Western and post-socialist women's movements. While Western women draw on an 'unbroken consciousness' in their aim to create the private or the PERSONAL AS POLITICAL (Einhorn, 1991), women in post-socialist societies often feel they have to 'catch up' on Western women's material benefits and at the same time argue Western women are patronising them. Creating a distinctly Eastern European form of feminism which moves away from seeing women only as 'victims' and also addresses the involvement of women in their own past and present is a task which is being pursued (Funk and Mueller, 1993; Renne, 1997). FS

Post-structuralism Supposedly referring to a form of literary and philosophical analysis that has come after structuralist analyses of language and society, post-structuralism has nevertheless taken on a wider meaning as a broad description for post-foundationalist philosophical positions. These positions share at least three common critical positions that Madan Sarup (1993) summarises as: the critique of the unified subject; the critique of HISTORICISM; and the critique of meaning. Nevertheless, post-structuralist positions range from Jacques Derrida's arguments about DECONSTRUCTION (e.g. Derrida, 1997), through Michel Foucault's genealogies of MODERNITY (e.g. Foucault, 1978) to Gayatri Chakravorty Spivak's feminist, Marxist and POST-COLONIAL negotiations with deconstruction itself (e.g. Spivak, 1990b). In this broad sense post-structuralism basically connotes some sort of rigorous questioning of the limits of narratives or theoretical explanations, a questioning which most commonly works by teasing out the moments where the NARRATIVE leaves something out or includes the seemingly unincludable.

Spivak's (1988a) deconstruction of the work of the SUBALTERN studies historians is probably the best and most geographical example of post-structuralism in operation in a feminist vein. In this work she shows how these historians effectively essentialise the identity of the subaltern in their otherwise valuable attempts to reinsert the VOICES and EXPERIENCES of peasant people back into the narratives of Indian history (a history from which imperial and elite nationalist historians had previously banished them). Spivak suggests that this act of writing the subaltern back into history constitutes 'a strategic use of positivist ESSENTIALISM in a scrupulously visible political interest' (1988a: 205), and to this extent she applauds the historians' work. At the same time, however, she deconstructs the way in which such an essentialised subaltern identity is predicated on a moment of incorporation-turned-exclusion. Specifically she argues that while the historians are 'scrupulous in [their] consideration towards women' (1988: 215) they simultaneously tend to overlook the role played by the concept-metaphor of WOMAN in their discourses about peasant insurgency. Woman thus becomes what Spivak calls 'the neglected syntagm of the semiosis of subalternity of insurgency' (1988: 217); or, in other words, discourses of Woman play an essential but unexamined role in the historiographical attempt to produce a new anti-colonial historical discourse about subaltern insurgency in Indian history. Spivak's specific post-structuralist argument about the constitutive exclusion produced by STRATEGIC ESSENTIALISM is of particular note to geographers in this case because she argues that in many cases it worked in tandem with a discourse about TERRITORIALITY. In the historians' narratives, she argues, social discourses about Woman were key to the organising of territorial KINSHIP networks and the resulting concept-metaphor of territoriality was in turn key to the organising of peasant insurgency. This feminist form of post-structuralist argument may seem a long way from either Foucault's study of shifting EPISTEMES or Derrida's deconstruction of writing, but, to the extent that Spivak herself makes clear her debts to both these authors *as* post-structuralists, her own work also points to the heterogeneity of post-structuralism *itself.*

Its broadness noted, post-structuralism must not be confused with the looser and less philosophically precise term of POSTMODERNISM. Not only does post-structuralism carry less social baggage (it neither refers to a postmodern historical epoch nor, still more loosely yet, to a postmodern *Zeitgeist* of ethical and analytical relativism), it is also distinct from postmodernism insofar as it does not include certain more relativist philosophies (of which Richard Rorty's pragmatism is probably the best example). Philosophical postmodernists such as Rorty (e.g. Rorty, 1989) do not generally engage with the problems surrounding the articulation of structural NARRATIVES of explanation and critique. They distance themselves in this way from CRITICAL THEORY and its call to praxis, and thus, unlike much post-structuralist scholarship, they tend to be indifferent to, or at least only ironic about, the more than personal political and ethical responsibilities that follow in the wake of the so-called 'critique of METANARRATIVES'.   MS

Poverty Poverty is a relational issue which refers to the uneven distribution of life chances and experiences amongst those people who have no or limited access to capital and cultural assets. Poverty is equated with DEPRIVATION and lack of social power, and thus with social and spatial inequalities (see UNDER-CLASS) (Glendinning and Millar, 1992; Golding, 1986). Women (and subsequently children) are overrepresented amongst the poor, are more likely to live in impoverished places and, as a result, are more likely to endure hardship and exclusion from mainstream activities (Oppenheim, 1993; Philo, 1995b). The feminisation of poverty is a global phenomenon and is connected to the near universal existence of the unequal sexual division of economic resources (see PATRIARCHY). The dynamics of poverty within a given territory reflect the gendered organisation of socio-economic systems and the gender ideologies built into the WELFARE STATE and LABOUR MARKET (Sainsbury, 1994).

Poverty tends to undergo periodic rediscovery. Studies of the Victorian city connected poverty to the exploitative class relations of capitalism, but remained silent about deprivation amongst women. In the 1970s welfare geographers linked poverty with social (in)justice and emphasised its stratification by class, 'race' and gender (Harvey, 1973). Recent studies address questions about quality of life: themes include poverty and social exclusion (Jordan, 1994; Smith, 1997); the resourcefulness of poor women (Kempson, 1996); and identification of new kinds of poor, and dangerous, places (Campbell, 1993). The extent of poverty amongst women may not be known because many of the social indicators used (income, unemployment levels and consumption expenditure) are GENDER-BLIND.   HC

Power In common-sense terms, power involves the capacity of A to make B do something that B would not otherwise do. While force or coercion may be important elements, the term power usually refers to a complex balance in which 'the ability to impose a definition of the situation, to set the terms in which events are understood and issues discussed, to formulate ideals and define morality' are important elements (Connell, 1987: 107). Social scientists have drawn up typologies of power which have been adapted by feminists to describe the variety of ways in which men dominate or exploit women. It can be readily shown, for example, that men have a near monopoly of AUTHORITY, force and coercion while women are largely restricted to 'weaker' forms of power such as influence and manipulation. Explanations as to why this is so tend to take for granted and reproduce current 'truths' about the fundamental bases of gender inequalities.

In both LIBERAL and MARXIST political theory power is intimately linked with authority or DOMINATION. In the liberal version the STATE is ideally meant to act as a neutral umpire, exercising legitimate power over its subjects in the interests of social order, with power spread widely enough to ensure that everyone's interests are represented. Since this is patently not the case, much empirical work has been devoted to mapping the concentrations of power that actually exist. Summarising these accounts, Stephen Lukes (1974) has identified 'three dimensions' of power. These move from situations of observable behaviour and overt conflict to the more complex ideological means whereby groups are able to control the agenda of politics and to shape subjectivities. In a manner that has struck chords with many feminists, Lukes asks, is it not the supreme and most insidious exercise of power to prevent people from having grievances by shaping their perceptions, cognitions and preferences in such a way that they accept their role in the existing order of things ... either because they can see or imagine no alternative ... or because they see it as natural and unchangeable ... or because they value it as divinely ordained? (1974:).

In Marxist accounts, the state is not even potentially a neutral umpire but controlled by the ruling class which uses it to dominate and exploit another CLASS which lacks power. Though STRUCTURALIST, Marxism tends to reduce power itself to little more than an effect of the ensemble of structures, most liberal and Marxist views stress the AGENCY of the individual or group even in a situation of structural constraint. Both tend to treat power as the possession of individuals or groups, a zero-sum game in which there are inevitably winners and losers. It is primarily repressive in its exercise, emanating from the top down, according to the classic model of sovereignty. Despite some attention to IDEOLOGY, power is usually assumed to rest on an economic or, in the case of gender relations, biological base.

Given the equation of power with sovereignty, much feminist attention has been directed to the power of the state in sustaining male domination. Some have argued not only that the state is in the hands of men but that it systematically pursues 'male interests' (MacKinnon, 1989). Gender divisions are here treated as analogous to Marxist class divisions, with men and women facing each other as competing blocs. Marxist feminists, unwilling to claim that women are a class, treated the power of men and the subordination of women as effects of imperatives outside the direct relationship between the two and emphasised the ideological apparatuses of the state. Liberal feminists identified the problem of women's confinement to the PRIVATE SPHERE as central to their low socio-political status and saw the problem in terms of how women could gain equal access to the public sphere.

At the same time feminists have long been concerned with understanding the operations of power in everyday life and are well aware that power does not reside only in the state apparatuses, however broadly defined. Some questioned whether feminism needs a theory of the state at all, for top-down theories did not account for the specificity of women's experience of oppression or the multiplicity of sites in which it takes place (Allen, 1990). In the 1970s many welcomed PSYCHOANALYSIS as a way of explaining how the ideology of MASCULINITY and FEMININITY constructs men and women as appropriate patriarchal subjects in capitalist society (Mitchell, 1975) but theories of ideology were also seen to have their limitations. As Gatens has pointed out, these theories are committed to a form of HUMANISM that puts the emphasis on the way in which a human animal is socially produced as a masculine or feminine SUBJECT and obscures the ways in which power takes hold of the body rather than merely conditioning the mind (Gatens, 1996: 66–7). As feminists contemplated the difficulty of entering the public sphere without denying their own CORPOREAL specificity, it became clear that it was not 'masculinity' as such but the male body and its culturally determined powers and capacities on which the liberal body politic is modelled (Pateman, 1988a). Feminists have become increasingly concerned with the operations of power at this micro-level of bodies and their capacities. For a variety of reasons they have been drawn to Foucault's work on the 'polymorphous' character of power relations and the 'biopolitics' that connect and consolidate these relations. For Foucault:

> Power is everywhere; not because it embraces everything but because it comes from everywhere ... Power comes from below; that is there is no binary and all-encompassing opposition between ruler and ruled at the root of power relations, and serving as a general matrix – no such duality extending from the top down and reacting on more and more limited groups to the very depths of the social body (1978: 93–4)

Foucault is not concerned with locating the bases of power or with the question of who has power and he insists that there is more to power than domination. Rather than being purely repressive, 'power produces; it produces reality; it produces domains of objects and rituals of truth' (Foucault, 1977: 194). As McHoul and Grace put it:

> He asks how power installs itself and produces real material effects; where one such effect might be a particular kind of subject who will in turn act as a channel for the flow of power itself. ... Power is not to be read, therefore, in terms of one individual's domination over another or others, or even as that of one class over others; for the subject which power has constituted becomes part of the mechanisms of power. It becomes a vehicle of the power which, in turn, has constituted it as that type of vehicle. Power is both reflexive, then, and impersonal. It acts in a relatively autonomous way and produces subjects just as much as or even more than subjects reproduce it (1993: 21–22).

This has provided a space for some feminists to consider how power constitutes certain types of bodies which in turn become vehicles for the transmission of power. Theorists of difference understand the body not as a biological constant but as a historical product and emphasise the productiveness of power as positive capacity as well as a site of subordination and inequality. While many feminists continue to work within the earlier frameworks, the main focus of feminist work on power has shifted decisively towards the discursive practices, including the biological ones, which constitute the body.   RP

Power geometry Doreen Massey (1991b) introduces the concept of power geometry to insist that although SPACE is fluid and always imbued with social meaning this does not mean that it is unstructured. She insists that not all people have access to the shrinking world described by theorists such as Harvey and Soja as TIME–SPACE COMPRESSION and instead argues that 'different social groups, and different individuals, are placed in very distinct ways in relation to these flows and interconnections' (1991b: 25). 'RACE', CLASS and GENDER all influence the ability of an individual or group to negotiate space, whether this be at the SCALE of cities, NATION-STATES or the globe.   JS
See also SPATIALITY.

Pragmatism Defined by Derek Gregory (1994: 471) in the *Dictionary of Human Geography* as 'a philosophical perspective which is centrally concerned with the construction of meaning through the practical activities of human beings'. This focus on issues of practical concern implies a shift of emphasis away from the more abstract and philosophical issues that occupy many philosophers. Pragmatism also has a specific meaning in linguistic theory where it is used to distinguish the actual everyday use of language (pragmatics) from its different systems of formal rules (syntactics and semantics). The turn to pragmatics reflects the general movement from STRUCTURALISM in the postmodern era with the recognition that the understanding and use of language and the reception of texts in different cultural circumstances is a crucial mediating element in the communication of ideas. This is a similar move to that in the social sciences where the growing emphasis on people's agency as well as the idea of SITUATED KNOWLEDGE is part of the move away from structuralist analyses of social relations.   LM
See also LANGUAGE.

Primogeniture Referring to the fact of being the first-born child, primogeniture in effect refers to the social arrangements of inheritance which privilege the eldest son in a nuclear family. Material property and titles held by the wealthiest classes, particularly in feudal and modern societies, were passed on to male heirs. Women were thereby excluded from social recognition and capital, as illustrated in Virginia Woolf's novel *Orlando*, when the eponymous hero wakes to find herself female and dispossessed of the family estate.   SAR

Private space/private sphere See PUBLIC–PRIVATE DIVISION

Privatisation The notion of PRIVATISATION has been used in a wide range of contexts, to refer to an array of policies which have acted to redraw boundaries between private and public, STATE and society. In many advanced capitalist nations, privatisation policies have reflected neo-liberal notions that the state should be 'rolled back' in favour of the market provision of services. Although the term implies a straightforward replacement of the state by the market, it is generally accepted that there are a number of dimensions to privatisation. While some authors limit the term's reference to the transfer of state assets to the private sector (e.g. through the sale of public enterprises), others include a broader range of policies, such as the deregulation of state activity, the introduction of user fees and the contracting-out of public sector services. Wider definitions emphasise the political imperative which lies behind the term: a view that the public sector should be disciplined by a 'free-market' ethos.

Feminists writing in the 1970s and early 1980s (prior to the sharpest attacks on public services by NEW RIGHT governments: see THATCHERISM) emphasised the patriarchal nature of the WELFARE STATE. Attention was drawn to the ways in which the state under FORDISM positioned women as dependent labour-market participants, as low-paid workers within the public sector and as submissive consumers of health and welfare services controlled by male PROFESSIONALS (Dale and Foster, 1986; Pateman, 1988b). It is now starkly apparent that women have also been disadvantaged by a variety of privatisation strategies, including, for example, the dismantling of benefit arrangements, the introduction of 'COMMUNITY CARE' and the contracting-out of local authority work forces (Escott and Whitfield, 1995; Walker, 1988; Wilson, 1987).   SR

See also POVERTY.

Production See MODES OF PRODUCTION/MODES OF REPRODUCTION

Professionalisation At first glance, the notion of professionalisation seems relatively straightforward: simply the process of becoming more like a professional. Here, professional is understood as the differently nuanced codes of conduct associated with experts who have educational experience and authority in specialist fields of knowledge, such as law, medicine or architecture (Corfield, 1995). So, for example, professionalisation of UK social work after World War II involved organising and elevating training programmes, standardised work practices, salaried occupational groups and a new vocabulary of professional language. However, writers such as Foucault brought to the fore questions of POWER and professionalisation (Burchell *et al.*, 1991). Foucault's work revealed how historically the institutionalisation of expertise in the form of the professions granted social groups powers to dominate and govern TERRITORIES and populations. His work provides a sophisticated analysis of the relations between power and knowledge through the study of DISCOURSES and it has been used by feminists and geographers to expose the social construction and privileging of 'professional expertise'. One aim has been to challenge dominating systems of knowledge (such as scientism). Another has been to scrutinise the politics of professionalisation and the ways in which struggles to resist such oppressive discourses face the risk of political co-option. The pressure on non-Western groups to codify and value their biomedical knowledge within Western scientific discourses is one such example (Shiva, 1993).   FG

Proletarianisation Derived from Marx's original distinction between the bourgeoisie (or capitalist class) and the mass of working people, or proletariat. The term has been used to refer to the deskilling and deterioration of pay and conditions within white-collar occupations such that they have become more similar in nature to manual work in factories. Whether or not such a *class*-based concept could be used to explain changes in the status of highly *gender*-segregated jobs prompted considerable debate in the 1970s and 1980s (see Lowe, 1987).   SR

Property, private The term private property is most commonly used in a relatively narrow way to refer to the private ownership of land. Yet property can be defined much more widely, encompassing a variety of personal possessions held by an individual. In this respect, it is important to recall the long history of women's positioning as the property of men. Historically, women have been required to surrender to men control of all property in the form of land and wealth. It was not until 1882 that married women in England, for example, were permitted to control their own property, and only in 1935 did single women receive the same right (for legislative details, see Lewis, 1984). Further, however, a belief that married women were the personal property of their husbands legitimised the physical possession of women's bodies.

Marxist-feminists writing in the 1970s were particularly drawn to Engels's (1985 [1884]) suggestion that the oppression of women was closely tied to the evolution of private property. It was thus possible directly to connect 'private property, responsible for class domination ... with gender domination' (PATRIARCHY) (Nicholson, 1986: 38). Other (primarily radical) feminist writers criticised Engels's account, arguing not only that patriarchy pre-dated class society but also that his conceptualisation failed to address 'the autonomy and persistence of patriarchy' (1986: 38).

Within geography, discussions of private property have been pursued in a number of different ways, although most rest upon a definition of property as land or territory. They include the following:

- *Private property and private space*: here, accounts of the 'privatisation of the architectural public realm' (Davis, 1990: 226) have documented the exclusion of particular social groups from spaces which were formerly defined as public.
- *Private property and home ownership*: the ownership of private property has been central to debates concerning the relationship between housing tenure and social class (Pratt, 1986; Saunders, 1984). Proponents of the incorporation thesis argued that private ownership of housing incorporates homeowners into the capitalist social order (see Pratt, 1989). There remains a question, however, as to how the position of women who are not defined as heads of household is to be viewed: are they necessarily incorporated in the same way as their male partners?
- *Private property rights and gender relations*: relevant work includes discussions (primarily within the context of the developing world) of the effects of changing land tenure arrangements upon gender relations (see, for example, Carney and Watts, 1990; MeizenDick *et al.*, 1997).   SR

Prostitution Prostitution (sex work) is a divisive subject among feminists. Some feminists focus on the exploitation of women and children which results from sex work (MacKinnon, 1989; Pateman, 1988a). Others believe commercial sex can be freely engaged in by adults and that it is a legitimate site of resistance against the dominant culture (Bell, 1994; Law, 1997; Rubin, 1984). In most countries the dominant IDEOLOGY denies the SEXUALITY of public places by imposing greater spatial restrictions on sexual minorities than on those who conform more closely to societal standards of 'proper' sexual behaviour

221

(Duncan, 1996b). These restrictions hide from public view, and thus privatise, the aesthetically and morally offending physical, psychological, medical and social problems surrounding the highly marginalised identities of prostitutes. Alternatively, they may segregate such behaviour in spaces where it is subject to SURVEILLANCE by the police. Prostitution can thus be constituted as a 'problem' not so much because of anything inherent to sex work itself but because of its marginalisation by dominant social institutions and ideologies, including, importantly, capitalism in which sexual labour is seen as ideally unpaid (Singer, 1993). To the extent that prostitution *is* a problem for prostitutes themselves, it is usually the manifestation of much more profound socially and politically structured problems such as POVERTY, unemployment, unequal class relations, PATRIARCHY and/or misogyny. Shannon Bell (1994) argues 'at the heart of prostitute discourse is a dichotomization of the prostitute body in terms of empowerment/victimization'. What is clear from the varied experiences of prostitutes themselves is that prostitution is a highly contested, plural subject position in which some play an empowering role as teachers, counsellors and healers, while others are abused, exploited and criminalised.   ND
See also PORNOGRAPHY.

Psychoanalysis Psychology and psychotherapy associated with FREUD and his followers. From the late nineteenth century this studied the mental processes operating outside of the realm of individual awareness.

FEMINISM has occupied a close but fractious relationship with psychoanalysis, ranging from appropriation to wholescale rejection. Psychoanalysis has provided a conceptual basis for feminist accounts of the GENDERING of SUBJECTIVITIES in PATRIARCHAL society. However, some feminists critique Freud for ESSENTIALISM and biological determinism. Lesbian theorists in particular have criticised psychoanalysis for assuming heterosexual relations in their models of the subject (see Butler, 1990).   JS
See also FREUDIAN THEORY; LACANIAN THEORY.

Public–private division The terms public and private both have a number of sometimes contradictory and sometimes overlapping meanings that reflect historical change in the institutional, social and economic contexts in which they have been used. For example, in the classical world the public sphere signified the power of citizens to debate and take part in political life whereas the sphere of the private signified lack of such power: 'the realm of necessity and transitoriness remained immersed in the obscurity of the private sphere. In contrast ... stood ... the public sphere as a realm of freedom and permanence' (Habermas, 1989: 3–4). However, in early industrial capitalism the private sphere came to be represented as the area of social life in which the individual could retain some freedom of individual expression compared with the power of economic relations and state authority in the public sphere. The changing meanings of the public and private sphere are reflected in the accretion of a variety of meanings of the words 'public' and 'private'. Thus, the term 'public' can refer to openness, as in 'public debate', to state ownership and authority, as in 'the public sector' and 'the public interest', to the representation of power and social significance, as in 'a public figure', and to the role of the citizenry as critics of political or moral issues, as in 'public opinion'. The word 'private' can denote those areas of life which should be kept hidden from 'public' view and an area of social life in which intimate emotions and relationships can be experienced and expressed. It can also be seen as the sphere of the market and of economic relationships, as in 'the private sector'.

Habermas (1989) suggests that in early industrial capitalist societies the private realm contained two separate institutional spheres – the FAMILY and the market economy and that the public realm also constituted two separate institutional spheres – the state and the arena of public political debate between citizens. He argues that in WELFARE STATE capitalism these distinctions within and between the public and the private are breaking down. The political role of the citizen has changed from one of active participation to being the recipient of political 'publicity' and social welfare. The separations between the family and the market and between the market and the state have also become blurred. In particular the role of the family as consumer and as recipient of consumer advertising has become more important.

Accounts such as Habermas's have been strongly criticised by feminists for ignoring the importance of the gendered and profoundly unequal social relationships implicit in the actual or conceptual separation of the family from the polity or from market relationships (Fraser, 1989). There are also feminist discussions of two particular versions of, or DISCOURSES, concerning the distinction between public and private which have political significance for feminists.

The first such distinction is the distinction between state ownership, viewed as being ownership by the 'public' as citizens, and ownership by non-state interests, either individuals, groups or firms. The second distinction is between social relations which are conducted in the realm of politics and paid WORK and which are represented as being conducted in the 'public' eye and those relationships and activities which are represented as being part of the private, domestic realm, concerned both with intimate 'family' relationships, childbearing, CHILDCARE and relations with kin and with the work done to maintain the home, such as unpaid housework, shopping and cooking. Both DUALISMS have been subjected to criticisms.

The distinction between public and private ownership groups together economic organisations owned and directed by the state under the heading of the 'public sector'. Examples in Britain and the USA are the postal service, state schools (public schools in the USA) and government bureaucracies. Private economic organisations or firms owned either by individuals or by shareholders are said to be in the 'private sector'. Organisations in the public sector provide 'public services' which are provided and consumed, at least partly, according to non-market criteria. Welfare public services such as state-provided schooling, health care, low-cost 'public' housing and financial support to low-income groups are often represented as being provided collectively in order to meet social objectives of equity and EQUALITY of opportunity. Such welfare services have been particularly significant to feminist political action both because these services are important to the well-being of many poor women and their children and because their organisation and delivery has been characterised by gendered assumptions concerning women's responsibilities for social relations and unpaid work in the 'private', domestic sphere so that the apparently gender-free notions of equality of opportunity and equity are argued to be gender-biased (Wilson, 1977). In many countries there has been a recent move to introduce more market-based mechanisms and criteria in the organisation of public services and the use of the non-profit sector and the private sector for their delivery (Wolch, 1990). These changes have blurred the line between the public and private sectors.

The distinctions between the private, domestic sphere and the public sphere of politics and market relations was of particular importance in the development of nineteenth-century and twentieth-century urban forms in industrialised, capitalist countries, underpinning and justifying the development of residential areas separated from areas of industry, commerce and political power (Davidoff *et al.*, 1976; McDowell, 1983). The private sphere was represented as the domain of women (but subject to the authority of the husband), in which care for others and care of the home were carried out. It should remain both spatially and socially separate from the public realm. The public sphere was represented as the domain of men. Feminists have argued that this conception has served to legitimate and reinforce women's OPPRESSION. For example, the notion of the private as a location in which the individual can express themselves 'freely' is one that ignores the significance of gendered power relations. Feminists also argue that this representation of separate spheres is false: for example, state power and market relations impinge strongly on the constitution of domestic social relations, and relations of intimacy are not confined to the private. SB

Q

Qualitative methods See FEMINIST METHODOLOGY

Queer/Queer Nation/queer theory Queer is a complex term having multiple origins and meanings; in the context of feminism it most commonly refers to the 'deconstruction' by literary critics, artists and, increasingly, social scientists, working in a POSTMODERN or POST-STRUCTURALIST framework, of oppressive BINARISMS, especially those related to GENDER, SEXUALITY and the sex–gender system (most notably the HOMOSEXUAL–HETEROSEXUAL binarism). Historically, it also reflects a strategic decision by some GAY, LESBIAN, BISEXUAL and TRANSGENDER activists to appropriate and redeploy a derogatory epithet in service of an anti-HETEROSEXIST political agenda and to reflect, celebrate and reproduce in language lived experiences of gender, sexuality and other axes of POWER and difference which are seen, from a queer perspective, to be much more diverse, fluid, fragmented and unstable than BINARY labels allow.

Queer Nation is the name of a loosely organised and ostensibly non-hierarchical political movement founded in 1990 by members of the activist organisation AIDS Coalition to Unleash Power (ACT-UP) and others specifically to fight HOMOPHOBIA and heterosexism. The goal was to adapt 'direct action' tactics and strategies which had proved successful in the political battle against HIV/AIDS to other issues, grounded in heterosexism and homophobia, that affect the lives of non-heterosexual people. Examples are same-sex kiss-ins in heterosexual environments, the 'outing' of privileged public figures who were known or widely rumoured to be gay or lesbian and the disruption of religious services to protest heterosexist church practices. The significance of the term 'nation' in the name is more rhetorical and symbolic than substantive, though some individuals (not necessarily associated with Queer Nation) have at times advocated some form of sovereignty within a territory for lesbian, gay, bisexual, transgender or other queer people. Queer Nation has been much less active in the late 1990s than it was in the early 1990s.

Like feminist theory, queer theory is a diverse body of literature and ideas with multiple roots. It is frequently seen as having emerged out of a literary criticism tradition within the field of gay and lesbian studies (see, in particular, the works of Eve Kosofsky Sedgwick, 1990 and Michael Moon, 1998). Now, however, it has spread beyond the humanities to engagements with the social and even, at times, physical sciences, where it has continued to evolve and mutate. Queer theory still tends very much to employ DECONSTRUCTIVIST methods of analysis and to be informed by postmodern and/or post-structuralist modes of thought (see Seidman, 1996). Perhaps most importantly, this includes a political commitment to destabilising binarisms and other oppressive categories of thought.

225

However, some queer theorists are now exploring the potential role of more traditional objects and methods of analysis (e.g. economics, statistics and quantitative methods) in such projects. While focused initially on deconstructing the homo–hetero binarism, for example, much queer theory today extends this mode of analysis to other categories, including, especially in feminist geography (see Gibson-Graham, 1996), 'the economy'.   LK

R

Race The word 'race' is often written in scare-quotes in social science litera-ture to indicate that it is a social construct. Scientifically there are no such things as discrete biological races. On the other hand, racialisation, the repre-sentational process whereby social significance is attached to certain biolog-ical and/or cultural characteristics, is real.

A key factor is the notion of inheritance of traits, which places women in a central position as the reproducers of races both biologically and culturally (Anthias and Yuval-Davis, 1992). It is this naturalisation of both race and GENDER which feminist geographers argue must be destabilised by the initi-ation of 'unnatural discourse' to challenge what is taken for granted (Kobayashi and Peake, 1994). 'Unnaturalisation' is necessary to change not only the structures of thought of dominant groups but also those of subordi-nate groups, for whom the experience of internalising racial inferiority is profoundly dehumanising (hooks, 1992b).

Because it is socially constructed, the notion of race has different meanings in different societies, drawing on the historical specificity of relations of POWER between different ETHNICITIES. There is thus a distinct geography of these meanings which must be taken into account when theorisations are used. In the USA, race is understood primarily as a black–white difference, particularly between collectivities linked by origins in slavery, that is African-Americans, and those of white European descent. However, there are strong regional variations, so that the largest collectivity of women racialised as non-white in California, for example, is Latina.

In Britain the black–white binary also dominates thinking about race, but black takes on a wider meaning reflecting diverse COLONIAL relationships. However, in Germany and areas of Europe occupied in World War II the notion of race includes a strong 'white' strand through its inextricable associ-ation with the Holocaust (Miles, 1993). Because of these sharply different contexts there are also incompatible and changing labels for racialised collec-tivities in different locations.

Notions of race tend to homogenise CLASS, gender and ethnic differences. In reality, where gender is not explicitly considered, masculine racial identities are privileged. One strategy adopted by feminist geographers is to make visible women's places in subordinated racialised collectivities (Spooner, 1996; Walter, 1995). Feminist geographers are also confronting more fully the mutual constitution of the different dimensions of social positioning, including race, class and sexuality, raising the issues of the primacy of the gender as a category and the ways in which it is conceptualised (WGSG, 1997). This may involve the recognition that in some cases gender is not the dominant form of difference for women. For example, black and Latina environmental activists in Los Angeles emphasise their place-based identities which involve community unity rather than feminist identification (Pulido, 1997).   BW

Racism Racism is a discourse and practice of inferiorising ethnic groups. It draws on a set of assumptions, images and practices which differentiate between social groups and construct some as inferior in essential ways. Racist practices may have both intended and unintended discriminatory effects (Anthias and Yuval-Davis, 1992). Racism occurs both within the state, through institutions such as education and the police, and within civil society, for example through youth culture and attitudes of neighbours. The question of POWER is central, distinguishing racism from xenophobia, where there is no capability to DOMINATE others.

Iris Marion Young (1990c) places racism among a set of oppressions which she describes as cultural imperialism. Dominant groups occupy unmarked, neutral and apparently universal positions, whilst subordinate ones are constantly reminded of their group identities and are regarded with aversion. Despised groups are thus imprisoned in their *bodies*, in opposition to the *reason* which defines dominant groups.

Physical visibility by skin colour is the most widely acknowledged marker of difference, but racism in Western societies against Jews, Irish people and travellers shows that many other signs which appear to naturalise inferiority are used (Cohen, 1988). The overriding emphasis on skin colour means that other forms of racism are often overlooked or denied.

Racisms are gendered in a number of ways. POST-COLONIAL deconstructionists argue that colonial others represent split-off parts of male, middle-class, white European selves, which had to be denied in order to maintain the myth of masculine independence (Pajaczkowska and Young, 1992). Colonised men were thus feminised because of their dependent status, and colonised women were doubly excluded. But there is deep ambivalence about these othered groups, who are also the objects of envy and desire (Brah, 1996).

The mutual constitution of race and gender can be seen in the ways groups are racialised as inferior through their representation as feminine, for example passive, sensual Orientals and emotional, irrational Celts. In the USA white racist discourse provides highly racialised and gendered representations of the urban UNDERCLASS as 'welfare queens' (African-American female welfare recipients) and 'urban gangs' (unemployed African-American male youths) (Gilbert, 1997).

Racism permeates the study of geography, including FEMINIST GEOGRAPHY (Rose, 1993a). For example, work that uses Marxist forms of analysis to focus on the HOME as a specific space of reproduction can render black women invisible by accepting a notion of 'universal womanhood' which erases the question of race. Thus anti-racist feminist strategies include research approaches which 'denaturalise' the position of dominant groups, or recentre subordinate ones. Social constructions of WHITENESS in childhood in different US locations have been placed under the spotlight (Frankenberg, 1993a and b; Twine, 1996).

Anti-essentialist approaches are also being developed in studies of women's HEALTH, linking racism with other forms of oppression focusing on class, AGE, sexuality and DISABILITY and also acknowledging the agency of subordinate collectivities and individuals (Dyck, 1995a). However, according to Iris Marion Young (1990c) the extreme difficulty in shifting racist patterns of thought, many of which are reproduced unconsciously, requires a more fundamental revolution in SUBJECTIVITY, affirming OTHERNESS in ourselves.　　　BW

Radical democracy See DEMOCRACY

Radical feminism Radical feminism emerged particularly in North America in the late 1960s and early 1970s. It grew out of a critique of radical left politics, and thus much early work reworked socialist ideas in relation to radical feminist theory. Radical feminist theorists argued that women are oppressed as women by men and this operation of male power cannot be reduced to other forms of power, such as the power of capital over labour (Firestone, 1970). Indeed, the male–female relationship is the paradigm of all power relationships (Millet, 1970) and male oppression of women takes precedence over other forms of oppression. Radical feminists focused attention particularly on SEXUALITY, on motherhood and the FAMILY and on VIOLENCE against women. Central to their arguments were that gender roles were socially constructed, not rooted in biological difference. Radical feminists focused particularly on motherhood, debating its replacement through reproductive technology (Firestone, 1970) or seeking to redefine motherhood free from patriarchal control (Rich, 1979). Radical feminists, such as Daly (1978) saw androgeny or radical lesbian feminist separatism as an alternative to the patriarchal sex–gender system. Radical feminist politics have centred particularly on campaigns against violence to women – setting up women's refuges and rape crisis centres – and on opposition to PORNOGRAPHY (Dworkin, 1981; MacKinnon, 1987) as well as focusing on women's health issues (Phillips and Rakusen, 1978) and the mobilisation of 'THE PERSONAL IS POLITICAL', particularly through consciousness-raising groups.

The influence of radical feminism can be traced through most other forms of feminism since the late 1960s, particularly in SOCIALIST FEMINISM and in

French feminism. Its legacy is also evident in feminism which posits a distinctive women's culture (Segal, 1987). Within feminist geography the legacy of radical feminism is less strong than that of socialist feminism but is evident, for example, in debates amongst feminist geographers about the significance of patriarchy (Foord and Gregson, 1986), in Rose's (1993a) challenging of masculinist geography and in ECOFEMINISM. CD

See also SECOND WAVE FEMINISM/SECOND WAVE WOMEN'S MOVEMENT.

Radical geography A general term used to subsume variants of critical geography, including socialist and feminist perspectives; it is associated in particular with the journal *Antipode*, which is a 'radical journal of geography'. This journal was founded in 1969 by a group of faculty and graduate students at Clark University, MA, at a time when anti-war, anti-poverty and anti-imperialist movements were important on US campuses, and to a lesser extent in the United Kingdom. The founders of the journal called for a radical change in universities, seeing them as institutions that stifled, rather than fostered, radical thinking (Stea, 1969). In the first issue Dick Peet listed the concerns of the NEW LEFT in the late 1960s – Vietnam, apartheid, Israeli occupations, poverty in cities, US imperialism and defence spending – which he argued geographers should be examining instead of the peopleless regional science that then dominated the discipline. Many of these issues identified in 1969 remain current for the Left and for that wider group of geographers, among whom David Harvey (1982, 1996) is pre-eminent, committed to a geohistorical materialist analysis of uneven development. The commitment of radical geographers to a materialist analysis was restated in an *Antipode* editorial marking 25 years of publication of the journal, which is now produced and marketed by Basil Blackwell publishers. The editors at that time, Dick Walker in the USA and Linda McDowell in the United Kingdom, argued there that they saw 'no reason for a radical new turn in radical geography ... [the aim] is to maintain *Antipode's* role as the principal and principled voice of the Left in geography' (1993: 1) while recognising that the forward edge of such work must include feminist, environmentalist and POSTMODERN work.

Although there seems to have been a decline in directly political aims of geographical scholarship, in the 1990s there is a strong group of scholars who would term their perspective 'critical' rather than radical to reflect the widening focus from strictly socialist or class-based critiques to include post-colonial work, for example, and post-structuralist critiques of METANARRATIVES. There is, however, some dispute about the continued assertion of the over-riding significance of CLASS-based inequalities by some radical geographers, who reject arguments about the contemporary significance of a politics of identity, or what Brah (1996) has termed a cartography of intersectionality (see, for example Harvey, 1996; Young, 1998). Many feminist geographers continue to feel marginalised in the debates of the Left.

A loose association of critical geographers based in the main on e-mail communication was set up in the mid-1990s as an alternative focus for those geographers who opposed the amalgamation of the Institute of British Geographers and the ROYAL GEOGRAPHICAL SOCIETY, and who espoused broadly critical political aims. LM

Rape The forcible or coercive, non-consensual penetration of the body. Rape is the ultimate performance of DOMINATION over the body space. In feminist theory, rape is conceptualised as an act of POWER and VIOLENCE rather than sex, as a political act, in most cases against an oppressed group. Throughout history and across space, rape has been used to control women (Brownmiller, 1975). It reinforces female subordination through social and spatial control, achieved through widespread fear as well as violence itself, and by inadequate state intervention to prevent rape. Rape is rarely reported, but feminist research has highlighted its commonness and countered other common images and myths: that women provoke rape through their behaviour and that men who rape are strangers. Popular culture (and many women's fear) identifies rape as a STRANGER DANGER, yet it is far more common in private spaces (Pawson and Banks, 1993; Valentine, 1989). The rape of men, which contradicts stereotypes about HETEROSEXUAL MASCULINITY, is similarly underreported. Rape laws are developed not to protect women, but to assure fathers' and husbands' control over women's bodies (Russell, 1992). Though rape within MARRIAGE has been criminalised in some countries, offenders are still unlikely to be prosecuted. Conflicts over rape reflect gender struggles; part of the backlash against feminism is the accusation of 'victim feminism', the idea that feminists have exaggerated the risks and heightened the fear of rape (Roiphe, 1993). RHP
See also ABUSE; HARASSMENT; VIOLENCE.

Rationality A term usually associated with ENLIGHTENMENT thought, especially liberalism, in which it is assumed individuals are able to make rational choices based on an evaluation of alternative evidence rather than being influenced by 'irrational' beliefs, myth or superstitions. Rationality is often associated with MASCULINITY and counterposed to the supposed irrationality or emotional character of FEMININITY. This BINARY distinction has been the focus of feminist criticism. An idealised notion of individual rationality is the basis of geographic location theory. Rationalist ideals also lie behind feminist arguments about equality. As Di Stefano (1990) explains:

> the rationalist position takes the Enlightenment view of rationality and humanism at its word and as its starting point. On this view, common respect is due to all people because they are rational. The human capacity for rationality is precisely what distinguishes us from the realm of NATURE which, not incidentally, is not accorded respectful treatment. Women have been unfairly excluded from the respect which they are due as human beings as the basis of an insidious assumption that they are less rational and more natural than men (1990: 67).

This association between masculinity and rationality or civilisation, and femininity and nature is a widespread binary distinction or DICHOTOMY which feminist anthropologist Shelley Ortner (1974) posited as a UNIVERSAL, although varied and varying, characteristic of human societies.   LM

Realism A philosophy of science which proclaims the existence of a 'multi-tiered' reality which exists outside human consciousness and which prioritises conceptual ABSTRACTION in constructing knowledge. Following Sayer (1984, 1992) the adoption of realism in geography has focused on research methodology rather than on refining realist debates (Pratt, 1995). For some, realism underpins an 'emancipatory social science' and a critical knowledge of the 'differentiations of the world' (Yeung, 1997).

Realist approaches, drawing on the work of Bhaskar (1975, 1979, 1986), see the social world as composed of complex events and phenomena which exist within 'open systems'. These events and phenomena are transformed by social actors whose actions, based on intentions, have causal properties. They are the 'objects of analysis' in social science and, within a realist ONTOLOGY, they structure scientific knowledge.

Realist methodology seeks to establish the nature of an 'object of analysis'. Sayer (1984, 1992) argues that this can be done by rational abstraction which isolates the components and relations which are *necessary* for the object to exist from those which are *contingent* and frame a particular temporal–spatial manifestation. This processes is distinguished from, yet draws on, pre-existing social knowledge. Many existing objects of analysis in social science are deemed to be 'chaotic' as their necessary components and relations cannot be clarified. The process of abstract (theoretical) research is presented as a prerequisite of concrete (empirical) research. In feminist geography a realist approach was used by Foord and Gregson (1986) to attempt a reconceptualisation of PATRIARCHY.

Some (Cloke *et al.*, 1991) have criticised realism in geography by claiming that it calls for an 'ultimate' truth or universal theory and thus totalises knowledge. Sayer (1992) on the other hand suggests that this is a misreading of Bhaskar and that realist philosophy recognises that all knowledge is partial.   JF

Reason See RATIONALITY

Refugee Defined by the United Nations Convention on Refugees (1951), a refugee is a person who 'owing to well-founded fear of being persecuted for reasons of RACE, religion, nationality, membership of a particular social group or political opinion is outside the country of his [*sic*] nationality' and is unable or unwilling (because of fear) to return. Just as migration is becoming feminised, so too refugee movements are highly gendered, structured by relations of POWER, GENDER and CLASS. Globally in 1992 there were between 17 million and 19 million refugees, of whom around 80% were women and dependent children, many of them in female-headed households. Another 12 million to 17 million people, again predominantly women and dependents, were internally displaced within their own country. Refugees are among the poorest of the poor, with the vast majority in low-income countries, such as Ethiopia, Malawi and Pakistan; overall, 33% of the world's refugees are found in Africa, 32% in the Middle East and 20% in Europe and North America. Gender-sensitive approaches to refugee flows recognise women's disproportionate involvement and the fact that disaster-type interventions can compromise women's long-term EMPOWERMENT in gender relations (Walker, 1994). By being pushed outside the social institutions that 'protect' disempowered groups, female refugees are subject to sexualised HARASSMENT and abuse from male combatants and camp administrators. In the refugee camps, women are further subject to structural constraints, such as loss of the domestic sphere in which they may previously have had authority, and loss of income sources. Moreover, women are often excluded from policy-making, planning and management in camps, despite their overwhelming numbers and efforts by the United Nations to provide guidelines for working with refugee women. Rather than the automatic distribution of resources to men, women's involvement in the design and implementation of HEALTH, education and income-generation projects in refugee areas has been shown to be beneficial. Waterpipe provision in African refugee camps recognises women's role in domestic tasks, while locating standpipes away from areas where sexual harrasment occurs. Additionally, despite the prevalence of RAPE and SEXUAL abuse that refugee women experience, family planning services and female health care are often perceived as luxuries rather than as rights.

Sexual VIOLENCE against women in former Yugoslavia in the 1990s raised issues of the gendered definition of refugees, particularly the question of whether refugee status could be granted on the grounds of gender-specific persecution of women. Linked to feminist debates around human rights and legal status, the UN Women's Conference raised the argument that rape, domestic violence and abuse could all be grounds for claims to refugee status. Although the UNHCR interpreted 'membership of a particular group' to include women in this light, few claims on this basis have succeeded (Beyani, 1995).   SAR

Regime A prevailing order or political system. The term is often qualified, to refer to a more specific condition. For example, Lipietz (1986) suggests that a 'regime of accumulation' is a sustained period of relative equilibrium or growth within a capitalist economy. Feminists have challenged Lipietz's ideas for trivialising gender (see McDowell, 1991b). They also propose that 'welfare regime' (Lewis, 1992) and 'GENDER REGIME' (Connell, 1987; Walby, 1997) capture the state of play of gender relations within an institution or across space more centrally.   RK
See also GENDER, PATRIARCHY.

Regime of truth See TRUTH/REGIME OF TRUTH

Region/regional geography A region is defined as a distinct, bounded area of the world. The study of regional geography 'organizes the knowledge of all interrelated forms of areal differentiation in individual units of area' (Hartshorne, 1939: 643). The debate between regional geography and SPATIAL SCIENCE revolved around whether regional synthesis was a worthwhile academic study, or whether this chorology – the division of global SPACE into a plethora of unique and unrelated chunks – doomed it to being a pre-scientific endeavour unable to identify general laws. This description of uniqueness could arguably explain nothing of the processes through which different PLACES, spaces and LANDSCAPES are produced and used.

Recently, some geographers have attempted to revive interest in regional geography. Agnew offered a reinvigorated notion of the region, produced not through areal *differentiation*, but through areal *variation*, which stressed connections (and thus extra-regional processes) as well as differences (Agnew, 1989). Entrikin (1991) sought to argue that alternative models of scientific causal explanation, which accepted a 'NARRATIVE' structure of explanation threaded through unique materials, would enable a rescuing of place (and hence also of 'region') from its supposedly descriptive ghetto. The emphasis of POSTMODERNISM on questions of DIFFERENCE (rather than UNIVERSALITY, SAMENESS, LAWS, etc.) has also led some to expect a revival of some form of regional geography (Gregory, 1989).   JS

Regulation School The work of the French Regulation School has sought to theorise transformations in post-war economy and society (Aglietta, 1979; Leborgne and Lipietz, 1990; Lipietz, 1987). Regulationists argue that although there may be periods of history during which the accumulation of CAPITAL proceeds in a relatively steady way, capitalist social relations inevitably fall into periods of crisis. Thus the economic crisis of the early 1970s led to a dismantling of both the FORDIST compromise between capital and labour and the link between wage growth and productivity gains. Declining levels of profitability and productivity made it increasingly difficult for the STATE to provide collective goods such as health, education and housing.

The utility of regulation theory in explaining prevailing arrangements under Fordism and contemporary economic restructuring has been questioned by feminists, who have pointed to the ways in which regulation theory constructs women as a secondary labour force/reserve army of labour, and defines their participation in the labour market through their role in the nuclear FAMILY (McDowell, 1991b). It appears particularly problematic to employ a regulationist perspective to explain the dramatic changes of the current era, when increasing numbers of women have been drawn into both full-time and part-time employment, often as sole household wage earners.

Difficulties in thinking about gender relations stem largely from regulation theory's MARXIST origins – commenting on this genealogy, Lipietz (1993: 99) has stated that the Regulation School was launched by 'rebel (economist) sons of Althusser'. Although Mahon (1991: 127) has suggested that 'the family [of those employing a regulationist approach] now includes some "daughters" too', the Marxist legacy has presented obstacles in thinking about GENDER relations. Despite a suggestion that spheres of production and reproduction (regime of accumulation/mode of regulation) are interconnected, regulationist accounts construct dualisms between the home/consumption and paid work/production. Such a perspective disregards feminist arguments concerning the importance of unwaged work in the HOUSEHOLD to the continuance and social relations of paid labour. Further, regulationists are often insensitive to the ways in which gender constructs SUBJECTS in the workplace (McDowell, 1991b).   SR

Relative/relativism/relativity An EPISTEMOLOGICAL and/or ONTOLOGICAL position which insists upon radical undecidability. Relativists believe that any claims to truth or representation are based purely in LANGUAGE or DISCOURSE. They also posit that there is no position outside of language, which implies that there is no way of evaluating the claims of competing discourses. A moral relativism emerges as it is impossible to judge the values and decisions of one CULTURAL, value or discourse system from another. As a result, relativism has offered a fundamental challenge to the UNIVERSALIST claims to truth or authority made by Western thought, and has also heralded the death of the self-knowing, unified SUBJECT as centre-point to this knowledge.

Relativistic approaches have emerged with the influence of postmodernism and post-structuralism in geography and related disciplines. Although relativism is in some ways allied to feminism in its challenge to any universalist claims by Western knowledge and in its fragmenting of the unified (masculinist) subject, it also challenges the efficacy of feminist POLITICS. Some feminists have suggested that, despite the radical propositions of relativism, it is in actuality a politically conservative position. As Fox-Genovese (1986: 121) has remarked,

> Surely it is no coincidence that the western white male elite proclaimed the death of the subject at precisely the moment at which it might have had to share that status with WOMEN and peoples of other RACES and CLASSES who were beginning to challenge its supremacy.   JS

See also POSTMODERNISM; POST-STRUCTURALISM.

Representation Refers to the ways in which meanings are conveyed or depicted. If GEOGRAPHY means 'writing the world', then representation – in this case 'writing' in its broadest sense – is one of geography's central concerns. Two broad approaches to geographical representation can be identified. The first claims that reality is present in appearance and that geographical observation can produce models and ultimately laws about behaviour in the human as well as the physical worlds. This approach has been associated with positivist models of spatial science and the 'quantitative revolution' and with ideas about the mimetic representation of a transparent, knowable world whereby a mirror can be held up to reflect reality. The second approach challenges the notion that reality is present in appearance and rather suggests that the process of representing the world is intimately bound up with the world that is being represented. In these terms, it is thought to be impossible to reflect reality through MIMETIC representation because the very process of representation is inseparable from the subject that is being represented. Such post-positivist approaches are often associated with qualitative ideas and methods of research which pay more explicit attention to the processes, practices and politics of representation.

Research within feminist geography has been particularly important in addressing the politics of representation and attempting to destabilise the unequal power relations between researcher and researched. Gillian Rose has argued that geographical representations of the world through, for example, FIELDWORK and LANDSCAPE descriptions reflect the masculinist production of geographical knowledge (1993a), and other feminist geographers have proposed different strategies of representation that can disrupt this masculinism. So, for example, feminist geographers have explicitly addressed the politics and ETHICS of representation, the importance of SITUATED and PARTIAL KNOWLEDGES, and alternative visions of 'the field' for geographical research. Work within both cultural and feminist geography has been sensitive to the difficulties of representing people and places in ways that avoid processes of 'OTHERING' or exoticising DIFFERENCE. The recognition of a 'crisis of representation' has been particularly acute in cross-cultural research because of the need to avoid academic 'tourism' or voyeurism (Clifford, 1988). Post-colonial approaches have highlighted the importance but difficulties of representation over space and time. A critical awareness of the politics of representation and the importance of situating knowledge provides one route through the impasse of wanting to represent people and places but the difficulties of doing so.

Alongside an increasing awareness of *strategies* of representation many geographers have begun to study different *forms* of representation. The rise of cultural geography has seen geographers studying a wide range of representations that include visual images, written texts and films (Barnes and Duncan, 1992; Cosgrove and Daniels, 1988; Gregory, 1993). Many feminist geographers and cultural critics have focused on the ways in which such representations are gendered as well as spatial (Bonner *et al.*, 1992). Important work here includes Griselda Pollock's spatial analysis of French Impressionism, in which she contrasts the interior, domestic and bounded spaces portrayed by female artists such as Cassatt and Morisot with the more public spaces of spectacle portrayed by artists such as Monet and Degas (Pollock, 1988). Pollock also discusses the position of the artist and the viewer, examining the spatial dynamics of visual representation in ways that parallel work on feminist film theory. A landmark text in this field by Laura Mulvey explores the different spaces of a masculinist filmic GAZE that exist on NARRATIVE, discursive and psychic levels (Mulvey, 1975). Subsequent work continues to suggest the inherent spatiality of a gendered gaze but has attempted to position female spectators in different ways (Bruno, 1993; Doane, 1982; Hansen, 1991). Feminist analyses of written texts have similarly addressed the spatiality of representation, focusing on different narrative spaces within the text and the POSITIONALITY of readers and writers. Within geography, research on gendered textual representations includes work on TRAVEL writing by women (Blunt, 1994) and juvenile fiction written in different ways for boys and girls (Phillips, 1997).

Over the 1980s and 1990s many geographers have shown a heightened awareness of the importance of a range of strategies and forms of representation and the ways in which such strategies and forms are intimately connected. This interest in representation has often been tied to POST-STRUCTURALIST and POST-COLONIAL approaches that examine discursive formations over space and time. Rather than separate 'representation' and 'reality', such approaches focus on the discursive constitution of the material world. Increasingly, representation is not seen as an unmediated process whereby a transparent reality is made known. Rather, many geographers recognise that processes of representation are inseparable from the world that is being represented. In this way, if geography means 'writing the world', geographers write different worlds in different ways.   AMB
See also TRUTH.

Repression For Freud, repression was a technical term that referred to the mechanisms by which the impulses, memories and painful emotions arising from conflicts between the pleasure and reality principles are thrust into the unconscious where they continue to influence individual EXPERIENCE and behaviour. It is a dynamic concept which treats the psychical apparatus as comprising forces which necessarily enter into conflict with each other. The organised ego is seen as the outcome of the successful control of the PLEASURE principle by the reality principle. At the ontogenetic level, repression takes two forms. First, there is primary repression, which refers to the non-admission at any stage of certain ideas to consciousness. It is this that gives rise to the unconscious itself which then acts in all subsequent repression, pulling elements back from consciousness. Secondary repression takes place on the border between the unconscious and the preconscious, viewed as systems. CATHEXIS (psychical energy) is withdrawn from the idea in the preconscious and it either retains its unconscious cathexis or gains a new burst. Repressed ideas do find their way to consciousness but in a disguised form, through dreams, symptoms and slips of the pen and tongue. What makes this possible is that the unconscious is governed by what is known as the primary process, notably the processes of condensation and displacement.

For Freud what was repressed was not drives themselves but the ideas that became attached to drives. The relation between the organic and the purely ideational elements was left unclear. Lacan emphasised the repression of signifiers, that is, the replacement of one element in the signifying chain by others, which then become elements in the conscious discourse of the subject.

At the phylogenetic (that is, genealogical history) level, Freud used the concept more loosely. He believed that civilisation itself is based on the repression of the sexual instincts and the sublimation of their energy. He also believed that the more complex a society becomes the greater amount of sexual repression is required, hence his famous saying that civilisation is premised on human unhappiness. Civilisation was under constant threat from within, for there was no guarantee that repression would be successful and always a danger that repressed instincts would break out in the forms of VIOLENCE, suicide, war and conflict.

All the critiques of SEXUALITY in capitalist society take as a starting point Freud's writings on repression and civilisation. The Freudo–Marxist tradition associated with such figures as Reich and Marcuse argued that systems of oppression and exploitation were built on repression. Sexual liberation was therefore a necessary component of class struggle. Foucault rejected what he called the 'repression hypothesis' that assumed that capitalist domination had been built on a steady increase of repression, starting in the seventeenth century and reaching its culmination in the nineteenth, and pointed instead to the growing obsession with sexuality and the production of new DISCOURSES and SUBJECTIVITIES. He rejected the idea that to struggle against repression is to strike a blow for freedom against power and argued instead that the repression hypothesis shared the same model of power it was criticising. He did not deny the existence of repression, but saw it as merely one effect of a complex set of mechanisms concerned with the production of discourse, POWER and knowledge.   RP

See also FREUDIAN THEORY; LACANIAN THEORY.

Reproduction See MODES OF PRODUCTION/MODES OF REPRODUCTION

Reserve army of labour See CASUALISATION

Resistance Resistance refers to 'any action, imbued with intent, that attempts to challenge, change, or retain particular societal relations, processes, and/or institutions' (Routledge, 1997: 69). It may involve the actions of individuals or groups (e.g. SOCIAL MOVEMENTS). Orthodox accounts establish resistance against a dominating POWER and most often juxtapose (open and confrontational) resistance to the STATE (in the form of rallies, strikes, uprisings and so on). Resistance can be violent (e.g. armed struggle) or non-violent (e.g. non-cooperation) in form. Although women have often been involved in violent struggle (e.g. wars of independence in Algeria and El Salvador, amongst others), feminist resistances have tended to adopt non-violent means.

Traditional explanations of resistance tend to construct a BINARY of opposing forces – between those of dominating power and those of resistance to it – constructing a central dialectic of opposed forces. However, POLITICS are never so clear-cut and, for example, there is evidence of domination operating within resistance movements (see, for instance, Third World critiques of Eurocentric assumptions in the feminist movement (Mohanty, 1992)). Theorisations of power since Foucault have challenged the binary of power and resistance. For Foucault, power is not held by particular people or groups and used to repress other identities and knowledges. Instead, power is diffused throughout society in a web of relations and is creative as well as repressive. Hence, all have power to act. This shifts attention from resistance as heroic acts of opposition to a dominating power to understanding it first as a form of power in its own right and second as operating throughout the practices of everyday life. In *Weapons of the Weak*, for instance, Scott (1985) argued that even the most powerless and MARGINALISED peoples have the ability to resist: not openly but through minor subversions and irritations that interrupt the easy running of the dominant system (e.g. 'losing' or breaking tools, sabotage or foot-dragging). From a feminist perspective this move represents a shift of attention from the traditional realm of politics – the PUBLIC sphere – to the PRIVATE, and so a valorisation of women's resistance within SPACES traditionally regarded as being outside politics.

Some have, however, pointed to the danger of seeing all acts as of equal importance in this diffuse understanding of resistance. The term 'resistance' loses explanatory utility if all acts could potentially be resistant. Literature influenced by POSTMODERN thought discusses the cultural expressions of resistance and TRANSGRESSION, that, although counter-hegemonic, fall short of open direct confrontations with the dominant and are embedded in the attitudes and lifestyles that exist in ordinary circumstances (see Hall and Jefferson, 1976). Although one must recognise that power is dispersed through society in a web-like structure, it is necessary to remember that power can accumulate in certain sites, so producing an uneven topography of power.

Others have questioned the desirability of theorising resistance, arguing that it represents an 'intellectual taming' (Routledge and Simons, 1995), or that 'to be discursively recognisable, may itself be a tactic which always already concedes too much' (Phelan, quoted in Rose, 1997: 186).

It is possible to see the different types of resistance articulated in each of the 'waves' of feminism. In the first wave, women resisted a PATRIARCHAL STATE which refused them the vote. SECOND WAVE FEMINISTS resisted MASCULINIST society which was seen to offer particular limited roles and possibilities to women. Third wave feminists have acknowledged the greater complexity of GENDER relations, not simply operating around a male–female binary but cross-cut by issues of RACE, CLASS and SEXUALITY. Feminist resistance is now articulated as a much more 'ambivalent' form of politics (Rose, 1991).   JS

Rights/equal rights/equal opportunities These underpin a particular strand of European–American feminism often identified as 'modern LIBERAL FEMINISM'. This form of feminism focuses on the acquisition of formal legal parity with men and dates back to the rise of liberals in the eighteenth century and to the emancipatory struggles for the vote in the nineteenth century. Liberal philosophers such as Locke and Rousseau rejected the assumed 'natural' order of power and authority held by the church and monarchy and introduced a concept of individuals as free and equal beings, EMANCIPATED from hierarchical bonds. 'Rights' became the expectations of individual citizens rather than their duty. Liberalism argued for a society based on a freely agreed and honoured 'social contract'. Mary Wollstonecraft (1792) first identified the social injustice between men and women in this form of CITIZENSHIP: women were excluded from the 'social contract' and therefore denied their status as individuals and access to the public domains of education, politics, PROPERTY ownership and inheritance. Feminist arguments suggested that women's failure to gain access to the social contract derived from pre-existing notions of 'rights' embodied in the 'conjugal and property rights' of men over women. The nineteenth century women's suffrage struggles for the vote and for access to education and employment were based on demands for equality of treatment with men and access to a liberal 'social contract'.

Equal rights is a term used by feminists to identify women's demands for an end to direct and indirect discriminatory practices in, for example, legal structures, workplace organisation and government legislation (see, for example, Peters and Wolper, 1995). Struggles for equal rights dominated the women's movements in Europe and North America in the 1970s, with key anti-discriminatory legislation being put in place (Sex Discrimination Act, Equal Pay Act, Equal Rights Amendment). Although many feminists supported these demands there was scepticism about both the possibility and desirability of equality with men. Cockburn notes that 'Since there is no place for women as women in men's social contract, for women to seek "equal rights" with men is to seek to be surrogate men' (1991: 21). Yet to argue for equality for women *as women* raises questions for many feminists about the universalising category 'WOMAN', which prioritises 'sameness' and denies DIFFERENCES between women (Evans, 1995).

Equal opportunities collectively represents a variety of strategies devised by organisations and workplaces, often through specialist equal opportunities working parties or pressure groups, to facilitate women's access to resources, services and employment. A key aim is ensuring that 'no doors are closed to women that are open to men' (Cockburn, 1991). This assumes that women's past history and present circumstances prevent them from taking up available opportunities or from competing on equal terms with men to take advantage of life chances. Special action is deemed necessary to support women, as a group, to move towards equality. Important action includes paid maternity leave, parental leave, special training programmes, childcare, return-to-work programmes and skills audits as well as policies to protect women from domestic VIOLENCE and to facilitate access to housing and women-centred health care. Such action has been successful in improving some women's employment experiences and expanding some HEALTH, care and housing services. Feminist critics of equal opportunities, such as Hakim (1996), point out the partial results of equal opportunities and its favouring of white middle-class career women.

Feminist geographers, working within this liberal tradition, produce work that documents women's different use of space and their unequal spatial constraints, particularly with respect to childcare and work opportunities (see, for example, Booth *et al.*, 1996; Bowlby, 1990; Little *et al.*, 1988; Tivers, 1984).

Liberalism bequeathed feminism a particular version of equality that was fundamentally bound by the forms of universalising law. Cockburn (1991) suggests, in societies where the law and processes of political representation formally govern social relations, equality for women may be a necessary step in advancing women's interests, but it is not sufficient in itself.   JF

See also BUILT ENVIRONMENT; CITIZENSHIP; COMPARABLE WORTH; ENLIGHTENMENT/ENLIGHTENMENT THEORY; WORK/WORK FORCE/EMPLOYMENT.

Royal Geographical Society See WOMEN AND GEOGRAPHY STUDY GROUP

Rural/rurality The term rural is often counterposed to that of urban in a binary or dualistic construction that emphasises geographical differences between the city and the countryside. However, it is now accepted that the social characteristics of urbanism are not necessarily restricted to those built-up areas that are defined as cities but may also include rural residents who are participants in generalised urban cultures or ways of life. Rural inhabitants may work in a town or city, do their shopping there, purchase national newspapers published in the capital city and have access to the same radio and television channels. And in a nation such as Great Britain, where a tiny proportion of the employed population works in agriculture, clear systematic differences between the CITY and the countryside are increasingly less evident than in most THIRD WORLD societies. However, despite these common features it is clear that location still matters and that rural residence has a common set of features. As David Harvey noted for the city in *Social Justice and the City* (1973), spatial location, as well as income, is associated with differential accessibility and proximity to goods and resources. For less affluent rural residents, the decline of publicly provided services, as well as commercial facilities, in villages has increased their isolation. Rural POVERTY is particularly marked for elderly residents, among whom women are a majority (Cloke and Little, 1990, 1997).

As well as investigating questions about rural living standards and lifestyles, including gender differences, rural geographers have also begun to examine the ways in which an idea of ruralness, or rurality, is a social construct that includes not only material differences between rural and urban areas but a set of ideas, beliefs and symbols that represent images of the countryside (Cloke *et al.*, 1994; Whatmore *et al.*, 1994). As feminist photographer Ingrid Pollard (see for example Kinsman, 1995) has documented, rurality in Great Britain is associated with a particular NOSTALGIC version of Englishness that excludes black British people.

There is also an additional distinctive focus in rural geography: investigations of the agricultural sector and food production, including studies of women's economic activities in the countryside (Whatmore, 1991). However, in this work the rural–urban distinction is increasingly blurred as the work uncovers the networks of connections that link the countryside and the city through patterns of ownership, production, distribution and CONSUMPTION of food and so might be more accurately seen as a specific example of economic geography rather than part of a distinctive rural subdiscipline (Cloke, 1990; Lowe *et al.*, 1994). LM

S

Safer cities A term used to refer to cities in which a variety of PLANNING and/or public transport initiatives have been introduced in response to male VIOLENCE to try to increase women's safety. In the United Kingdom a government Home Office scheme backed such measures in a number of major urban centres. Initiatives introduced under the Safer Cities scheme included 'safe women's transport', additional lighting in car parks and subways and closed-circuit television (CCTV). Safer Cities initiatives have been criticised as concentrating on physical responses to violence against women. JL
See also BUILT ENVIRONMENT; CITY.

Same/sameness Refers to the centred constitution of the 'SELF' against a more marginalised 'OTHER'. The 'same' often appears to be a transparent norm against which the 'other' is marginalised and defined. But, as Edward Said has shown in the context of ORIENTALISM, the 'same' and 'other' are mutually constituted in relational terms (Said, 1978). The power relations between the 'same' and 'other' are unequal and the 'same' is often perceived as transparent and taken for granted against which the 'other' is defined as different. So, for example, masculinity, WHITENESS and the WEST may all be constructed in transparent terms in the context of masculinist, white supremacist and Western power relations. An increasing amount of work has begun to destabilise such notions of the 'same' by revealing the power relations at the heart of its constitution (Hall, C., 1992; Young, R., 1990). At the same time, many critics have been keen to destabilise the spatial as well as the social DICHOTOMY between the 'same' and 'other'. Homi Bhabha imagines a 'third space' beyond the limits of the hegemonic 'same' and its 'other' margins (Bhabha, 1994). Teresa de Lauretis has written about a 'space-off' beyond the coordinates of 'same' and 'other', describing it as

> the elsewhere of discourse here and now, the blind spots, or the space-off, of its representations. I think of it as spaces in the MARGINS of hegemonic discourses, social spaces carved in the interstices of institutions and in the chinks and cracks of the power–knowledge apparati (1987: 25).

In geography, Gillian Rose has written about 'paradoxical space' in an attempt to destabilise the spatial opposition of the 'same' and 'other' (Rose, 1993a). AMB

Scale A technical term, referring to the representation of area. At its most basic, scale is used to interpret CARTOGRAPHIC representations of any area. In recent years, however, the production of scale has been scrutinised as a political–economic process (see Herod, 1997; Smith, 1990; Swyngedouw, 1997). Rather than taking particular scales of social action for granted (such as the nation-state or local wage bargaining) these scholars are pointing to the way in which scale is produced as a result of conflict and struggle. Taking the example of HOMELESSNESS, Smith makes a fourfold case for understanding the production of scale:

> First, that the constructon of geographical scale is a primary means through which spatial differentiation 'takes place'. Second, that an investigation of geographical scale might therefore provide us with a more plausible language of spatial difference. Third, that the construction of scale is a social process, i.e., scale is produced in and through societal activity which, in turn, produces and is produced by geographical structures of social interaction. Fourth, and finally, the production of geographical scale is the site of potentially intense political struggle' (1993: 97).

This approach has allowed geographers to address the ways in which processes operating at different scales relate to each other, and Swyngedouw (1997) has coined the term 'glocalisation' to highlight the simultaneous importance of local and global developments. Doreen Massey's (1994c) work has also been particularly influential in this regard. In unpacking the interconnections between different places and people she argues that a 'global sense of PLACE' illuminates the links between everyday life and various social relations which are 'stretched out' across space, sometimes at global dimensions. Likewise, some feminists have argued for new transnational political connections, uniting women across national boundaries (see Grewal and Caplan, 1994; Mohanty *et al.*, 1991).

In recent years feminist geographers have also highlighted the importance of the BODY as a new scale of analysis, drawing attention to embodiment as a factor in social relations (see Longhurst, 1995, 1997). JW
See also GEOGRAPHY; SPACE.

Science/the science question The orthodox definition of science is that it is the systematic study of the nature and behaviour of the material, physical and social worlds, based on observation, experimentation and measurement. Science promises universal truths: a pure, technical, objective and RATIONAL knowledge free of cultural, political or social bias. However, feminists and POSTMODERNISTS have been critical of the primacy of scientific explanation in modern thought. Now, as Demeritt (1996: 484) has explained, there are two clearly defined and apparently irreconcilable understandings of science: it is both 'celebrated as our guiding light in the wilderness and exorcised as the cause of our expulsion from the garden'. Feminists and postmodernists have argued that far from offering the TRUTH, scientific knowledge is a social construction and so only one of a number of competing explanations for the world and its workings.

Harding (1986) suggests that there are three different feminist critiques of science. The first she identifies as 'feminist empiricism' which sees only 'bad science' as the problem. These LIBERAL critiques have pointed to discrimination in disciplines of science which have hindered the promotion of women. This critique argues that if sexism could be removed from science then scientific knowledge would be more objective.

Harding's second position is 'STANDPOINT THEORY'. Arguing from a Hegelian position, Harding argues that men's dominant position in social life results in partial or perverse understandings, whereas women's subjugated position provides the possibility of more complete, less perverse understandings (Harding, 1986: 26).

Both of these positions imply that OBJECTIVITY cannot be increased by value neutrality. Instead, it is a commitment to 'anti-authoritarian, anti-elitist, participatory and emancipatory values and projects that increase the objectivity of science' (Harding, 1986: 27). The viewpoints of the subjugated are less attached to positions of domination and so offer this greater objectivity.

Feminists have, however, questioned the existence of 'a' feminist standpoint to argue that this position is cross-cut with issues of RACE, CLASS and ETHNICITY. Haraway (1988: 584) goes further to suggest that the 'standpoints of the subjugated are not "innocent" positions' and have to be contextualised just as do the dominant positions of the scientists. In these critiques it has been argued that questions of gender are inherent to scientific DISCOURSE. Like other theorists of the sociology of science, feminists have been interested in what is counted as science and why, as well as in the exploration of the ways in which scientific discourses have been influential in constructing rather than reflecting BINARY distinctions between bodies and ideas about male and female biological characteristics. Science is based upon the same binary logic that can be found throughout modern Western ENLIGHTENMENT thought. Merchant (1980) and others have examined the masculinist/heterosexual erotic language of science, especially notions of seduction and rape. For example, Francis Bacon, the 'father of the experimental sciences' argued that the scientist should 'make no scruples of entering and penetrating into these holes and corners, when the inquisition of truth is his whole object ... [because] nature betrays her secrets more fully [when subdued]' (quoted in Demeritt, 1996: 488).

Other feminists have criticised the masculinist GAZE of the disembodied scientist who 'has excluded himself from "NATURE"' (Haraway, 1989: 159). Here the mind–body dualism reappears, setting the male scientist outside of what he is seeing. This critique has been levelled at the geographical gaze on the LANDSCAPE or at objectifications of social life in CARTOGRAPHY (see Rose, 1993a). Others have looked to the inherent drive of science to control nature.

Merchant (1980) has charted the rise of modern science and the modern scientist as Priest. She argues the culture–nature dichotomy parallels and reinforces the dichotomy of man–woman and suggests that there are important connections between the oppression and domination of women and the exploitation of nature.

Harding's third perspective, feminist postmodernism, challenges the premises of science more fundamentally. Postmodern positions challenge the very existence of such Enlightenment concepts as reason, progress, emancipation and science, arguing that these are only some amongst many possible stories of reality. Alternative explanations are silenced by scientific ones which have great POWER as totalising explanations. Postmodern feminists argue that as feminism seeks to contextualise knowledge it is not compatible with scientific reasoning.

Harding poses 'the science question' as whether there should be an attempt to produce a feminist science or whether feminists should persist with their critiques of scientific knowledge. Haraway (1988: 585) argues that science 'has been utopian and visionary from the start; that is one reason "we" need it'. Haraway argues against rejecting science in favour of a relativism of subjective positions, instead insisting on the importance of maintaining a sense of objectivity, not from a godlike position that sees all, but from the viewpoints of the subjugated.

There have been a number of attempts to assert that geography is a science, most notably by SPATIAL SCIENTISTS. Others have questioned the possibility of using science to study social phenomena, even arguing that geography needs to be viewed as an art (Meinig, 1983) and so must adopt techniques and theories from the humanities (Duncan and Ley, 1993).   JS

Second wave feminism/second wave women's movement These terms are used to refer to the resurgence of political activism and feminist writing from the late 1960s onwards in the USA and Great Britain. It was termed the second wave to distinguish it from turn-of-the-twentieth-century 'first-wave' struggles for the vote (Banks, 1981). Some of the initial political campaigns of the second wave were related to the BODY and to women's rights over their own bodies. One of the earliest demonstrations in the USA, for example, was the 'No More Miss America' demonstration which was the demonstration that earned feminists press coverage as bra-burners. In fact, as Susan Bordo recalls,

> 'no bras were burned at the demonstration, although there was a huge 'Freedom Trash Can' into which were thrown bras, along with girdles, curlers, false eyelashes, wigs, copies of the *Ladies' Home Journal*, *Cosmopolitan*, *Family Circle* etc. The media, sensationalizing the event, and also no doubt influenced by the paradigm of draft-card burning as the act of political resistance par excellence, misreported or invented the burning of bras. It stuck like crazy glue to the popular imagination (1993: 19–20).

A similar media image dogged early feminist events in the United Kingdom too, events such as the demonstration against Miss World in 1971. As feminists claimed that 'THE PERSONAL IS POLITICAL', campaigns around issues such as pregnancy, ABORTION and maternity leave became part of the body politics of the early years. An important publication was the Boston Women's Health Collective self-help manual for women, which was published in both Britain and the USA. Other campaigns in these years included the Working Women's Charter Campaign, which began in 1971 in Britain, the same year as the first women's liberation conference, held in Oxford.

One of the key aspects of the women's movement in these early years was its non-hierarchical organisation, and its insistence that politics also involved 'the problems of everyday life, the problems about which women talk most to other women', as Sheila Rowbotham, Lynn Segal and Hilary Wainwright (1979: 13) noted in their assesment of the first decade of the second wave. During this decade, small groups of women, often meeting in each other's homes, and sometimes referred to as consciousness-raising groups, were an important part of the growth of feminist organising. In the next decade, however, anything approaching a formal movement began to fragment, as BLACK FEMINISTS and lesbian feminists criticised the movement for its implicit biases, often choosing to organise separately rather than to change the women's liberation movement from within. The dominance of SOCIALIST FEMINISTS also led to antagonisms with RADICAL FEMINISTS and others more interested in single-issue campaigns, such as the GREENHAM COMMON women's campaign against US nuclear bases in Britain or women interested in cultural issues rather than economic questions.

By the beginning of the 1990s a new wave of younger women either rejected the label feminist altogether or challenged what they saw as the puritanical, or dowdy, image of that generation of women who had been active in the late 1960s and early 1970s, emphasising the power and energy of what they saw as a 'new feminism' (see, for example, Roiphe, 1994; Walter, 1998; Wolf, 1994). For the former group, whose political training had been in left-wing or Marxist organisations, the individualistic politics of the 1990s seemed antithetical to feminist demands (Oakley and Mitchell, 1997). The theoretical shift over the same decades also emphasised the differences between women rather than their commonalties and, at times, seemed to whip the rug from under FEMINISM as a movement depends on the identification of joint interests. However, the continuing inequalities between the status of women and men, albeit in many different ways in different parts of the world, give continuing reason for feminist organising. In 1996, for example, the United Nations Fourth World Conference on Women was held in Beijing, with delegates from almost every nation in attendance.   LM
See also FEMINISM; SUFFRAGE.

Second World See WORLD: FIRST, SECOND, THIRD, FOURTH

Segregation: social, spatial, occupational Social segregation refers to the presence of social divisions between different groups within a population, whereas spatial segregation is used to characterise the patterning of social segregation across SPACE. Geographers' considerations of both social and spatial segregation date back (at least) to the work of the Chicago School sociologists, who drew upon processes of plant ecology (including notions of 'invasion' and 'succession') to model human activity within the city. As a result, social segregation generally has been used with reference to urban residential segregation, with an early focus upon the 'ghettoisation' of ethnic minority groups within the city. The legacy of the Chicago School's highly problematic attitudes towards 'racial' assimilation has been a strong one; however, more recent accounts have been concerned to understand the intersections between segregation and racist and discriminatory practices within labour and housing markets (see Anderson, K., 1991; Western, 1993). Further, it is now recognised that both social and spatial segregation emerge from processes and practices of inclusion and exclusion (see Sibley, 1995a).

Occupational segregation refers to the clustering of particular groups of workers within different occupations. It has been used most frequently to describe the enduring tendency for women to work in different occupations from men, although it is also important to consider processes which divide workers of different ETHNICITIES (see Hiebert, 1997).

Feminists often distinguish between horizontal segregation (the extent to which men and women are concentrated in different jobs) and vertical segregation (which refers to differences within occupations in respect of pay, skill, status, promotion prospects, etc.). Interestingly, debate about the reasons for continuing occupational segregation can be seen as being implicitly geographical in that arguments revolve around where the causes of occupational segregation are located: in the home and within the family, or through workplace-based practices (Hanson and Pratt, 1995: 4–8).   SR
See also COMPARABLE WORTH; LABOUR: DOMESTIC, SWEATED, WAGED, GREEN, SURPLUS; WORK/WORK FORCE/EMPLOYMENT.

Self Refers to that which is intrinsic to a person, one's nature, character or physical constitution or appearance, considered as different from some other. Used in this way the concept is associated with the ENLIGHTENMENT view of the SUBJECT as the RATIONAL, autonomous individual who pre-exists in society as an embodiment of some universal. As a consequence of classifying phenomena as natural and 'normal', it has been suggested that dichotomies have been created between subject and object, self and other and that these binary opposites have been offered in dominant theories of knowledge and identity (Haraway, 1989, 1991). These DUALISMS, together with their spatial counterpart PUBLIC–PRIVATE, have become an important structuring principle upon which characteristics commonly associated with MASCULINITY and FEMININITY are arrayed and forms of POWER exercised. Thus public reason has been associated with a supposedly neutral observer whereas those marked by variations from the posited 'norm' derived from sex, skin colour or other differences have failed to qualify for full participation in the liberal democratic model.

Critiques of this rationalist position emphasise the idea of self as a social construction which cannot exist without society. Early twentieth-century theorists and social psychologists interpreted self-identity as a creative and reflexive process formed through social experience involving reflection and reaction from others and interpreted through the lattice of self-perception. Varied formulations of 'self' which explicitly bring together mind and body, imagined and 'real' have also emerged within PSYCHOANALYSIS (Pile and Thrift, 1995). The geographical encounter with these related disciplines, and with POST-STRUCTURALISM in particular, has given rise to the concept of the geographical self whereby human subjectivity is made in spatio-temporal relation to human and non-human others. Pile (1996) surveys early geographical theorisations of the self. Like Sibley (1995a), he suggests that psychoanalysis and the workings of the unconscious are important in shaping the thoughts, feelings and actions which define, maintain and reproduce the relationships between self and other.

Within a rapidly growing literature on the geographical self three strands can be identified (Matless, 1997): one which addresses the formation of self within the context of broad spatial relations, a second which focuses on issues of conduct and a third which is concerned with the decentred self. Examples include how gendered IDENTITIES have been and are shaped within specific places or sites by the local internalisation of senses of the NATIONAL, IMPERIAL and GLOBAL (Blunt and Rose, 1994). Formations of the geographical self through conduct which is conformist or transgressive have been linked to the role of institutionalised geography in encouraging particular modes of social and environmental behaviour associated with the production of gendered identities and the practices of CITIZENSHIP (Bell and McEwan, 1996; Maddrell, 1997). Equally, the shaping of the geographical self through movement has found expression in the gendering of TRAVEL and exploration (Blunt, 1994; Mills, 1991). Feminist concerns with the decentring of the human subject as opposed to the centred human have found common ground with recent interest in non-human subjects and ecological relations. Here the deconstruction of the human arises in part from implicating the formation of the human self in non-human ecological processes (Soper, 1996).   MB

See also NATION/NATION-STATE/NATIONALISM; NATURE; OTHER/OTHERNESS.

Semiotics Originating in structural linguistics, semiotics is the study of culture as a system of signs and symbols (or, more generally, as a process of signification). The linguistic roots of semiotics are based in quite abstract and formal distinctions between signifier and signified, for example, or between *langue* and *parole* (written and spoken language). Through the work of Roland Barthes and Jacques Derrida, the scope of semiotic analysis (or semiology) has been expanded to cover the production and interpretation of signifying systems in art and architecture, ANTHROPOLOGY and literary analysis.

Beginning with theories of LANDSCAPE iconography (Cosgrove and Daniels, 1988), geographers have applied linguistic and visual metaphors to the interpretation of 'landscape as text' and as a 'way of seeing'. The idea of CULTURE as a signifying system has been employed in the field of urban semiotics, approaching the CITY as a system of meaningful signs and symbols (Gottdiener and Lagopoulos, 1986). The intertextual links between different modes of REPRESENTATION (landscape painting, gardening, photography, cartography, etc.) have also been explored (Daniels, 1993), drawing on literary notions of discourse, metaphor and TEXT (Barnes and Duncan, 1992).

While geographical research on processes of signification has been focused on different ways of reading the landscape, linking modes of representation to their wider social formations, it has been criticised by feminist geographers for failing to interrogate the extent to which landscape painting and related modes of representation embody a masculinist way of seeing (Rose, 1993b). As a result, feminist geographers have been particularly interested in the work of feminist artists and critics who have attempted to subvert the 'masculinist GAZE' (Nash, 1996b; Rose, 1995c). PJ

Sense of place See PLACE

Separatism Given the recognition of the regulations of TIME and SPACE which have often separated the social worlds of men and women, feminism has been particularly concerned to break down the exclusions between men's and women's worlds, challenging such distinctions as 'PUBLIC' and 'PRIVATE' and increasing access for women to male-defined institutions. However, there has also been a recognition of the value of separate spaces, not as sites of confinement for women but as spaces for empowerment and the building of strong female relationships, exemplified in consciousness-raising groups or women-only social events. Separatism was most strongly advocated by some RADICAL FEMINISTS and lesbian feminists, who argued that 'COMPULSORY HETEROSEXUALITY' (Rich, 1980) and the workings of heteropatriarchal society were at the root of all women's oppression. Separation from men and from all the institutions, activities, roles and relationships of patriarchal society was seen as one possible solution (Frye, 1983), ranging from political separatism to complete separatism. The latter was explored in experiments in separatist communal living such as the 'lesbian lands' established in the 1970s in the USA (Valentine, 1997).

Separatism has also been debated amongst groups of feminist geographers, most notably in the organisation of projects within groups such as the WOMEN AND GEOGRAPHY STUDY GROUP (WGSG) of the Royal Geographical Society/Institute of British Geographers (WGSG, 1997: 3; Silk, 1995). While recognising the dangers of exclusions and ESSENTIALIST identity politics, feminist geographers have acknowledged some of the benefits of developing separate spaces for the development of female solidarity and empowerment. CD

Sex/sexuality Sex has a complex relation to at least three other terms: WOMAN, GENDER and sexuality. The Victorians had a habit of referring to women as 'the sex' in a way that conflated all these terms. As Simone de Beauvoir observed, 'she is simply what man decrees; thus she is called "the sex", by which is meant that she appears essentially to the male as a sexual being. For him she is sex – absolute sex no less' (1949: 16).

For much of the twentieth century 'sex' was used to describe sexual difference in a way that took for granted that social and psychological differences flowed naturally from the facts of biology. Little distinction was made between MALE–FEMALE, man–woman or MASCULINE–FEMININE. The emerging schools of sexology and psychoanalysis tended to treat sex as a powerful physical drive which had to be tamed by social forces. The term 'sexuality' emerged in the late nineteenth century to refer to patterns of behaviour associated with sexual expression and has steadily taken on broader meanings relating to REPRESENTATION, IDENTITY and DESIRE. While 'sexuality' was recognised to have diverse expressions it was assumed to be based on LIBIDO or 'sex drive' and hydraulic images abound. 'Sex' was also seen as a term linking sexuality and gender in the sense that being a 'real' man or woman implied an engagement, or at least a capacity to engage with, normative HETEROSEXUAL relations. Homosexuality was often recast as the 'masculine' in one partner reaching out to the 'feminine' in the other.

Since the 1970s there have been numerous attempts to disaggregate these terms. GAY theorists (see Plummer, 1981) challenged the assumption that gender and sexuality are closely linked and insisted that the two should be treated separately. As gender and sexuality came to be more carefully theorised the links of both of them to the concept of sex were also problematised. Feminists challenged the use of the term 'sex' as a blanket category to refer to men and women and reduced it to a biological marker of sexual difference and distinguished it sharply from the socio-cultural meanings attaching to gender. It was argued that, far from following 'naturally' from biological differences, gender differences involved an exaggeration of those differences and imposed a simplistic binary on a population whose psychological characteristics and social capacities are in reality much more varied. Sex role theory played down the importance of biological differences and emphasised the importance of socialisation in the creation of men and women.

In the area of sexuality, the emergence of labelling theory, social scripting theory and symbolic interactionism also problematised the notion of a powerful 'sex drive' existing outside the social, and argued instead that 'sex' involves a socially constructed set of meanings and behaviours for which the appropriate scripts have to be learned. In an influential work Gagnon and Simon (1974) specifically referred to 'conduct' rather than behaviour to indicate the importance of the social, not just in shaping a powerful and pre-existing drive but also in actually constituting it. Ken Plummer (1975) similarly emphasised the wider socio-historical formations which generate the identities which people assume and specifically highlighted the importance of stigmatic labelling in creating the separate worlds of sexual deviance. Both works stressed that sexual development should not be seen as something relentlessly unwinding from within but as something constantly shaped through encounters with significant others.

These ideas were extended by those who have emphasised the discursive construction of both sex and sexuality. Foucault (1978) demonstrated how DISCOURSES on sexuality came to be of central importance in the control and regulation of populations. Sex, he believed, is a historical construct rather than a universally natural property or biological given. Thus he challenged the traditional notion of sex as an instinctual drive or force whose expression might be intrinsically liberating. Not only did he question the validity of Freudo–Marxist assumptions about sexual repression but also he claimed those arguments were formulated within the same discursive and strategic limitations as the power they want to attack. Far from sexuality being repressed, new sexualities are constantly being produced through discourses which are played out directly on the bodies. 'Sexuality' is set up as the key to identity, and people actively seek to live up to the 'norms' thus created in order to produce their 'true' selves. In this account 'sex' is no longer perceived as the biological category underpinning sexuality. On the contrary, 'sex' is itself produced as an apparently natural force by the machinery of sexuality. In the most famous example, it has been pointed out that relations between women were unable to be seen as sexual at all until the discursive category of LESBIANISM became available.

Feminist thinking about the sex–gender relation has followed a similar path. Feminists drew heavily on the work of Robert Stoller (1968), an analyst who in theorising TRANSSEXUALISM first drew the distinction between sex as a biological component and gender as the social element in sexual difference. The separation of sex and gender was important in the early days of the women's movement because it served to make clear that 'sex roles' and psychological characteristics do not automatically arise from the 'facts' of sexual difference and do not necessarily take a binary form. But in the 1980s 'the sex–gender distinction' came under attack for treating the relation between the two as arbitrary and for ignoring the importance of the lived experience of the body. Moira Gatens (1996) argued that the sex–gender division mirrors the BODY–mind split and that sexual difference was reduced to mental conditioning. But biological differences are not immutable and at very least there needs to be an interactive relationship between sex and gender. What is downgraded in our culture is not 'femininity' but femaleness. Pateman stressed that 'men do not exercise power as, or over a "gender" but over embodied women. Men ... exercise power as a sex, and wield sexual power' (1988a: 402). While it is acceptable for men to incorporate elements of 'femininity' into their identities and even achieve 'genius' status for so doing, masculinity is not socially valued if it is attached to female bodies. The body politic was seen to take the male body as the norm and allow women to participate only at the price of denying their sexual specificity and behaving as 'honorary men'. The emphasis on changing gender ideology fails to address the ways in which women are either reduced to bodies (in ways that echo de Beauvoir's observation) or denied bodily specificity. As a result of this many feminists have been less enamoured of 'gender' and returned to the term 'sex' (e.g. Probyn, 1993). This should not be interpreted as a return to essentialism, for sex is not being used here to refer to BIOLOGY but to the importance of bodily difference. In this view, bodies have a meaning only to the extent that they are already cultural products, and biological science is part of discourse rather than of the world of 'nature' that it purports to describe.   RP

Sexism Discrimination against women on the grounds of sex; a summary term that refers to masculinist assumptions of female inferiority, sometimes also referred to as male CHAUVINISM, as well as to discriminatory actions.   LM

Sex tourism Predominantly refers to people travelling to purchase sex. It is mostly young women, many driven by economic desperation to enter prostitution or sold into it as children, who are the 'objects' of the sex trade. Many feminists argue that central to sex tourism is the deliberate objectification of women by governments and the private sector, with its success depending on normative masculinist and racialised notions of sexuality (Brock and Thistlethwaite, 1996; Enloe, 1989, 1993; Truong, 1990). As countries such as Thailand, South Korea, the Philippines, Indonesia and Sri Lanka rely ever more heavily on the tourist trade for economic development, the offering of sex tours to international businessmen from Japan, Western Europe, North America, the Middle East, China and Australia has become a key way of attracting foreign currency (see Seager's, 1997 map.) Many destinations of sex tourism grew out of 'rest and recreation' sites for First World militaries in developing countries, though sex tours in North America and Eastern Europe are also on the increase. While some may argue for the benefits of PROSTITUTION to women (Chapkis, 1997), the global sex trade is considered largely coercive and a crime against women and children by many international feminist organisations (e.g. End Child Prostitution in Asian Tourism, and International Federation of Women's Travel Organisations). These groups are working to address the economic and health risks associated with sex tourism, such as the threat of sexually transmitted diseases and HIV and AIDS

A large network of local and international commercial, entertainment and governmental entities directly or indirectly support profitable trade in sex beyond female and child prostitution, however, from gay–lesbian holiday package tours to sex tours on the Internet.   KM

Sexual abuse See ABUSE; VIOLENCE

Sexuality See SEX/SEXUALITY

Sexual violence See ABUSE; RAPE; VIOLENCE

Shopping Often subsumed within wider debates about CONSUMPTION (Jackson and Thrift, 1995), shopping is a more specific term encompassing all of the activities that go on before, during and after the exchange of money for goods. Shopping involves processes of appraisal, comparison, purchase and use – or sometimes (as in the case of 'window shopping') only some of these processes. In most cases, the process of shopping stretches back in time before the moment of exchange and often extends beyond the point of sale into cycles of use and re-use as consumers incorporate goods into their everyday lives (Mackay, 1997).

Inspired by Walter Benjamin's analysis of nineteenth-century shopping arcades (Buck–Morss, 1989), much of the geographical literature on shopping has focused on the history of the department store and on North-American-style 'mega-malls' (particularly the West Edmonton Mall in Canada). These studies have tended to emphasise the recreational nature of 'lifestyle shopping' (Shields, 1992), where consumers actively manipulate their IDENTITIES through the purchase of particular commodities, rather than the more mundane world of the high street and the supermarket. In all of these contexts, however, shopping remains a highly gendered activity, performed predominantly by women within the domestic context of the household and strongly influenced by ideologies of the HETEROSEXUAL nuclear FAMILY (Miller *et al.*, 1998).

Feminist research on shopping has focused on demonstrating the unacknowledged LABOUR involved in this skilled social accomplishment (often dismissed as a 'labour of love'). As well as highlighting practical issues, such as improving access for women with small children, feminists have drawn on literary and historical sources to demonstrate the masculinist nature of the consumer's objectifying gaze (Bowlby, 1985). Feminists have also explored the hitherto neglected spaces of the car boot sale and the thrift or charity shop (Gregson and Crewe, 1994), 'second-hand worlds' that are just as strongly marked by traditional gender roles and ideologies as the more fully researched world of the high street and the mall (Morris, 1988).

Recent research on shopping includes empirically grounded studies of 'the shopping experience' (Falk and Campbell, 1997) and theoretically informed work within the 'new retail geography' (Wrigley and Lowe, 1996).   PJ
See also CITY; FLÂNUER/FLÂNEUSE.

Sisterhood The language of sisterhood was important for 1970s feminism, particularly radical feminism, evident in the common slogan 'sisterhood is powerful' (Morgan, 1970). Suggesting that women were universally oppressed by male POWER, the language of sisterhood emphasised that differences between women were less important than the common cause of women against men. It is clear that dissent around this theoretical position and the issue of differences between women was raised in the early period of RADICAL FEMINISM (Evans, 1995: 17) but was most strongly expressed in the writings of BLACK FEMINISTS in the early 1980s, who argued that 'it is very important that white women in the women's movement examine the ways in which racism excludes many black women and prevents them from unconditionally aligning themselves with white women' (Carby, 1982: 232). Such critiques were taken up more broadly as feminists recognised the need to address the DIFFERENCES between women and were also supported by POST-STRUCTURALISTS' critique of overarching theories and the recognition of multiple discourses and identities. Although the language of sisterhood is seen as outdated for contemporary feminists, 'sisterhood' should be acknowledged as a term which was important in mobilising women and encouraging solidarity. It can be seen as a term which is not necessarily about an ESSENTIALISED definition of the category 'woman' but as a political project of alliances which might be forged across differences (hooks, 1986). One example of this is the women's group Southall Black Sisters (SBS, 1989) a campaigning group against domestic VIOLENCE which strongly promotes links between women across different 'ETHNIC' and 'cultural' backgrounds. CD

See also SECOND WAVE FEMINISM/SECOND WAVE WOMEN'S MOVEMENT.

Situated knowledge Associated primarily with Donna Haraway's (1991) revisioning of a critical feminist objectivity, situated knowledge refers to a form of EPISTEMOLOGY attuned to the profoundly social, RELATIONAL and contextual underpinnings enabling all knowledge production in diverse global situations. As a non-masculinist and yet also non-idealist understanding of epistemology, Haraway argues that

> Situated knowledges require that the object of knowledge be pictured as an actor and agent, not a screen or a ground or a resource, never finally as slave to the master that closes off the dialectic in his unique agency and authorship of 'objective' knowledge' (1991: 198).

The resulting sensitivity to EPISTEMIC VIOLENCE (i.e. the ways in which universalistic and godlike knowledge claims have been implicated in relations of domination) reflects the critical implications of this account, and it is these implications that Haraway's historicisation of primatology has documented in such devastating detail (1989). Nevertheless, her argument about the situatedness of knowledge production is equally illustrative of CRITICAL THEORY'S anticipatory and socially transformative moments. Indeed, while her account of primatology examines how it was previously shaped by profoundly sadistic as well as RACIST and PATRIARCHAL familial ideologies, she also shows how feminist researchers have been able to rework and reconstruct new, non-violent approaches to primate research. Haraway thus insists that: 'We need the power of modern critical theories of how bodies and meanings get made, not in order to deny meanings and bodies, but in order to live in meanings and bodies that have a chance for a future' (1991: 187). For feminist geographers these arguments have underpinned the project of reflexive revisioning and reconstruction of previously patriarchal methodologies such as FIELD-WORK. More than this, they also point to the significance of other geographical relations in which popular as well as elite and academic knowledge formation is situated, and some of these relations, including the 'PUBLIC and

PRIVATE' (Marston, 1990; Sparke, 1996; Staeheli, 1994) and 'work and home' (Pratt and Hanson, 1995), are already being unpacked by geographers concerned with producing situated feminist knowledge. MS

Social construction of reality According to a social constructionist view, gender is not biologically determined but is achieved through a process of socialisation into gendered roles and expectations. These norms, far from being rigidly determinate or narrowly defined, are usually quite fluid and open to degrees of resistance and negotiation by groups and by individuals. The wide individual variation in hormone types and levels, the existence of TRANSSEXUALITY and the fact that some biological influence and limitations can be overridden by social, cultural and technological factors indicate that SEX as well as GENDER can be seen as socially and historically produced. To the extent that gendered behaviour (gendered practices and PERFORMANCES of identity) is socially produced, it is also shaped by cross-cutting influences of RACE, CLASS, generation, ETHNICITY and other relations of POWER. Although there is broad agreement among many feminists on the idea of social construction, the exact meaning of the social construction of reality and its compatibility with materialist perspectives has been an issue of debate. Certainly a belief in social construction need not entail an idealist epistemology and ontology as some have assumed. Thus it can be argued that social construction and materialism can be quite compatible. Several closely related versions of the thesis serve as useful starting points: symbolic interactionism, Berger and Luckman's social constructionism, critical or scientific REALISM, Gramscian and other non-reductionist Marxisms and Giddens's structuration theory. Each of these theories posits a mutually constitutive or dialectical relation between social, economic and political structures and individual agency. While they tend to place different degrees of emphasis on STRUCTURES or AGENCY, all these perspectives see the relative balance as an open question to be investigated empirically. ND

Socialism Socialism as a set of writings, ideas and practices stresses ideals of equality and justice, usually on the basis of common ownership of the means of production. Particular forms include utopian socialism, democratic socialism practised in many parliamentary industrial states and communism (founded on Marxism–Leninism) (Smith, G.E., 1994). Early socialist writings addressed the so-called 'woman question': Karl Marx declared 'the status of women provides a barometer for the level of humanity attained in a society'; August Bebel wrote, 'There can be no EMANCIPATION of humanity without the social independence and EQUALITY of the sexes' (cited in Einhorn, 1993: 17). The emancipation of women from the HOME and entry into paid employment represented a common cause between socialist and feminist campaigners. However, the balance between the relative significance of PATRIARCHAL and CAPITALIST structures of oppression has been subject to constant debate, played out, for example, in the women's sections in trade union movements and socialist political organisations.

In the form of socialism implemented across Eastern Europe and the USSR (communism, or state socialism) considerable attention was given to the full involvement of women in society through employment. High levels of female labour force participation were achieved not with part-time employment (as in many Western countries) but with full-time employment (Bridger *et al.*, 1996). In recognition of the difficulties which women faced in combining home and WORK, states provided widespread CHILDCARE, legislated for equality of the sexes and in many cases improved access both to family planning and to benefits for families with children. However, the conditions under which such participation was achieved remained difficult for many women, particularly in situations of poor access to consumer goods and shortages of even basic goods (Corrin, 1992).

Although the achievements for women and the effects these had on changing gender identities must be recognised (Attwood, 1990), both internal and external critiques emerged which argued that STATE socialism, while of emancipatory intent, was fundamentally a patriarchal ideology. It was the state which declared women and men equal. However, women's 'emancipation' was based on a dual definition of women as workers and MOTHERS and on an unequal involvement of women in the labour force. State social measures designed to facilitate the combination of domestic and work roles were aimed at women rather than men and reinforced gendered definitions of domestic and child-rearing work as 'female' in a situation where many household tasks required hours of additional work (Bridger *et al.*, 1996; Einhorn, 1993). Women's employment was often highly concentrated in less-well-paid sectors and grades of employment while senior school pupils, for example, were often guided into highly gender-specific forms of future employment (Winkler, 1990). This 'double burden' (Corrin, 1992) coincided with exhortations for women and men to engage in social and political work beyond the home (Tóth, 1993). Participation in formal representation was much higher than in most Western states, for example, but again the conditions of entry to such spheres were far from equal, with few women in senior posts.

Such experiences evoked a variety of responses. In the mid to late 1980s, for example, some otherwise 'reformist' moves towards liberalisation and openness incorporated neo-conservative 'solutions' to the multiple demands on women:

> it was precisely because women suffered wage discrimination and bad conditions and because child care, health care and public services were all inadequate that [...] 'we must rethink the stereotypes that have developed and realise that, for the future of the country and of socialism, the most important form of creative work for women is the work of motherhood' (cited in Bridger *et al.*, 1996: 24).

Many women apparently desired such an 'escape' from multiple burdens and particularly looked forward to more Western-style consumerism as a solution to their difficulties. However, many women were sharply aware of their own capabilities and preferred to challenge the unequal implementation of these policies. They formed women's grass-roots organisations which contributed to the mass movements of the late 1980s and feminist movements in POST-SOCIALISM. (Hampele, 1993). FS

Socialist feminism Founded on the belief that the oppression of women is intimitately connected to CAPITALISM. While acknowledging the importance of PATRIARCHY, socialist feminists argue that in capitalist society GENDER divisions are used and upheld in the interests of the ruling class. In particular, it is argued that the oppression of women ensures that they are a source of cheap and flexible LABOUR and that women's domestic labour services male workers and secures the next generation of workers. Socialist feminists tend to subscribe to either 'dual systems theory' or 'single or unified systems theory' in explaining the position of women. Dual systems theory suggests that capitalism and patriarchy are two separate systems that interact with each other whereas single systems theory suggests that capitalism and patriarchy are inseparable (see Tong, 1989; Walby, 1990).

Socialist feminists have thus been concerned with the articulation of class and gender relations and, in contrast to Marxist feminism, they have not prioritised one set of cleavages over the other (see Foord and Gregson, 1986; Gregson and Foord, 1987; McDowell, 1986). Moreover, socialist feminists attach considerable importance to organising working-class women into trade unions, using class as a mechanism to bring women together to highlight gender inequality in the workplace. Issues that are of particular importance include equal pay, maternity rights, employment benefits and access to promotion (see TRADE UNIONS).

During the 1980s feminist geography was strongly influenced by socialist feminism, and scholars looked at the way in which capitalism and urbanism intersected with gender (see McDowell and Massey, 1984; Massey, 1983; Nelson, 1986). More recently, however, feminist geographers have been attracted to POST-STRUCTURAL perspectives that explore the social construction of gender, and the gendered nature of theory and the academy.   JW
See also RADICAL GEOGRAPHY.

Social movements Social movements were popularised by Castells's (1978) theorisation of the CITY as a spatial unit of COLLECTIVE CONSUMPTION (by which he was referring to STATE provision of housing, transportation, education, recreational facilities and so on). He claimed tensions and contradictions surrounding the uneven provision of these goods and services led to urban struggles, which transformed into urban social movements when they had the ability to promote structural change. A necessary condition for this was the linking of issues of collective consumption to those of PRODUCTION and the wider working-CLASS movement. He has been criticised for positing too mechanistic a link between crises in state provision and urban struggles, for supposing that crises in the realm of collective consumption would necessarily lead to anti-capital alliances (Short, 1993) and for his lack of consideration of the gendered nature of such movements (Fincher and McQuillen, 1989; Mackenzie, 1989). Castells (1983) has since revised his rather economistic notion of urban social change (rooted in French political experience) to incorporate the realms of POLITICAL self-management and cultural IDENTITY as integral to urban EXPERIENCE and mobilisation.

Urban social movements are apparently being displaced (in terms of analytical interests at least) by new LOCAL–GLOBAL modes of communication collectively referred to as new social movements. These comprise networks of collective actors who have common interests and a common identity, with their prime source of POWER being the threat of mobilisation and the ability to open up new arenas for political DISCOURSE, primarily within, although not restricted to, the sphere of CIVIL SOCIETY (Wekerle and Peake, 1996). Painter (1995) claims that in geography, outside consideration of urban social and nationalist movements, there has been little regard for new social movements, but critical geographers are currently expanding their horizons in this area of inquiry (see Meono-Picado, 1997; Peake and Trotz, 1998).

Encompassing both progressive and regressive POLITICS new social movements rally around the rights of WOMEN, blacks, people of colour, LESBIANS and GAYS and cover issues of peace, DISABILITY, religious funda-mentalism, ethnic chauvinism, the ENVIRONMENT, ecology and animal rights. Mouffe (1992) sees them as movements in which SUBJECT positions previously considered apolitical have given rise to new political subjects and spaces of mobilisation. Pringle and Watson (1992) claim that because Mouffe (like Foucault) treats GENDER IDENTITY solely as a subject position within a discourse then women's movements are only one among many new social movements. The extent to which feminists agree that alliances between new social movements can work in women's interests hinges on the degree to which they allow for the celebration of private differences in public while not constituting a rainbow coalition in which already constituted differences prevent a redefining of common interests. Undoubtedly, the ability of these movements to engage in struggles over the production of new meanings and collective identities, their potential to create linkages, however *ad hoc*, disrupt hierarchies and BOUNDARIES and create their own (confusing) local–global spaces is shaping the politics of the twenty-first century.   LP
See also ECOFEMINISM; GREENHAM COMMON.

Space Although one of the central concepts of human geography, space is notoriously hard to define and has been central to many of the most heated debates in the discipline's history. Space has been theorised in abstract, material, relational, metaphorical and imaginary forms.

Questions of space have not always represented a self-conscious focus for geography. REGIONAL GEOGRAPHY had an inherent understanding of space as a patchwork of distinct assemblages of material phenomena ('areal variation' or 'areal differentiation') but did not study space *per se* (the classic work of regional geography is Hartshorne, 1939). SPATIAL SCIENTISTS, on the other hand, examined the role of space as a key variable in shaping the organisa-tion of human activity. From an examination of spatial patterns, generalisa-tions and predictions (laws) could be formulated about the geometry of social life. However, it can be argued that despite their apparent central focus upon space, spatial scientists actually reduced space to a transparent and homogenous backdrop to their neo-liberal calculations of friction of distance, location and so on. Soja pointed out that this reduction of society to geometric spatial relations, and the assumption that spatial patterns were evidence of spatial processes, represented a crude 'spatial fetishism' (Soja, 1980). This has further implications for geographical knowledge: space here is an arena within which objects interact under the watchful eye of the spatial scientist. Rather than understanding the observer as caught up in the processes under investigation, 'he' is sovereign and detached (see SUBJECT/SUBJECTIVITY; GAZE; VISION/VISUAL/VISUALITY).

Marxist geography introduced an interest in space as part of society rather than as a passive backdrop to it (what Soja (1980) termed the 'sociospatial dialectic'). Harvey (1973) regarded the spatial organisation of society, particularly of the city, to be fundamental to the workings of capitalism (a process he named historicogeographical materialism). Lefebvre (1991) argued that space was produced through three different processes which he named 'spatial practice' (which embraces the 'production and reproduction, and the particular locations and spatial sets characteristic of each social formation' (1991: 33)), 'representations of space' (the abstract space of state plans, economic analysis and so on) and 'representational spaces' (those everyday representations of spaces 'linked to the clandestine or underground side of social life' as articulated in cultural products such as art and poetry, often produced as resistance to representations of abstract space (1991: 33)). The concentration on capitalism has led figures such as Harvey and Soja to consider CLASS to be the main components of a person's access to spatial resources (e.g. their ability to traverse space or own land). Massey (1991b) and others (e.g. Kaplan, 1987) have suggested that other aspects of subjectivity such as 'race' and gender also influence a person's ability to traverse space.

A number of geographers have suggested that in modern social and political thought space has been equated with stasis and reaction in contrast to time, which was seen to drive social process and transformation. However, the rise of feminist, POST-COLONIAL, POST-STRUCTURAL and POSTMODERN epistemologies has led to critiques of universalising NARRATIVES grounded in temporal processes and instead to an increased interest in space as a set of concepts embodying notions of DIFFERENCE. The popularity of the work of Michel Foucault, both within geography and beyond, reinforces this interest in the centrality of space to understanding the organisation of society and POWER relations. Some geographers have been critical of this recent interest in space, especially by theorists working outside the discipline, arguing that the engagement is with spatial METAPHORS, which renders space innocent and ignores the relations of power which always run through material (what might be considered as 'real') spaces. Smith and Katz (1993: 75) argue that spatial metaphors are problematic, 'in so far as they presume that space is not'.

FEMINIST GEOGRAPHERS have focused upon the ways in which space is gendered. There have been three main feminist interventions in understandings of space and society. First, geographers studied the 'geography of women', basically considering how women utilise space differently from men. The second type of engagement between feminism and geography involved an examination of the ways in which gender relations are expressed in space. Many have argued that the division of social life into PUBLIC and PRIVATE spheres is replicated in space. Public space, the space of transcendence, production, POLITICS and power, is the sphere of men, whereas private, domestic space, the space of reproduction, is women's space. Linda McDowell (1983) has shown how the development of the modern city has reinforced this division through suburbanisation which has increased the distance between public and private. Women in public space are often regarded as 'out of place', especially when this is at night: women are sometimes held to be partially culpable when attacked at night because of their failure to realise that their being in public space after dark (especially if wearing 'revealing' clothing) was tempting attack (see Pain, 1991; Valentine, 1989). Women and girls are socialised into believing that such spaces are dangerous to them, and as such should be avoided. Studies of non-Western societies have found similar results (see Spain, 1992; but see Sharp, 1996). Political geographers have also demonstrated the replication of this gendering of space at the STATE and international level.

The third intervention in the theorisation of space by feminists has been more fundamental in that feminists have suggested that the very conceptualisations of space entertained by geographers are MASCULINIST. The drive to demarcate spatial relations and spatial identities has been argued as a masculinist practice at odds with a feminist definition of space as fluid, uncontrollable and, ultimately, unknowable (Rose, 1993a, 1996a). Rose, (1996a) suggests that the common insistence that only material social space is real, constructs a distinction between real and non-real space. Through a masculinist practice of power this distinction is created as a HIERARCHY, which is gendered: the real space is 'simultaneously concrete and dynamic', signifying the masculine, whereas non-real space is 'simultaneously fluid and imprisoning, but always engendered as feminine' (Rose, 1996a: 59). JS
See also GEOPOLITICS; SPACES OF EXCLUSION; SPATIALITY.

Spaces of exclusion Exclusion occurs as a result of processes in which a powerful human grouping more or less consciously seeks to distance itself from other less powerful groupings (e.g. white people distancing themselves from BLACK people, men distancing themselves from women). The processes, which lead to exclusion can be helpfully conceived of as both social and spatial processes; hence the phrase 'spaces of exclusion'. Feminist writers have long been concerned with exclusion in relation to many aspects of women's lives; but particularly in relation to male-dominated public spaces which often result in women's exclusion in private spaces such as the home.

Recently, theories of exclusion have become arguably more sensitive to issues of difference and broader in their understanding of who excludes, what can be excluded and how this happens. Sibley (1995a) has been instrumental in bringing feminist and PSYCHOANALYTIC theories about the self to bear upon considerations of social and spatial exclusion. Sibley has argued, using Kristeva's (1982) writings about processes of abjection, that exclusion can be conceived of as manifestations of psychic and bodily rejections of the 'OTHER'. This rejection is learned within Western society in childhood through (among other things) the literal distancing from excreta and the subsequent symbolic separations of purity and impurity. This basis of (adult) social rejection is realised in space by the erection of boundaries and BORDERS (both symbolic and material) which stereotype and separate objects and peoples:

> stereotypes play an important part in the configuration of social space because of the importance of distantiation in the behaviour of social groups, that is, distancing from others who are represented negatively, and because of the way in which group images and place images combine to create LANDSCAPES of exclusion (Sibley, 1995a: 14).

In this way we can see how writers such as Sibley spatialise the exclusion endured by different groups of people such as gypsies, ethnic minorities and those with mental and physical impairments.   HP
See also BOUNDARY; MARGINS.

Spaces of representation/representational space Terms used by the Marxist philosopher Henri Lefebvre to refer to SPACE as it is directly lived, embodying images, symbols and imaginary elements. In his book *The Production of Space* he suggests that spaces of representation are linked to 'the clandestine or underground side of social life', to the space of 'inhabitants' and 'users' as well as the critical work of artists and some writers and philosophers (1991: 33, 39; the English translation refers to the term as 'representational spaces'). This space is subjected but at the same time it is a realm of counterspaces and spatial representations that contest dominant practices and spatialities. Lefebvre relates spaces of representation to two other moments: spatial practices, which refer to the routines, sites and interactions through which social life is produced and reproduced; and representations of space, which refer to space as it is conceived by scientists, planners, urbanists and others and which have a regulatory and dominating role being 'tied to the relations of production and to the "order" which those relations impose' (1991: 33). Together the concepts are dialectically interlinked, forming a triad that has been employed by a number of theorists addressing the social constitution of space–time (Harvey, 1989) or 'social spatialisation' (Shields, 1991). But Lefebvre attaches particular political importance to spaces of representation in struggles against the colonisation of everyday life and capitalist 'abstract space'. This sense of a political project gives the notion of spaces of representation much of its current critical edge as suggested by recent accounts of spatial politics and artistic practices drawing upon Lefebvre's work (Deutsche, 1997) and as explicitly asserted by Edward Soja (1996), who highlights the radical openness of the concept and suggests that it approximates what he defines as third space.   DP

Spatial division of labour See DIVISION OF LABOUR: DOMESTIC, INTERNATIONAL, GENDER, SPATIAL

Spatiality The social production and meaning of SPACE. Recognising this sociality of space and spatial relationships resists the tendency of spatial science to reduce space to geometric relations (but see POWER GEOMETRY) and draws attention to the fact that social life necessarily happens in certain spaces and PLACES. This reinforces the idea that space is thoroughly embedded in social relations rather than being a mere backdrop to them. Edward Soja (1980, 1985, 1989) has termed the inseparability of space and society the 'sociospatial dialectic'. MARXIST GEOGRAPHERS, for example, argue that different modes of production involve different spatialities (see Harvey, 1982; Massey, 1984).

For Soja (1989: 132) spatiality also has existential origins, in the distinction between 'being-in-itself (the being of non-conscious reality, of inanimate objects, of things) and being-for-itself, the being of conscious human persons'. To be human, then, does not simply involve quantifiable spatial details such as distances and areas but also their transformation 'through intentionality, EMOTION, involvement, attachment' (Soja, 1989: 133).

Gregson and Lowe (1995: 225), however, argue that there is an over-identification of spatiality with the production of time–space (especially of CAPITALIST industrial production systems). They wish to extend the use of the term to examine other forms of spatiality such as that of the 'day-to-day social reproduction' within the HOME.   JS

Spatial science A branch of the discipline developed in the 1950s and 1960s that sought to describe and explain the spatial organisation of society by formulating and testing scientific laws (Berry and Marble, 1965). The use of scientific methods allowed geography to be reconceptualised as a discipline in distance (Watson, 1955) and to claim a place in the academic division of labour. Spatial science emerged during the quantitative revolution, draws on the philosophy of POSITIVISM and is associated with the development of specialisms within the discipline.

The main emphasis was on formulating spatial models with predictive powers. Examples include studies of migration (Wolpert, 1967), locational decisions (Getis, 1963) and land-use patterns (Garrison and Marble, 1957). Work in this tradition stressed the applied nature of geographical research and the capacity of a common scientific method to draw together physical and human geography. According to Nystuen (1965) these spatial studies were united by their emphasis on three core geographical phenomena: direction, distance and connection.

In the 1970s spatial science was criticised for prioritising theoretical models of artificial space that bore little relation to the 'real' world (Balchin, 1972). More recently, critiques have focused on its implicit geometrical determinism and trivialisation of human AGENCY (Gregory, 1981). Feminists have added weight to these critiques by pointing to the androcentrism of spatial science (Monk and Hanson, 1982).   HC

Standpoint theory One of the difficulties in defining standpoint theory is that the term 'standpoint' groups together a number of different feminist epistemologies which, while they all share a perspective which privileges women's 'ways of knowing' and recognise the gendered construction of knowledge, vary in how this is theorised. Thus, while some versions of feminist standpoint theory have been criticised for being ESSENTIALIST more recent versions have developed a less universalist and more relational concept of standpoint theory.

The starting point for standpoint theory is a critique of the 'RATIONAL' disembodied knowledge of the ENLIGHTENMENT, a 'view from nowhere' which is in fact a masculinst discourse (Lloyd, 1984). Standpoint theorists argue that what has counted as knowledge in modern, Western cultures is tested only against a limited or distorted social experience – that of men. If the experiences arising from the activities assigned to women, analysed through feminist theory, are used then a more complete and less distorted starting point can be developed. Indeed, it is argued that the view of women as an excluded and subjugated group is more 'inclusive and critically coherent than that of the masculine group' (Di Stefano, 1990: 74).

This position is exemplified in the work of Hartsock (1987) who develops what she defines as a feminist materialist standpoint. Hartsock argues that because of the structual differences between the positions of men and women 'the lived realities of women's lives are profoundly different from those of men' (1987: 158). Women have a deeper or more rooted view of the world because of their relationship to the material social relations of reproduction and childrearing. Hartsock draws on the work of OBJECT RELATIONS THEORISTS such as Chodorow (1978), Gilligan (1982) and Dinnerstein (1976) who argue that women's identity is constructed relationally rather than, as in the case of men, through the separation of self from others.

Harding (1986, 1991) has also been influential in developing feminist standpoint theory. Working within a critique of modern SCIENCE, she argued that the value of standpoint theory is that it both challenges the implicit bias within 'rationalist' science and offers an alternative, and preferable, perspective:

> In claiming that inquiry from the standpoint of women (or the feminist standpoint) can overcome the partiality and distortion of the dominant androcentric/bourgeois/Western sciences, it directly undermines the point-of-viewlessness of objectivism while refusing the relativism of interpretationism (1990: 97).

Yet Harding also acknowledges the dangers of universalising knowledge and essentialised identities inherent within standpoint theory and argues that the construction of knowledge must recognise differences between women. She suggests that a 'political and epistemological solidarity' is required within which the possibilities of the 'permenanent partiality of a feminist point of view' (1986: 193) can be explored. Harding's ideas are developed in Donna Haraway's (1988, 1991) theorisation of 'SITUATED KNOWLEDGES' which she defines as a 'doctrine of embodied objectivity' or a 'feminist objectivity'. Recognising that all knowledge is embodied, Haraway is nevertheless wary of the dangers of romanticising or essentialising the viewpoints of the oppressed. Instead she argues that all standpoints are partial perspectives, offering situated accounts. However, the viewpoints of the less powerful are of particular interest because they are less attached to axes of domination.

While few feminist geographers might locate themselves as standpoint theorists, partly perhaps because of its links with RADICAL FEMINISM, McDowell (1993a) suggests some of the possibilities of standpoint theory for geographers and outlines areas of feminist geography which she sees as taking a standpoint epistemology.   CD

State The state is a territorially embedded, institutionalised series of discursive networks of POWER relations, usually centralised and dominant yet contested. Through various organisational forms – such as welfarist, corporatist, regulationist – and its monopoly of the legitimate use of violence over a specific TERRITORY – it attempts to exercise sovereignty over citizens. Thus it includes such institutions as the government, armed forces, civil service and local councils. In the late twentieth century the NATION-STATE is its most common form, rendering recent pronouncements about its withering away rather premature, although city-states, such as Singapore, still appear on the world map. The state's spatiality is manifest in various ways: it shares frontiers with other states, it can have numerous subdivisions, extreme regional differences and sometimes non-contiguous boundaries between the state and the nation, plus the state apparatus for monitoring and governing the population has to be spatially dispersed (see LOCAL STATE) (Johnston, 1993). Geographers have traditionally studied spatial phenomena such as state BOUNDARIES whereas state theories have traditionally been aspatial, but the two have been brought together through the concept of uneven development (Smith, 1990) and the understanding that all states operate in particular cultural contexts.

Different components of the state can have different interests, making it difficult to identify the state as a monolithic unity with a single purpose and set of interests. Nevertheless, and despite many categorisations, its functions have generally been divided between those of legitimation and accumulation (of capital), including regulation that governs the social relations of production. Beliefs about the degree to which the state is autonomous from classes and interests vary between theorists of different political persuasions (see PLURALISM and MARXISM).

The state received a great deal of attention in the 1970s both from feminists and from Marxists. Feminists, though, have held apparently contradictory views on the centrality of the state in supporting, constituting and legitimating women's SUBORDINATION. In some places, such as Australia and Scandinavia, feminist analyses have focused on how WOMEN have been able to develop state FEMINISM, manifested, for example, through femocrats, equal opportunities, affirmative action and ministries for women, although others have seen these interventions as increasing the threat of co-option which is never too far away from even the most successful women's organisation. In other places, such as Britain and Canada, the emphasis has been on the ways in which state policies have maintained women's dependence, although whether the major beneficiaries of such practices are capital – through control of the REPRODUCTION of LABOUR power manifested through both the sanctioning of MOTHERHOOD and population control policies – or men – through the state's regulation of male VIOLENCE, SEXUALITY, procreation and MARRIAGE – has been a major point of contention between socialist and radical feminists (Barrett, 1980).

Women's subordination by the state is also being questioned by postmodern and POST-STRUCTURALIST feminists such as Pringle and Watson (1992) who claim the state and feminist POLITICAL strategies need to be reviewed in light of the rejection of abstract categories and overarching deterministic frameworks (which have endorsed the notionview of the state as inherently masculinist and the arbitrator of public PATRIARCHY) (Walby, 1990). The diversity among states has also led to post-colonial feminists arguing that much Western feminist theorising of the state has ignored the experiences of THIRD WORLD women (Rai, 1996). Given the enormous material and ideological resources under its control it appears the state will continue to be a major site of day-to-day struggles into the next century, not only in terms of economistic concerns but also for those who do not conform to dominant notions of SEXUALITY and CITIZENSHIP, that is, around issues that contain disciplinary and gendered subtexts. LP

Stonewall A New York City gay bar raided by police in June 1969, while patrons mourned the death of camp ICON Judy Garland; resulting street demonstrations are now commemorated as Gay Pride observances; often regarded (largely erroneously) as the beginning of the contemporary gay–lesbian civil rights movement in the West. The name was adopted by a gay rights organisation in Britain in the 1980s. LK

Stranger danger The notion that violent crime occurs largely in PUBLIC places between strangers. The moral panic of 'stranger danger' reflects a tendency to identify and locate risk, crime and criminals as social and spatial 'others'. Most often associated with fear of crime amongst women, children and older people, who, statistically, are all at far greater risk of violence and abuse in the home from people well known (the 'spatial paradox'). RHP
See also CRIME; FEAR OF CRIME; VIOLENCE.

Strategic essentialism See ESSENTIALISM

Structural adjustment Structural adjustment involves the reorganisation of an economy in order to promote export-led growth. Since the 'THIRD WORLD' debt crisis of the 1980s IMF (International Monetary Fund) credit has largely been conditional upon the implementation of Structural Adjustment Packages (SAPs). While these measures were designed to ensure that 'poor countries' paid debts owed to Western banks, many of the characteristics of SAPs mirror neo-liberal policies in the 'First World' (e.g. widespread privatisation, cutbacks in the public sector and the general 'rolling back of the state').

Much controversy about structural adjustment has focused on the social costs of SAPs and their impact on women's daily lives (Elson, 1991). Not only are SAPs linked to the feminisation of poverty but also specific groups of women bear the brunt of adjustment when economic crisis forces them to participate in diverse survival strategies. Consequently, a triple burden of community management is added to double roles in PRODUCTION and REPRODUCTION (Moser, 1993a). As Radcliffe and Westwood (1993) have shown for Latin America, some of these survival strategies evolve into SOCIAL MOVEMENTS while others remain short-term responses to crisis. Moser (1993b) argues that 'Third World' women are mistakenly seen as endlessly flexible because policy-makers assume women's unpaid activities in households and communities will provide cushions for economic adjustment. Diane Elson has shown how the conceptual tools underpinning structural adjustment reflect 'male bias' (Elson, 1995). Such feminist critiques have helped to discredit the shock adjustment tactics of the late 1980s, while promoting calls for 'development with a human face'. NL

Structuralism A widely used term in the social sciences and humanities referring to explanations and analyses that focus on the abstract or formal structures that are the invisible logics underpinning social organisation and language. Thus Marxism is a structuralist perspective in which the underlying structures that determine the operation of different modes of production may be theorised by intellectual effort but whose existence is neither visible nor measurable, although the effects are empirically evident. Thus the appropriation of surplus value is a structural concept whereas the exploitation and emiseration of the working class may be measured through empirical social research methods such as social surveys. The abstract formal structuralism of the French marxist theorist Louis Althusser had an impact in urban studies and economic geography in the 1970s, especially in the work of Manuel Castells on cities. In a more general sense the turn towards structuralism was an important part of the challenge to the POSITIVISM of SPATIAL SCIENCE at that time. In British geography, REALISM became a more significant influence than the formal structuralism of Althusser throughout the 1980s.

Although it was predominantly the political economy versions of structuralist analysis that influenced feminist geographers in Britain in the 1970s and 1980s – for example, discussions of PATRIARCHY and CAPITALISM in the dual and single systems debate – a structuralist approach is also important in other fields. It has been significant in the development of linguistic theory, for example, where analysts such as Ferdinand de Saussure and Noam Chomsky have isolated the formal rules of LANGUAGE, and here geographers James and Nancy Duncan (1988) drew on Saussure in their exploration of a textual approach to LANDSCAPE analysis. Structuralism is also particularly significant in French philosophy where, for example, the work of Roland Barthes in literary theory, Claude Levi-Strauss in ANTHROPOLOGY and Jean Piaget in psychology are important parts of twentieth-century intellectual history. These versions of structuralism have had an impact on more recent feminist analysis, especially in the turn towards gendered symbols and representations rather than material social relations. However, the POST-STRUCTURALISM of Foucault in particular but also the ideas about DECONSTRUCTION in the work of such theorists as Derrida have largely replaced purely structuralist analyses as these approaches seem more in tune with the contemporary geographical focus on difference and unevenness and on context-dependent analyses of thought and action.   LM

Subaltern Subaltern studies emerged out of Indian historiography and represented a turn to consider the significance of dominated groups in shaping history – a history from the bottom up (see Guha and Spivak, 1988). Moving beyond a historiography which charted the colonised and coloniser as the key social forces, subaltern studies looked instead at the dynamics of DOMINATION and OPPRESSION within both COLONIAL and POST-COLONIAL societies in terms of a wider diversity of exploitative and dominating relations and elite groups. Although emanating from a strong movement in post-independence India, subaltern studies has been associated with the social history movement and Marxism and has had substantial impacts around the world (e.g. Bozzoli and Delius, 1990). Gayatri Spivak, a major figure in post-colonial and POST-STRUCTURALIST feminism, has also been associated with subaltern studies. The focus of attention has been on the agency and creativity of the subaltern in responding to situations in which they are relatively powerless (see, for example, Scott, 1986) and in recovering the actions and voice of the subaltern in history. Spivak (1988a) has pointed to the difficulties in recovering women's VOICES in history, the exemplary case for her being that of widow's self-immolation, sati. Others are more optimistic about the possibilities of recovering subaltern women's voices from the historical record. Women's position as subaltern points to the importance of recognising that many different forms of power relations shape positions of subalternity, and that a subaltern position is often in a contradictory relation to POWER, as subaltern men, for example, may be in a position of power in relation to some women. JR
See also AUTHORITY.

Subject/subjectivity 'The subject' is a term routinely used to refer to the individual human being/agent, accenting both physical embodiment and the range of EMOTIONAL–mental processes through which it thinks its place in the world. All forms of geographical thinking presuppose some theory of subjectivity. Recent debates over the nature of the subject have taken a number of forms, but perhaps the most significant has been that between humanists and post-(or anti)humanist positions, the latter being associated with STRUCTURAL and POST-STRUCTURAL thinking. The humanist tradition in Western thought imagines the subject as being a self-knowing, bounded and unique individual. The subject is contained within the BODY, which captures it and demarcates it as separate from what is outside and OTHER (Pile and Thrift, 1995: 44). This humanist position sees 'man' to be at the heart of the world and the basis for knowledge (this privileging of the unitary gendered subject has attracted critique from feminists).

Alcoff (1988) suggests that it was FREUD'S initial problematising of the coherent subject that allowed later rejections of it. PSYCHOANALYTIC conceptions of the unconscious undermined the subject from any position of certainty. Now that the subject is no longer regarded as entirely self-aware and self-fashioning, it becomes necessary to take seriously the many external forces and processes which construct the subject in everyday life. Feminists have used psychoanalysis to challenge the gendered subject to reveal 'the fictional nature of the sexual category to which every human subject is none the less assigned' (Rose, J., in Alcoff, 1988: 430).

Structuralist thought similarly questions the AUTHENTICITY and ability of individuals to act and think independently and self-consciously. Althusser (1971) suggested that the idea of internalised subjectivity is an IDEOLOGICAL construct which is created in CAPITALIST regimes through 'ideological state apparatuses', most importantly the school and the family. Such institutions teach humans to understand themselves as subjects of dominant ideologies.

Post-structural theorists also reject the notion of the Cartesian (centred, knowing) subject. To post-structuralism, the 'self-contained, AUTHENTIC subject conceived by humanism to be discoverable below a veneer of cultural and ideological overlay is in reality a construct of that very humanist discourse' (Alcoff, 1988: 415). Perhaps most familiar and influential to feminist theories of the subject is the work of Michel Foucault (see especially Foucault, 1977, 1978). For him, subjectivity is an epiphemonenon of DISCOURSE: there is no ontological SELF, but rather a sense of selfhood is an effect of discourse, and a location within networks of POWER–knowledge. In his earlier work on the subject, represented in his *Discipline and Punish* (1977), Foucault focuses on attempts to 'produce' docile subjects through spaces of disciplining and surveillance (primarily schools, hospitals and prisons). In later works, such as the *History of Sexuality* volumes, he studies the 'technologies of the self' through which individuals are taught to adopt – and importantly, to want to adopt – certain forms of subjectivity (e.g. sexual identification). For Foucault there is no subject prior to knowledge, power and discourse, instead subjectivity is a product of these. For example, discourses about HOMOSEXUALITY as an IDENTITY pre-exist homosexual subjects, rather than vice versa. As a result, there can be no knowable subject which exists outside of history (and GEOGRAPHY): always the subject must be historicised. Foucault proclaimed the 'death of man', by which he meant that emerging trends of Western thought at the end of the twentieth century (including post-structuralism), have de-invested 'man' as a stable category. Now that it has been decentred, the subject cannot serve as an absolute ground of knowledge, as a trascendental historical figure or as an ultimate justification for MORAL theories.

This has implications for politics. Whereas for such theorists as Hegel and Marx subjectivity was a necessary aspect of achieving freedom, for Foucault subjectivity represented enslavement to self-identity.

Although very influential, some feminists have drawn limits to the collaboration that a feminist politics can have with Foucault. His theorisation of subjectivity is considered by some as being too passive. As Linda Alcoff suggests, Foucault ignores the fact that, some of the time, 'thinking of ourselves as subjects can have, and has had, positive effects contributing to our ability effectively to RESIST structures of DOMINATION' (Alcoff, 1990: 73). Other feminists have reacted more powerfully to postmodern and post-structural pronouncements of the 'death of the subject', wondering whether this had occurred just when the male, white, subject might have had to share its status with those formerly excluded from subjectivity (see Fox-Genovese, 1986; Mascia-Less *et al.*, 1989). Alcoff has argued that the post-structural position on the subject reproduces a similar effect to humanism: humanists argued that issues of 'RACE', CLASS and GENDER were irrelevant to questions of truth and justice because underneath we are all the same. Post-structuralists argue that 'race', class and gender are constructs and so have no relevance to questions of truth and justice, and so once again, underneath we are all the same (Alcoff, 1988: 420–1).

Feminist film theorist Teresa de Lauretis has gone beyond LANGUAGE and discourse as the sole source and locus of meaning to offer a more active notion of subjectivity. She regards habits and cultural practices as crucial to the construction of meaning and argues that gendered subjectivity is a construct which is formulated in a non-arbitrary way through a matrix of habits, practices and discourses. This allows the articulation of a gendered subject which is fluid rather than being pinned down. Further, the female subject is a site of DIFFERENCE, 'an identity fractured by the gap between the conscious and the unconscious, between language and DESIRE, between discourse and its excess' (Rose, 1995c: 333). De Lauretis argues that 'subjectivity' is central to feminist politics in two senses (Rose, 1995c: 332): first, as an analysis of how women are subject(ed) to masculinist definitions of femininity and, second, as a search for women's resistance to this subjection and the establishment of subjectivity on women's own terms.

In a similar vein, Judith Butler foregrounds the PERFORMATIVE, arguing that if there is no 'being' behind identity, then the 'doer' is merely a fiction added to the deed, and 'the deed is everything' (1990: 25). For Butler the performance of identity establishes the fiction of a subject who is performing. This allows a possibility for POLITICS and for transformation, for if there is always a compulsion to repeat, 'repetition never fully accomplishes identity' (1990: 24). Furthermore, identity is only secure when performed 'correctly', and, as Butler's work has demonstrated, this offers a great potential for subversion. However, a question arises as to who is directing the performance if there is no subject: the notion of agency is diffuse.

The theorisations of subjectivity by the likes of Foucault, de Lauretis and Butler have been central to feminist geographies of the subject. In different ways each considers SPACE to be central to subjectivity, most importantly in resisting the humanist image of the master subject as offering a 'view of everywhere from nowhere which hopes to construct a transparent space in which the whole world is visible and knowable' (Rose, 1995c: 335). Metaphors of MARGINS (hooks, 1991), decentring and fragmentation suggest an adoption of an alternative understanding of the subject which allows for the creation of 'SITUATED KNOWLEDGE' (Haraway, 1988).

Importantly for geographers, the concept of performance requires an understanding of space: it is important *where* subjectivity is performed as this influences the outcome and meaning of the performance and so the meaning and effect of subject identity (a point noted by Goffman in his 1969 work on the 'presentation of self in everyday life'). The spatiality of subjectivity offers a key issue for FEMINIST GEOGRAPHY.   JS

See also MASQUERADE; SOCIAL CONSTRUCTION OF REALITY.

Subjugation Subjugation, applying both to people and to places, implies being conquered, being brought into a state of subservience or submission. While SUBJECTIVITY implies the possibility of agency, action and authorship, their putative opposite, subjection, has been analytically passed over in favour of the more popular term disempowered. Subjugation has seen a revival of interest, though, through the work of Foucault on subjugated knowledges, which he defines as knowledges that have been, 'disqualified as inadequate to their task' (Bell, 1993: 89, quoting Foucault). In feminist circles Haraway (1991) has further problematised the notion of discourses of RESISTANCE, revealing that there is no one standpoint of the subjugated; knowledge is not transcendent but partial and located, what Haraway (1988) terms 'situated knowledge'. Feminist geographers' recognition of the contextual embeddedness of knowledge has further revealed the geographic nature of positionality (see Jones *et al.*, 1997; Rose, 1997a).   LP

Subordination A general term associated with being of inferior importance or submissiveness to authority or discipline. GENDER subordination has been defined by Pearson *et al.*, (1981: x) as a term 'which conveys the general POLITICAL character of male–female relations, while reserving concepts such as PATRIARCHY and EXPLOITATION for historically specific forms these relations may take'. Studies of the underlying causal forces of women's subordination have focused on the HOUSEHOLD, the LABOUR MARKET, SEXUALITY, and the STATE with an emphasis on women's WORK and the sexual DIVISION OF LABOUR. There has also been a recognition of ideological aspects of women's subordination such as the enduring encoding of NATURE as the (heterosexual) FEMALE BODY (Merchant, 1995).

In late CAPITALISM theorisations of the relation between subordination and domination have become less programmatic and increasingly spatialised. Elizabeth Grosz claims,

> The subordinated are implicated in power relations even if they are not directly complicit in them: they are implicated in the sense that, as a mobile set of force relations, power requires the structural positions of subordination, not as the outside or limit of its effectivity, but as its internal condition, the 'hinge' on which it pivots' (1995: 210).

The subordination of women is now increasingly realised not through structural analyses of 'objective' interests but in terms of their common subject positions, or what Pringle and Watson (1992: 68) refer to as their 'discursive marginality'. New understandings and forms of subordination have also met with new forms of resistance such as new SOCIAL MOVEMENTS.   LP

Suffrage/suffrage campaigns These are general terms that refer to campaigns for women's formal EQUALITY with men, especially the right to vote. In Britain, a distinction was made in the early twentieth-century movement between women who were prepared to use violent means to achieve CITIZENSHIP rights – these women are termed suffragettes – and supporters of suffrage, who condemned violence. This distinction took a personal form in the renowned Pankhurst family where Emmeline Pankhurst and her daughter Christabel launched a militant suffragette campaign in 1905, whereas her second daughter Sylvia remained wedded to the peaceful aims of the Women's Social and Political Union that her mother had founded in 1903. Early campaigns for the vote are sometimes referred to as the first wave women's movement to distinguish them from the resurgence of women's political activism in the second part of the twentieth century.   LM

Surplus labour See LABOUR: DOMESTIC, SWEATED, WAGED, GREEN, SURPLUS

Surplus value The capital extracted by employers when they employ workers. The difference between the value of the goods produced by any worker and the wages paid to them amounts to surplus value. Surplus value thus goes towards the profit made by any employer. Although traditionally applied to blue-collar manufacturing work, there is no reason why the same arguments cannot be applied to service employment. Feminists have used the term to refer to the unpaid DOMESTIC LABOUR performed by women for men (see Delphy and Leonard, 1992).   JW

Surveillance A term from Foucault's work about POWER in modern societies which refers to the ways in which individuals internalise the dominant relations or norms of a particular discursive regime through the action of what Foucault termed capillary power or biopower in contrast to power that is imposed from above. The term is also used to refer to the increasing introduction of technologies such as closed-circuit television (CCTV) in the surveillance of public and quasi-public spaces such as streets and town centres, shopping malls and leisure centres.   LM
See also DESIRE; PLEASURE; POWER.

Sweated labour See LABOUR: DOMESTIC, SWEATED, WAGED, GREEN, SURPLUS

Symbolic order See LACANIAN THEORY

Symbols, symbolism Drawing on anthropological work about the cultural significance of material artefacts and cosmological symbolism, geographers have had a long-term interest in interpreting the signs and symbols of the material LANDSCAPE, including the iconography of CITIES in different cultural contexts. While geographers such as Paul Wheatley (1971) have approached Chinese and West African cities as systems of culturally meaningful signs and symbols, others have attempted to relate landscape symbolism (in textual, visual and cartographic forms of representation, for example) to their underlying social formations (Cosgrove, 1985b). Feminist geographers and other social scientists have extended these ideas to include the way that PATRIARCHAL social relations are embodied in landscape symbolism, criticising earlier work for its neglect of GENDER relations and inequalities (Rose, 1993b; Spain, 1992).   PJ

T

Territory Power is implicit in the notion of territory, which describes a portion of land controlled by some form of authority. Rights to territory can be defended by force, and the desire to expand territory can be justified as a natural entitlement. The SCALE varies from a nation–state to an individual's personal space, each contributing to senses of identity derived from shared space.

Gendered DISCOURSE is used to assert control over territory at the scale of the NATION–STATE. Colonisers have frequently represented conquered territories as feminine (Nandy, 1983). In reasserting independence in post-colonial eras, territories may be redefined as masculine. Catherine Nash (1993, 1996a) shows that Gaelic nationalist reclaiming of Irish territory from Britain in the early twentieth century was accompanied by constructions of hypermasculinity and hyperfemininity.

Territories may simultaneously be identified at a variety of local scales, including workplaces, leisure spaces, neighbourhoods and streets. Phil Cohen (1993) argues that territorialisation 'privatises' public spaces by making them exclusive preserves, and may also extend domestic order to outside locations. Historically, spatial dominance has been exercised by adolescent and adult males, but claims are now being made by women and girls. These local forms of territoriality are associated with class and ethnic subordination which restrict movement away from home.

At night city streets become masculine territory in which women are 'out of place' (Valentine, 1989). Saunderson (1997) showed that although Belfast women she interviewed identified strongly with the city, all felt in danger of sexual attack at night, although only 20 per cent feared sectarian violence.

BW

Text/textual/textuality/intertextuality A text refers to the spaces and cultural practices of signification, such as written words on a page, maps, paintings, landscapes and social institutions. This inclusive definition of text reflects critical hermeneutical and post-structural approaches that reject naive REALISM (see Eagleton, 1983). Because of the discipline's history as a descriptive science, human geographers uncritically used a naive realist view of LANGUAGE until roughly the mid-1980s, assuming that words correlated exactly to the objects and ideas they signify (Barnes and Gregory, 1997). More recently, scholars have questioned the production of geographical texts as mimetic ways of representing 'reality' (Barnes and Duncan, 1992; Cosgrove and Domosh, 1993; Duncan and Ley, 1993; Harley, 1989; Jackson, 1989; Pratt, G., 1992), as masculinist (Deutsche, 1991; Massey, 1991c; Rose, 1993a) and as historically excluding women (Domosh, 1991; Mayer, 1989; Monk, 1983).

The history of textual analysis is associated with hermeneutics, or the study of interpretation, a method dating back to Wilhelm Dilthey's interpretations of biblical texts in the nineteenth century. Dilthey described a 'hermeneutical circle' through which meaning was produced through the interactions between a written passage, the larger whole of the text, and the intellectual framework of the interpreter. By the twentieth century, philosophers Paul Ricoeur and Hans-Georg Gadamer expanded the meaning of text to include all cultural practices. Ricoeur argues that the text model is a good paradigm for the social sciences because cultural productions, like written texts, are always (re)interpreted by individuals and groups in new situations. This negotiated process of signification renders the meaning of social action unstable and open to a wide range of interpretation. Anthropologists such as Clifford Geertz (1973) have adopted this interpretive model to define culture as text that is read and produced by individuals in a society. Similarly, geographers have adopted this approach to examine the signifying process of LANDSCAPE as a text (Barnes and Duncan, 1992; Duncan, 1990) or have challenged the model's usefulness (Mitchell, 1993; Peet, 1993).

In the early twentieth century, linguistic STRUCTURALISTS such as Ferdinand de Saussure were critical of naive realism and problematised the relationship between the signifier (sound or image) and the signified (meaning). Saussure also distinguished between everyday language, *la parole*, and the deeper, underlying set of rules that govern signifiers, *la langue*. More recently, POST-STRUCTURALISTS have challenged Saussure and argue that meaning is not determined by the structural relationships among signifiers. Texts emerge from the interplay of signs, and meanings are continuously being produced through the interaction of texts with other texts (intertextuality) rather than between a text and 'reality'. Jacques Derrida's method of DECONSTRUCTION attempts to undermine traditional binary hierarchies (e.g. speech–writing, life–death, soul–body, inside–outside, master–servant), reveal the contradictory meanings of texts and examine the inconsistent logic of the language use. Derrida (1967 [1973]) has argued that the most encompassing BINARY is MASCULINE–FEMININE, a hierarchy well examined by Hélène Cixous (1986). Another post-structural approach that has had a more influential role in geography is discourse analysis. According to Michel Foucault, discourses are social networks of POWER – ensembles of concepts, statements and social practices that make the world knowable and meaningful. Although discourses universalise particular viewpoints, as partial knowledges they are always open to contestation.

Feminist and post-structuralist geographers examine how texts and discourses naturalise GENDER, SEXUALITY and RACE (Bell and Valentine, 1995; Berg, 1998; Brown, M., 1995b; Duncan, 1996a; Kobayashi and Peake, 1994), while at the same time caution that post-structural approaches may reinforce ANDROCENTRIC viewpoints and may not lead to changes in the material conditions of women and peoples of colour (Bondi, 1990b; Kobayashi, 1993).   KT

See also NARRATIVE.

Thatcherism A term used to describe the set of characteristics associated with the radical right Conservative Governments in the United Kingdom between 1979 and 1990 – the chief attribute of which was, according to the British sociologist Stuart Hall, authoritarian populism or an appeal to the apparently selfish and individualistic interests of the majority of the population. These years were marked by growing inequalities in income and wealth, a decline in the provision of public services such as the health service and the privatisation of many formerly public-owned utilities such as gas, electricity and water companies and the railway network. Despite Margaret Thatcher – often referred to as the Iron Lady – being the first woman Prime Minister of the United Kingdom, her premiership was marked by growing gender inequalities, especially marked for working-class women.   LM

Third space A number of social scientists and cultural critics, including geographers, have recently explored the weaknesses of DUALISTIC EPISTEMOLOGIES as bases for theorising one's own positionality, analysing difference and promoting strategies of resistance. It is alleged that supposedly neutral dualisms such as PUBLIC–PRIVATE, MALE–FEMALE and right–wrong can become the hidden means by which dominant assumptions are perpetuated (Pile, 1994). In this context there has been increasing interest in Homi Bhabha's, search for a 'third space' (Bhabha, 1990, 1994). Although Bhabha has used the term in a variety of ways (Bhabha, 1994: 36–7) geographers have focused particularly on third space as a location of knowledge and RESISTANCE. The strengths of the term are then seen as lying in the fact that it elaborates the 'grounds of dissimilarity' on which dualisms are based; acknowledges that there are spaces beyond dualisms; and accepts that third space itself is fragmented, incomplete and the site of struggle for meaning and REPRESENTATION (Pile, 1994: 273; Law, 1997: 109).

Ed Soja champions 'third space' as a means of theorising our appreciation of space in the contemporary world, exploring late-twentieth-century Los Angeles and rethinking issues of temporality and historicity. 'Third space' is forwarded as a tentative and flexible term which attempts to capture a constantly shifting milieu of ideas, events, appearances and meanings (Soja, 1996). In his exuberance, Soja arguably claims too much for this concept; the rather grand pretensions to move beyond all conceivable dualisms occasionally verges on the ridiculous (Barnett, 1997).

The critique of Soja's 'third space' should not be confused with Gillian Rose's attempt to explore the applications and limitations of Bhabha's theorisation (Rose, 1995d). Rose argues that Bhabha's 'brave new HYBRID world' is, perhaps, gendered in both its writing and its practice. She refers, in particular, to a lack of attention to the ways in which powerful discourses are reproduced and an undertheorisation of the social and CORPOREAL in Bhabha's work (Rose, 1995d: 372).

Certain geographers have shown, however, that Bhabha's work might indeed provide the basis for an understanding of how MARGINAL groups contest and rework identities and spaces. Lisa Law, for example, in a study of SEX TOURISM in the Philippines, uses third space in two ways: to reconceptualise the researcher's own positionality and to theorise the bar as a negotiated space where Filipino women actively reinvent themselves (Law, 1997). In a similar vein, Steve Pile has drawn on Sarah Radcliffe's work in Latin America to reveal how 'marginal' groups can create alternative geographies which bring together space, politics and hybrid identities. The women's resistance movement in this country recombined elements of public and private discourse and space to create an elision, a third space (Pile, 1994). Certainly, this represents an exciting area for feminist geography, but whether or not the approach of Law and Pile escapes the rather uncritical optimism of Bhabha, and points the way towards a more genuinely social and embodied application of third space, is an open question.   CJ

Third World See WORLD: FIRST, SECOND, THIRD, FOURTH

Time It is widely recognised that notions of time and ways of valuing time are socially constructed (Adam, 1990, 1995; Elias, 1992). Two aspects of the measurement of time are considered particularly important: the measurement of *duration* and the delineation of *past*, *present* and *future*. The ways in which time is both measured and experienced varies between societies, social groups and individuals and can be measured and experienced by the same person in different ways. For example, time can be measured through clocks or calendars, seasonal change, religious and public festivals, life-course events, body rhythms and bodily ageing. Time can also be conceived of as a social, group or individual resource for which claims can be made, contested, negotiated and competed over.

Industrial capitalism involves increased reliance on clock time and the exchange of time spent in production for wages. This has resulted in the time of those who could exchange their labour for money being seen as more valuable than the time of those who cannot do so. Furthermore, whereas the timing of paid work is task-led and (fairly) predictable, the timing of caring activities does not match the time demands of paid work and is less predictable than paid work time. This creates difficulties, exacerbated by an unequal division of domestic labour, for women attempting to combine paid work with either child or elder care, as is shown by a variety of time-budget studies (Davies, 1990; Hochschild, 1990).

Geographers have been interested in the relation between time and space and in the consequences of the observation that it takes time to traverse space (Carlstein *et al.*, 1978; Harvey, 1990). In particular, feminist geographers have engaged critically with the approach of time-geography initiated by Hagerstrand (1973) to explore the ways in which women negotiate, manipulate and experience the conflicting time–space demands of caring and paid work (Dyck, 1989, 1990; Rose, 1991; Tivers, 1985).   SB

Time–space compression This term derives from the work of David Harvey (1989), who has developed Marx's ideas to suggest that technological developments, improved transport and enhanced communications have allowed CAPITALISM to conquer SPACE and TIME, to make production and CONSUMPTION more immediate and to facilitate increased capital accumulation. Harvey illustrates his contention with a striking image of the earth shrinking in relative size throughout the development of industrial capitalism.

However, Doreen Massey (1991b) has argued that the shrinking of space is a very uneven process and one primarily enjoyed by privileged, white men. She argues that for many MARGINALISED people, the world is as large and inaccessible as ever. More than this, Massey argues that the increased mobility of some groups (most usually those already in positions of POWER) can actively reduce the mobility of others (as, for example, when there is a reduction in investment in public TRANSPORT).   JS
See also POWER GEOMETRY; SPATIALITY.

Tourism Anglophone studies of tourism are gaining momentum as increased GLOBALISATION has made tourism one of the fastest, if not the fastest, growing industries in the world. As massive flows of travellers connect people and places rapidly and in new ways, human geographers and others study its meaning as a cultural phenomenon and in terms of large-scale patterns of PRODUCTION and CONSUMPTION. Recent studies examine the changing nature of tourist destinations, and reconceptualise what constitutes a tourist site. Tourism is viewed as an economic development strategy in many places, manifested as cultural repackaging of the past through museums, countryside resorts, craft and heritage industries, and in creating spectacle of experience in theme parks. The AESTHETICISATION OF EVERYDAY LIFE is demonstrated as planners, retailers and media moguls produce 'simulated' tourism in large-scale urban developments, shopping malls, restaurants, and via global film and television (Urry, 1995). The many social and environmental costs and benefits of tourism are also of concern to geographers, such as the exploitation of local cultures and people for outsider consumption as well as contentious claims to authenticity and meanings of place (Kincaid, 1988). 'Ecotourism', a form of small-scale tourism sold as an experience of a local ecology, has developed as a reaction against the negative environmental impacts of development.

Tourism, like other forms of 'displacement' (Kaplan, 1996), is always a gendered, classed, racialised and sexualised process, so an understanding of these social constructs is necessary to understand tourism itself. FEMINIST GEOGRAPHY critiques of tourism patterns and processes have focused primarily on three aspects: the ways in which gender difference situates women as workers in feminised job categories, the ways in which social forces position women materially and discursively as particular kinds of consumers of tourist sites and the ways social forces position women as producers of cultural knowledge about them, primarily in the form of written TEXTS.

New forms of exploitation of women are evident in the rising tourist trade. While SEX TOURISM has attracted the most concern, women's roles in the organised and regulated service economy of tourism has also received attention (Enloe, 1989). These roles rely extensively on gendered and racialised notions of appropriate work for local people, especially in developing countries, and by defining 'women's work' as unskilled or low-skilled.

A burgeoning feminist critique of European and North American women's travel and travel writing has appeared in geography (Domosh, 1991), although the distinction between travel and tourism in this literature is itself contested as an artificial demarcation between high and low cultural forms. Fussell (1980), for instance, argues that travellers are distinctive by virtue of their quest for self-enrichment whereas tourists are mass consumers. Blunt (1994) and Kaplan (1996) take issue with such categorisations, and this seems especially relevant since most women's access to travel was severely limited until transportation technologies, the construction of new tourist sites such as national parks in the US and Canadian West and the advent of professional tourist agencies, such as Thomas Cook's in 1850s Egypt, made travel respectable and safe for affluent women.

Scholarship on the cultural production of tourist destinations, in the forms of textual NARRATIVES and also photographs, drawings and collectibles, concentrates on the meanings that women have attached to tourist settings and processes and how they represent(ed) themselves and other people and places within them (Blunt and Rose, 1994; Morin, 1998). Difference based on the social identifiers of GENDER, RACE, CLASS, CULTURE and NATIONALITY take on special significance in historical works, especially as many tourist destinations were established within the context of Euro-American COLONIALISM or IMPERIALISM.   KM

Tourist gaze While the unmarked, unlocated masculine 'gaze from nowhere' has been widely criticised in feminist studies and feminist geography generally (Haraway, 1988; Rose, 1993a; see GAZE), the subjective tourist gaze has come under scrutiny as tourists visually and textually view people and places outside of their everyday experience. Urry (1990) has closely examined the tourist gaze as the fundamental consumption activity of tourism. Feminist geographers explore the many social and material constraints that enable or limit the gaze (Blunt, 1994); they have called attention to the positioned and embodied ways of seeing and evaluating tourist sites, which change with the viewer, over time, and with respect to the social and spatial contingencies of the specific location.   KM

Trade unions Associations of wage earners who organise to protect and improve their living and working conditions. Historically rooted in the friendly and mutual societies of the pre-industrial era, trade unions have played a key role in the evolution of twentieth-century capitalism. Since the 1970s, economic restructuring, new management practice and changing societal norms have prompted trade union decline across the Western world (see Martin *et al.*, 1996). As a result trade unions are now looking to rebuild their organisations and refocus their culture away from traditional associations with male, white, manual workers towards women, part-time employees and service-based staff (see Wills, 1998). While women have always played a key role in trade union history (e.g. the Bryant and May match factory strike, London, in 1888; Rose, 1996b), they have rarely reached high union office, their needs have traditionally been ignored and, in some cases, they have met discrimination from their male brothers in the union movement (see Boston, 1980; Cockburn, 1983). As the FEMINISATION OF THE WORK FORCE continues, and trade unions target women for membership and high office, it is likely that women will come to play a greater role in trade unionism in the future. In the contemporary labour market, the very future of the labour movement cannot be secured without greater participation by women.

Levels of unionisation amongst women have been suppressed in some developing countries, as cheap, female labour is used to attract foreign capital to export processing zones (EPZs) (see Enloe, 1989).   JW

Tradition Associated with 'a general process of handing down', according to Raymond Williams (1983: 319), 'but there is a very strong and often predominant sense of this entailing respect and duty'. Tradition is bound up with MEMORY and, far from being inert, is a process entailing not only repetition over time but also forms of invention, adaptation and re-creation, as well as inclusions and exclusions that construct a 'selective tradition': 'an intentionally selective version of a shaping past and a pre-shaped present, which is then powerfully operative in the process of social and cultural definition and identification' (Williams, 1977: 115; see also Hobsbawm and Ranger, 1983). The simultaneously spatial and temporal formation of traditions is discussed by Wills (1995, 1996) in her studies of the geographies of trade unionism, and, within the discipline of geography itself, attempts to destabilise 'the geographical tradition', with its exclusions and 'paternal lines of descent' (Rose, 1995a), are proceeding. Interest in radical as well as reactionary aspects of traditions – the difficulty of making clear distinctions between the two is frequently noted – is apparent in other recent feminist and post-colonial writings where the maintenance of a sense of the past and history has been connected to RESISTANCE, the work of 'organic intellectuals', and basic struggles for survival (e.g. Collins, 1992). General theories of tradition, 'post-traditional society' and 'DETRADITIONALISATION' are also currently being developed (Giddens, 1995). These seek to move beyond simplistic oppositions between tradition and MODERNITY, a BINARY formulation that has been increasingly criticised, including in relation to its use in studies in non-Western societies by some feminists who, according to Ong (1988), have consequently constructed linear either/or arguments about the destruction of 'traditional customs' that obscure the social meanings of change for the people involved. DP

Traditional societies A widely used, although problematic, term which refers to societies apparently untouched by progress, development or modern life. By denying history and social struggles, the term 'traditional' implies static, unchanging relationships between women and men. A traditional society is often attributed with specific gender relations, usually involving the containment and oppression of women. The politicisation of the term 'tradition' has been analysed by feminist theorists of MULTICULTURALISM and POST-COLONIALISM (e.g. Yuval-Davis, 1997) who point to the way in which conservative gender relations are buttressed by reference to 'tradition'. As Giddens (Beck *et al.*, 1994: 105) summarises, 'up to and beyond the threshold of modernity, gender differences were deeply enshrined in tradition and resonant with congealed power'. SAR

transgender A contested and problematic term referring to a variety of conscious, unconscious, intended, unintended, acknowledged or unacknowledged TRANSGRESSIONS against gender and sexual norms and/or the sex–gender system. The self-styled transgender political activist Leslie Feinberg (1992) uses the term to refer to any refusal to be bound by the culturally constructed BINARISM of 'MALE–FEMALE'; others, including some TRANSSEXUALS struggling to secure their eligibility for funding for gender reassignment therapies (including surgery), argue that the term should be more narrowly construed as referring to a medical and/or psychological *condition* for which treatment or intervention is required and should be provided. Still others view it as a more general social–psychological orientation in which individuals do not necessarily identify with their ascribed (and socially constructed) GENDER (as opposed to their anatomical 'SEX'), but do not necessarily feel a need to 'correct' the situation either (or even to challenge binary notions of gender).   LK

Transgression A form of RESISTANCE that operates through going beyond or overstepping an accepted limit. Such actions challenge the 'normal' characteristics of things. QUEER politics have made much use of the idea of transgression, especially in exploring and challenging the definitions of SEX and GENDER identities. For geographers it is important to recognise that the PLACE where transgression is enacted affects its POLITICAL impact.   JS
See also TRANSGENDER; TRANSSEXUAL.

Transitional space A term developed by the child development theorist and pioneer of object relations theory, D.W. Winnicott. Defined as a space where the child experiments and plays with connections between the external world and an internal conception of the self, the term connotes a place to which the child brings objects, cultural practices and self images, and where s/he is able to play with/experiment with these in developing knowledge of self and other, inner psyche and external world. Contrasting with Freudian and Lacanian analyses of psychic development, the term has been taken up by Jane Flax (1993) and by Aitken and Herman (1997) in human geography, who draw connections with Lefebvre's notion of trial by space.   NG

Transport Feminist studies of transport have generally focused on inequality in access to transport between men and women. Research has identified women's relatively poor access to private transport and, in particular, the monopolisation of the household car by adult males. Poor access to private transport increases women's dependency on public transport which is frequently expensive, inconvenient and inappropriate to their needs.   JL

Transsexual A contested term referring to certain conscious, unconscious, intended, unintended, acknowledged or unacknowledged transgressions against gender and sexual norms and/or the sex–gender system; most commonly used to reference individuals whose anatomical and/or chromosomal traits lead to their cultural construction as either 'male' or 'female', while their individual psychological and/or emotional identification is with the culturally constructed 'opposite sex'. However, some individuals prefer the term 'TRANSGENDER' to describe these circumstances.

Transsexual is distinguished from transgender, then, in various ways, depending upon the interests and politics of those doing the distinguishing. The medical establishment, along with many self-described transsexuals, tends to distinguish it by its grounding *in*, rather than against, binary conceptions of SEX and GENDER. Hence, by this definition a transsexual is a 'man' presently or formerly 'trapped in a woman's body' (or in the process of having his sex and gender 'reassigned'), or a 'woman' presently or formerly 'trapped in a man's body' (or in the process of having her sex and gender 'reassigned'). Transgender, by contrast, refers to a medical and/or psychological *condition* which must be 'corrected'. Others, however, distinguish the term transsexual simply by its grounding in the concept of sex rather than gender, which may or may not involve (though it typically does) a *binary* conception of sex grounded, in some sense, in anatomy. LK

Travel/travellers/travel writing Refers to movement over SPACE and TIME and written representations of experiences away from home. Some writers position travel on a continuum between exploration and TOURISM, suggesting it is a more individual than a collective process that is often associated with self-development away from 'home'. Spaces of HOME and away are more blurred than distinct through travel because the very idea of home can only be imagined from a distance and necessarily changes on return (Van den Abeele, 1992). In recent years, an increasing amount of critical attention has been paid to the ways in which travel is bound up with the production of knowledge, POWER relations and IDENTITY formation. The imperial development of disciplines such as ANTHROPOLOGY and GEOGRAPHY was closely linked to imperial travel and the immediacy and authority of observation associated with 'being there' and viewing 'other' people and places at first-hand. Several writers have focused on travel as a gendered practice, tracing discourses of heroic MASCULINITY that shaped imperial exploration (Driver, 1992; Pratt, M.L., 1992) and the ambivalent place of white women such as Mary Kingsley and Isabella Bird who travelled in the context of imperialism (Blunt, 1994; Mills, 1991). Research on British women travellers suggests that they were empowered to travel and to transgress in the empire while away from the feminised DOMESTICITY of living at home. Such women travellers were among those who were nominated as the first female Fellows of the ROYAL GEOGRAPHICAL SOCIETY in the 1890s. In contrast, other writers have shown the ways in which travel can be forced as well as voluntary, pointing to the movement of slaves, REFUGEES, asylum seekers and the dispossessed away from home and experiences of displacement, exile and alienation (Bammer, 1994; Broe and Ingram, 1989). In both cases, experiences of travel are closely shaped by relations of power over space.

Travel has come to be a metaphorically as well as a materially significant term. In cultural criticism, travel has been seen as a 'translation term', reflecting cross-cultural dialogue and the production of knowledge over space, and time has been interpreted in terms of 'travelling theory' (Said, 1983). Another important area of work has addressed the spatial constitution of SUBJECTIV-ITY in terms of travel, suggesting the mobility of anti-essentialist subject positions over space and time (Kaplan, 1996; Robertson *et. al.*, 1994). Writers such as bell hooks and James Clifford have cautioned against celebrating travel in uncritical and ungrounded terms, highlighting diverse experiences of travel. As hooks writes, 'from certain standpoints, to travel is to encounter the terrorizing force of white supremacy' (hooks, 1992a). James Clifford proposes the use of the term 'travel' in cultural studies

> precisely because of its historical taintedness, its associations with gendered, racial bodies, class privilege, specific means of conveyance, beaten paths, agents, frontiers, documents, and the like. I prefer it to more apparently neutral and 'theoretical' terms, such as 'displacement,' which can make the drawing of equivalence across different historical experiences too easy (1992: 110). AMB

Travelling theory Travelling theory has two incarnations. First, Edward Said uses the term in exploring how theories and ideas 'travel' (Said, 1983). He shows that when theories, such as Foucault's theories of power or Lukacs idea of reification, are transferred through time, and from person to person, the radical potential and explanatory power of that theory or idea changes correspondingly. Travelling theory also refers, however, to something mobile in the nature of theory. This has been associated with the rise of travel as a metaphor and as a means of theorising the disruption of place within disciplines such as ANTHROPOLOGY and GEOGRAPHY (McDowell, 1997b: 136).

The work most immediately relevant to feminist geographers has related to travelling theory in its second incarnation. James Clifford places travelling theory at the centre of POST-COLONIAL criticism and the reconfiguration of anthropology as a discipline. He shows how the metaphor of travel assists in de-essentialising researcher and subject and exposing the unacknowledged relationship of POWER and control which characterised post-colonial encounters (Clifford, 1992). Drawing on her own travel experiences, bell hooks has described how travelling can disrupt reifications of 'the OTHER' and provide insights into one's own and others' social locations (hooks, 1991a).

Within geography, and social science more generally, TRAVEL and travel writing has become a significant and exciting new area of emphasis (Blunt and Rose, 1994; Pratt, M.L., 1992). In her analysis of the travels and travel writing of Mary Kingsley, Alison Blunt shows how imperial women's travel was distinctive with respect to how they travelled, how they wrote about their journeys and how their writings were received (Blunt, 1994: 10). The act of making women visible within history is taken by Blunt and others as an opportunity to remake the categories through which the past is discursively constituted. Mona Domosh takes advantage of this spirit of DECONSTRUCTION and reassessment in attempting to outline a feminist historiography of geography through a focus on the contibution of female travellers to geographical knowledge (Domosh, 1991).

Although these efforts have gone a considerable way towards freeing the idea of travel from its MASCULINE associations, a number of critics see such metaphors of mobility as being tainted by their links with GENDER, CLASS and RACIAL forms of privilege or DOMINATION (Clifford, 1992; Pratt, M.L., 1992: 242). For Janet Wolff, the men/travel association is the other half of a historically disempowering women/home IDEOLOGICAL construction (Wolff, 1993).

Geraldine Pratt and Janet Wolff also have a more fundamental difficulty with accepting mobility metaphors. These authors argue, in different ways, that an emphasis on fluidity and movement runs the risk of reproducing pretensions of an ungrounded, detached GAZE: a view from nowhere. Pratt then builds on Nancy Fraser's work in highlighting the ultimately disempowering implications of an overemphasis on the fluidity and destabilisation of identities. She argues that the stability and reification of some identities have to be acknowledged if hard political questions regarding class, gender and racial difference are to be addressed (Pratt, G., 1992: 242). It may be that other spatial metaphors – such as BORDERS and borderlands – are ultimately more suggestive and less potentially disingenuous, than the second sense of travelling theory in advancing our thinking on subjectivity, categories and positionality (Pratt, G., 1992; Wolff, 1993).   CJ
See also MARGINS.

Truth/regime of truth Foucault uses the term regimes of truth to refer to 'the ensembles of rules according to which the true and the false are separated and the specific effects of POWER attached to the true (Foucault, 1980: 132). There are some similarities here with sociology-of-knowledge contentions regarding the conditions under which knowledge is produced and recognised as such. However, Foucault also challenged any assumption that IDEOLOGY can be demystified or that the undistorted truth can be attained. He brings to our attention the complex network of disciplinary systems and prescriptive technologies through which power operates in the modern era, particularly since the normalising disciplines of medicine, education and psychology have gained ascendancy. For example, an immense strategy for producing truth has been constituted around SEXUALITY: 'We demand that sex speak the truth . . . and we demand that it tell us the truth . . . about ourselves' (Foucault, 1978: 69). In psychoanalytic discourse, sexual desire takes on the capacity to become the most revealing sign of our truest, deepest selves.

Foucault is not concerned with uncovering 'the truth' but rather with discovering how certain DISCOURSES claim to speak the truth and thus can exercise power in a society that values this notion of truth. He argues that making a claim to be a SCIENCE is an exercise of power because other knowledges, by comparison, are accorded less status and can exercise less influence. Foucault showed that truth does not exist outside power. Each society has its own 'regime of truth': the types of discourse accepted as true, the procedures that make it possible to distinguish between truth and error. In place of 'organic' intellectuals producing the truth, he believed it was the task of the

'specific' intellectual to work upon the particular regime of truth in which he
or she operates.   RP
See also PARTIAL TRUTH.

U

Unconscious See FREUDIAN THEORY

Underclass A controversial term which is used to apply to those who are
socially and economically marginalised and excluded from mainstream
society. There is considerable debate about the political value of this term as
those living in POVERTY often do have some casual relationship to wider
society.   JW
See also CASUALISATION; SPACES OF EXCLUSION.

Unemployment The term unemployment refers to the condition of those who
are without paid WORK. Traditionally, women's unemployment merited little
attention, as men were expected to provide for their household. As women's
labour-market participation increases, however, they are more likely to register as unemployed if they are made redundant. In many societies, however,
it is difficult to calculate rates of unemployment when paid work is often
casual, interrupted or in the informal sector.   JW

United Nations The United Nations (UN) was created by the Bretton Woods
agreement in 1944. Its international mission to maintain political and military
stability means that, as well as providing international policing, UN agencies
also address the consequences of conflict and poverty. These consequences
are often gendered. Therefore in the 1970s the UN launched the first Decade
for Women (1976–85), named 1975 the INTERNATIONAL YEAR OF THE
WOMAN and made INTERNATIONAL WOMEN'S DAY an official UN holiday.
UN-sponsored international conferences have been organised in Mexico
(1975), Nairobi (1985) and Beijing (1995), to highlight worldwide gender
inequalities. Other gender successes of the UN have been the promotion of
offices for women within governments and international agencies and the
development of policies to target vulnerable groups.
   The UN was first influenced by Women in Development (WID) strategies
which aimed to incorporate women into DEVELOPMENT processes by making
their roles more visible. Recently, however, WID approaches have been
challenged by Gender and Development (GAD) analysis which problema-

tises POWER relations and highlights DIFFERENCE between and among genders. The role of the UN in promoting gender-sensitive development, however, has been controversial. Some feminist academics have suggested that WID has exaggerated the numbers of women in certain vulnerable categories (e.g. female heads of households) and has made other women invisible (Varley, 1996). Activists have criticised large UN international conferences for being merely public relations exercises (Pietila and Vicker, 1994). In particular, the Fourth World Conference in Beijing was attacked for being held amidst protests concerning Chinese human rights abuses.   NL

Universalism The term universalism refers to conceptual claims, modes of reasoning and forms of knowledge that purport to be universal in scope. Classic examples include many of the claims of, and about, SCIENCE: for example, gravitational theory is couched in terms that assume it applies equally in all times and all places, and the methods of science are widely presented as universally applicable. Many concepts used in the social sciences, including within feminist thought, also carry universal claims. For example, the notion that GENDER is a form of social division that operates in all cultures is a universal claim, even if it is acknowledged that its parameters and content vary from place to place and through time. So too is the idea that all women have something in common, whether that 'something' is understood in terms of BIOLOGY, psychology and/or cultural experience.

Universalism is problematic for feminism: ideas that are at least potentially universal in scope are often useful but carry attendant dangers. This can be illustrated in relation to the idea that women have certain things in common. On the one hand it forms a basis for the development of a political movement of women: what brings women together to resist oppression based on experiences common to women. But on the other hand, presumptions about what women are like homogenise diverse experiences and generate stereotypes that feminists have been keen to contest. Related to this, feminists have criticised many conceptual claims about 'people' on the grounds that they refer implicitly to men or to masculine positions or experiences and are therefore *false* universalisms (Gilligan, 1982; Lloyd, 1984).

The double-edged quality of universalism has been felt sharply within SECOND WAVE FEMINISM. Identifying experiences shared by women, especially experiences generally regarded as 'personal problems', was very important in the emergence of the WOMEN'S MOVEMENT. But early enthusiasm produced over-extended claims about women and over-simplified conceptions of 'SISTERHOOD'. Critiques advanced by WOMEN OF COLOUR, by Third World women, by working-class women and by lesbians together suggested that such claims generalised from the experiences of white, Western, middle-class, heterosexual women (Carby, 1982; *Feminist Review*, 1987; hooks, 1982; Mohanty, 1988; Snitow, *et al.*, 1984). In so doing they constituted false universalisms, just like some of the claims about 'people' these feminists had themselves criticised.

It is possible to identify two broad positions taken by feminists in relation to the issue of universalisms, which often co-exist in the form of an oscillation between the imperatives of EQUALITY and DIFFERENCE (Alcoff, 1988; Snitow, 1990). The first is clearly illustrated by LIBERAL FEMINISTS who follow the tenets of liberal HUMANISM and LIBERAL THEORY in arguing that similarities among people far outweigh the significance of differences (whether of gender, CLASS, CULTURE, ETHNICITY or whatever) (Okin, 1989). This implies that universal statements about HUMAN NATURE are possible. Consequently false universalisms are criticised on the basis of their inaccuracy. RADICAL FEMINISM has produced a variant of this in which it is assumed that gender constitutes a fundamental cleavage in human nature, across which it is not possible to make universal claims. However, it is assumed that on each side of this cleavage fundamental similarities operate that are sufficient to support general claims about 'women' and about 'men' (Firestone, 1970).

The second position contests the whole strategy of universalistic modes of reasoning and is more concerned with deconstructing or destabilising universalisms than with improving their accuracy. This perspective is allied closely with POST-STRUCTURALIST thought and fosters multiple knowledges rather than singular, overarching claims about the world (Hekman, 1997; Scott, 1988). Linked to the critique of universalisms is a critique of DUALISMS, which take the form of mutually exclusive and exhaustive CATEGORIES, such as the division of the human population into two genders, one masculine and one feminine. This position has become influential within feminist geography where it is used to criticise the false universalism of much human geography, to discuss tensions between universalism and relativism in relation to feminist politics and to open spaces for the production of knowledges that destabilise, disrupt, resist or subvert universalism (Bondi and Domosh, 1992; McDowell, 1991a; Rose, 1993a). LB

Utopia/utopian thought/utopian communities The concept of utopia refers both to somewhere good and to nowhere. The double meaning is contained within the word itself, which plays on the Greek compounds *eu-topos* (happy or fortunate place) and *ou-topos* (no place). Interest in utopias and utopianism runs through much feminist thought, surfacing in fictional portrayals of ideal societies, in the establishment of experimental COMMUNITIES, in anticipations of a better future world in critical theory and in a variety of imaginings, plans and poetic longings. Indeed, feminist movements in North America and Western Europe have been an important force alongside other 'new' political movements in countering pronouncements by many commentators that utopia is coming to an end, by engaging creatively with utopianism and stimulating new approaches to the field especially since the 1960s, a decade that according to Jameson (1988: 75) saw 'the reinvention of the question of Utopia'. Utopian communities aim to transform spatial structures as well as social relations, and feminist studies have considered how utopian socialist communities as well as more recent radical and counter-cultural examples have addressed issues of GENDER, SEXUALITY and domestic arrangements in the creation of new environments (Hayden, 1976; Taylor, 1983). Research has also explored utopian visions developed by early feminist organisations, including proposals for the design of HOMES and CITIES, with one study taking as its epigraph lines from Susan Anthony in 1871: 'Away with your man-visions! Women propose to reject them all, and begin to dream dreams for themselves' (Hayden, 1981; see also Anderson, 1992).

An interest in the dreams women 'dream for themselves' has similarly underpinned feminist reconsiderations of 'utopia'. Utopia is often defined as a literary genre, involving the depiction of an ideal polity. Feminist critics have challenged existing histories of literary utopias by uncovering a neglected tradition written from the perspective of women's desires (e.g. Albinski, 1988; Bartkowski, 1989). In returning to texts such as Christine de Pizan's *The Book of the City of Ladies* of 1405, they have contested standard definitions of what constitutes 'a utopia', seeking to rewrite the meaning of the concept along with histories of its literary expression. Visions of gender relations within literature by men have also been critically examined, exposing the interweaving of reactionary and radical elements in different utopias, while the expansion of feminist utopian fiction especially in the 1970s has stimulated debates about the distinctiveness of its forms and strategies (prominent examples include Gilman's *Herland* of 1915, and more recently Gearheart, 1978; Piercy, 1976; Wittig, 1971; as well as 'anti-utopias' such as that in Atwood, 1986). Rather than viewing utopia as a closed, fixed form or a blueprint to be realised, a number of feminist critics have recently been reconceptualising 'the utopian' in more fluid, open and partial terms as 'an *approach toward*, a movement beyond set limits into the realm of the not-yet-set' (Bammer, 1991: 7). This approach has been important in areas of feminist theory, including within FRENCH FEMINISM (see Whitford, 1991). It typically gives a central role to *desire*, and it is opening up new paths in feminist utopian thought that are multidirectional, open-ended, TRANSGRESSIVE and not afraid of ambiguity or flux (Sargission, 1996; see also Reichert, 1994).   DP

V

Violence Violence may be physical, sexual, verbal, emotional or representational and may include interpersonal and institutional actions not necessarily construed as CRIME. It is central to feminist conceptualisations that violence occurs along a wide spectrum, thus boundaries with ABUSE and HARASSMENT are blurred, though violence is most frequently used to describe extreme events which lead to physical harm. Violence is a social and political act, both determined by and central to the continuation of POWER relations of GENDER,

SEXUALITY, age, 'RACE' and NATIONHOOD, in individual, local and international contexts. The incidence, nature and policing of violence are profoundly conditioned by the social structure of SPACE.

Feminists have radically redrawn understanding of sexual violence against women and children. Some studies suggest it is experienced by most women over the life course (Kelly, 1987); this includes what others distinguish as forms of harassment. Most sexual violence takes place within the home. Because of the nature of private space, where traditionally men's rights of authority and privacy have been legally enshrined, 'aggressive forms of misogynous MASCULINITY are often exercised with impunity' (Duncan, 1996b: 131). The HOME also provides the main site in which domination based upon age and generation finds expression in a high incidence of violence against children and older people. Feminists have strongly criticised the emphasis of crime prevention and policing strategies on violence in PUBLIC space; although in many countries this is changing, support services for survivors of domestic violence are generally underresourced, prosecution rates are low and the problem remains largely hidden.

Racist violence and harassment are also endemic to most societies. In some localities where public RACISM is prominent, it provides another dimension of fear which consolidates spatial boundaries and control. HOMOPHOBIC violence on grounds of actual or apparent sexual orientation also appears to be increasing, as alternatives to heterosexuality become more visible (Herek and Berrill, 1992). Attacks and routine harassment effectively police sexual behaviour, especially of those who are seen to transgress the gendered, HETEROSEXUAL nature of many public and private spaces (Namaste, 1996). The threat of violence is thus central in maintaining the social and spatial order of the city. The avoidance of violence is routinised in most people's everyday lives (Stanko, 1990), and expressed in their use of space, most sharply in women's concerns about RAPE.

While men's violence to women has been the main focus of feminist research (and some suggest this is the most common form of violence), the power relations involved in violence are complex, multiple and contradictory rather than fixed and predictable. Women do not live outside PATRIARCHAL ideology and practice, and some perpetrate abuse against the people they are most likely to have control over, especially children and older people within the home (Kelly, 1991).

Violence is also a feature of organisations, for example, in achieving interests of powerful groups in the workplace, and is perpetrated by institutions, including the state (see Fawcett *et al.*, 1996).   RHP
See also ABUSE; CRIME; FEAR OF CRIME; HARASSMENT.

Vision/visual/visuality Of the five senses, vision is a privileged way of experiencing, understanding and knowing about the world in modern Western societies. As the popular adage demonstrates, 'seeing is believing'. The act of looking, however, is not a neutral one, and is tied to specific histories of social practices, institutions and POWER relations (Berger, 1972; Foucault, 1971; Gaines, 1988). The 'normalising gaze' associated with vision as a system of knowledge is a kind of SURVEILLANCE that makes it possible to classify, judge and punish individuals (Foucault, 1977). This Western obsession with visual difference is tied to the history of market capitalism – a political economy historically structured by the production and consumption of cultural REPRE-SENTATIONS of OTHERS (King, 1997).

Scholars examine the problems associated with a visual theory of knowl-edge by analysing the technologies and representational forms of art, photog-raphy, cinema and science. Donna Haraway (1991) has demonstrated how the scientist distances himself from the objects he sees and studies to produce knowledge. This illusion of knowing and seeing the whole world while remaining distanced from it is called the 'god-trick' by Haraway. Through this 'view from nowhere' scientists claim objectivity and UNIVERSALITY, while denying the origins of science and its historical ties to militarism, CAPITAL-ISM, COLONIALISM and male supremacy. Similarly, anthropologists and post-colonialists argue that the taxonomic imagination in the West is a visual form of knowledge that results in the racialisation and systematic exploitation of Others (Clifford, 1988; Fabian, 1983; Haraway, 1984).

Geographers examine how SPACE has been imagined and controlled by visual forms of knowledge. Gillian Rose (1993b) argues that the masculinist gaze of the SPATIAL SCIENTIST treats space, and the things and peoples in that space, as trans-parent objects of study. This all-seeing and all-knowing geographer can be thought of as a surveyor, hero and sovereign in the history of geographic thought (Barnes and Gregory, 1997). In each of these roles, the geographer looks at the world from a privileged position of power. The surveyor maps landscapes, peoples and objects; the colonial hero explores new and exotic places for Western audiences; and the sovereign helps locate, administer and defend the spaces of nation-states. Historically, the maps, photographs and landscape representations produced by geographers have created an IMAGINATIVE GEOGRAPHY, a way of seeing the world through the eyes of white men (Ryan, 1994).

To challenge masculine claims to universal seeing/knowing, feminists have explored different methods of situating knowledge, such as POSITIONALITY and reflexivity. Yet Rose (1997a) suggests that these methods continue to treat vision as a form of power/knowledge. The illusion of a researcher who can fully know herself and situate subjects on an even terrain of power is for Rose a 'goddess-trick': 'I want to suggest that both these tactics [positional-ity and reflexivity] work by turning extraordinarily complex power relations into a visible and clearly ordered space that can be surveyed by the researcher' (1997a: 310). The visualised space of SITUATED KNOWLEDGES is contradictory according to Rose because the researched are always separate and different from the researcher. Moreover, rendering oneself 'transparent' through reflexivity assumes that relations between individuals (and their positions of distance/difference) can be mapped on a visible landscape of power. Nast also criticises reflexivity when it is theorised as a mirror because such an approach 'depends upon a very particular viewing context, one that speaks of privilege and bodily/spatial control' (1998a: 100). KT

Voice A central concern of FEMINIST and POST-COLONIAL theory is to give a voice to the experiences and histories of those formerly excluded from, and silenced by, academic theories and histories. QUALITATIVE METHODS are adopted in order to destabilise the central authority of the researcher–author and to leave space for the voices of the research participants.

However, in an influential paper, Gayatri C. Spivak (1988b) questions the possibility of recovering a SUBALTERN voice without producing an 'ESSENTIALIST fiction' of the MARGINAL. What she means is that 'one cannot construct a CATEGORY of the "subaltern" that has an effective "voice" clearly and unproblematically audible above the persistent and multiple echoes of its inevitable heterogeneity' (Ashcroft *et al.*, 1995: 8). bell hooks also fears the possibility of tokenism in this gesture. She claims that rather than allowing the voices of the marginalised to challenge the authority of dominant representation, all too often the power of representation still lies with the dominant:

> No need to hear your voice. Only tell me about your pain. I want to know your story. And then I will tell it back to you in a new way. Tell it back to you in such a way that it has become mine, my own (1991: 343).

Others have questioned the importance of having a voice, and instead suggest the POLITICAL effectiveness of silence (Miller, 1990). JS

Voyeurism From PSYCHOANALYSIS, voyeurism represents the active DESIRE to look. Pleasure is derived from this unseen looking.

Many FEMINISTS have been influenced by Laura Mulvey's analysis of cinema. Her celebrated paper on voyeurism in the cinematic tradition, 'Visual pleasure and narrative cinema' (1989) explains the role of cinematic techniques in maintaining a particular relationship between audience and film. Mulvey argues that film narrative constructs a viewing space – a variation on the theme of HETEROSEXUAL MASCULINITY – for the audience which directs viewers towards one rather than other interpretations of the screen action. Similar arguments have been used for other AESTHETIC productions including art, and also for hegemonic readings of the LANDSCAPE within geography (Rose, 1993b). JS
See also GAZE; VISION/VISUAL/VISUALITY.

W

Waged labour See LABOUR: DOMESTIC, SWEATED, WAGED, GREEN, SURPLUS

Wealth/wealth distribution The concentration of resources, power and social status amongst people who have good access to CAPITAL and cultural assets. Access to capital is spatially and temporally uneven and is intrinsically related to capitalist development processes (Gibson-Graham, 1995). The most straightforward indicator of wealth is position in the labour market, although alternative measures such as inheritance, consumption patterns and health are also important (Green, 1997). Growing numbers of women have been drawn into paid employment (FEMINISATION OF THE WORK FORCE) and have some independent access to economic resources, but wealth is stratified by gender as well as by other factors such as 'race', age and sexuality.

At a global level women are underrepresented amongst the wealthiest sections of the population and are subsequently more vulnerable to POVERTY and exclusion. In 1980 a UN report summarised this state of affairs: 'Women constitute half the world's population, perform nearly two thirds of its work hours, receive one tenth of the world's income and own less than one-hundredth of the world's property' (see McDowell and Sharp, 1997: 319).

The seemingly universal discrepancy between the work that women do and the poor reward they receive for it is explained by the pervasiveness of patriarchal social relations (Hartmann, 1979). Gendered DIVISIONS OF LABOUR ensure that the majority of women work in jobs dominated by members of their own sex, often for low pay and for little status (EXPLOITATION). SOCIALIST FEMINISTS have shown that access to paid employment is not a guarantee of wealth for women living and working in the global economy (Mitter, 1986; Pearson, 1986). This is because inequalities between women and men operate at the very heart of contemporary economic processes themselves (Jensen, 1989; McDowell, 1991b).

This global picture hides significant spatial variations in the position of women in the distribution of income and wealth. The ways in which women are drawn into paid work vary within and between nation-states, depending on factors such as the level of development, the nature of the economy and the role of the welfare state. In Britain, for example, the national rise in female employment is associated with the emergence of a service-based economy and the needs of capital for a low-wage and flexible labour force, best served by employing women on a part-time basis (POST-FORDISM). For many (working-class) women then, employment is not so much a passport to wealth as a way of preventing poverty (McDowell, 1991b; Walby, 1997).

The most recent phase of capitalist accumulation has involved some reconfiguring of the gender relations of wealth, particularly in North America and Western Europe. Growing numbers of women have moved into professional jobs and benefited from recent rounds of capitalist accumulation. Studies emphasise the connections between the increased wealth of these middle-class women and the regional geography of employment (Perrons, 1995), socio-economic polarisation (Fainstein *et al.*, 1992) and housing practices.   HC

See also MARXIST GEOGRAPHY.

Welfare state A welfare state is a social system in which the government assumes primary responsibility for the welfare of its CITIZENS by protecting minimum standards of income, nutrition, health, housing and education and in which access to these benefits is assumed to be a political right. The term is commonly used to describe the twentieth-century political economy of industrial democracies of Western Europe, the USA, Canada, Australia and New Zealand. Welfare state programmes have been informed by the economic theories of John Maynard Keynes, who advocated government intervention to increase employment. Welfare states can also be seen as the expansion of government to fill a void created by the separation of the workplace and the FAMILY. They are characterised by a high degree of institutional involvement in family life. As such they regulate families, enforcing specific norms of MOTHERHOOD and GENDER relations in the family.

The British welfare state has it roots in the Poor Laws initiated at the end of the sixteenth century and later modified in 1834. Owing in part to the advocacy and philanthropy of middle-class women's organisations, relief for the poor from both the state and private charities expanded throughout the late nineteenth and early twentieth centuries. It was in the wake of World War II, however, that the national programmes that constitute the backbone of the British welfare state were established. Programmes such as the National Health Act of 1946, the National Insurance Act of 1946 and the 1948 National Assistance Act established a network of institutions and programmes designed to provide social and economic protection. In the 1980s and 1990s, Margaret THATCHER's Conservative Government reduced, although did not eliminate, many aspects of the British welfare state.

In the USA, the policies and institutions of the welfare state first emerged in local and state policies at the turn of the century. By World War I, a patchwork of policies were in place to protect, educate and socialise children, support widows with children and provide health services to the very poor. As in Britain, many US programmes were pioneered and supported by middle-class women (many of whom were feminists) concerned with the immediate economic needs of poor women and children. The New Deal policies of Franklin Roosevelt in the 1940s marked the transition of the federal government to a welfare state, with federal programmes such as Social Security Insurance (SSI) and Aid to Families with Dependent Children (AFDC) creating a national safety net. Since the 1940s the US welfare state has provided two different types of social support programmes: social insurance programmes (such as unemployment insurance and SSI), which provide a level of support to all citizens, with the amount of support tied to employment-based contributions; and public-assistance programmes limited to those with economic need, especially female heads of households and their dependent children. During the 1960s the US welfare state expanded further with the 'war on poverty', the political demands of women and people of colour, and the growth in unwed motherhood. In the 1980s and 1990s, welfare state programmes have been reduced and privatised, and the very notion of the welfare state has come under sustained attack from conservative critics.

Feminist scholars have brought to light the contradictory implications of welfare states for women. Because welfare state programmes mediate between the PUBLIC and PRIVATE sphere they embody many of the larger social contradictions about 'women's place'. Thus welfare state programmes in the USA and Britain recognise and support women's economic dependence and central role in the private sphere, but they also contribute to the perpetuation of women's economic dependence and exclusion from the public sphere. Because of this contradiction, socialist feminists argue that social welfare policies have also regulated the contradiction between the need for women as low-wage workers and the need for their unpaid DOMESTIC LABOUR by channelling some women into homes and others into the labour market. Women have also dominated the work force of the welfare state, encouraging many feminist scholars to view welfare state programmes as the PROFESSIONALISATION of women's caring work. Changes in the organisation and delivery of social services have affected women as workers, and efforts to reduce the welfare state generally increase the workload and reduce the pay of social service workers. Means-tested welfare programmes also affect women by distinguishing between the deserving and the undeserving poor, reinforcing divisions among women based on CLASS, RACE and marital status.   DM

West, the/Westernisation Despite its common associations with geographical position, 'the West' is an inaccurate descriptive term. Although the term originated in Western Europe and came to include the USA, it tends to exclude areas within the Western Hemisphere (such as Latin America) and may now also include Japan. It may prove more fruitful to examine 'the West' as an *historical* construct (Hall, S., 1992: 276–7). A 'Western' society tends to be one that is developed, urbanised, capitalist, secular and modern. The rise of 'the West' may then be related to the latter Middle Ages in Europe and specific processes of economic, political, social and cultural change. In this sense, the term 'Western' is almost synonomous with the term 'MODERN'.

Edward Said has explored how the historical dominance of the West – meaning here Europe – was sustained by the construction of an irrational, exotic, erotic, despotic, heathen Orient which thereby secured the West as rational, familiar, moral, just and Christian (Said, 1978). 'ORIENTALISM' is viewed as a discourse in which the West's knowledge of the Orient is bound up with its economic and political dominance. The currency of the East–West dualism is therefore related to the global history of economic injustice and the specific goals and requirements of IMPERIALISM (Jacobs, 1996a).

One of the most telling criticisms of Said's 1978 work is that it ignores internal divisions within 'the West' (Lewis, 1996; Young, R., 1990). A number of feminists and feminist geographers have turned their attention to women's involvement in the imperial cultural production of the 'East' and 'West' (see Blunt and Rose, 1994; Lewis, 1996; Mills, 1991). Reina Lewis, for example, argues that women's differential access to forms of imperial production resulted in an Orientalist GAZE that was less pejorative and absolute than the discourse emphasised by Said (Lewis, 1996: 4).

Running parallel to, and influenced by, such efforts to disaggregate 'the West' historically are attempts to undermine the BINARY construction upon which the term rests. This is related to at least two, connected, cultural, social and informational shifts in the contemporary world. First, there has been a rise in the visibility and extent of African, Asian and Latin American DIASPO-RAS within North America and Europe. This is often referred to as an aspect of 'globalisation' and summarised by the idea of 'the rest' in 'the West' (Hall, 1993) or the 'Third-Worlding' of the West (Koptiuch, 1996). This opens up the possibility of research into how diasporic communities rework individual, regional or national identities in their new locations. Janice Boddy, for example, shows how Sudanese women refugees in Toronto have utilised established binary constructions of 'the West' and TRADITION so as to refashion their personal and national identities (Boddy, 1995).

Second, there has been an expansion of informational and media networks into what has been classically termed the 'THIRD WORLD': a process often conceptualised as a form of 'Westernisation'. This might be best theorised as a dialogue between a European, or North American, 'presence' and 'indigenous' consumption and expression whereby cultural identity is consistently renegotiated (Hall, 1993). The task for a feminist geographer here might be to explore in which conditions these articulations operate to reinforce or break down established gender hierarchies (Chowdhry, 1994).   CJ

Whiteness Following debates about the social construction of 'RACE' in human geography and across the social sciences there has been a growing recognition that these arguments apply to dominant groups as well as to so-called 'ETHNIC minorities'. The analysis of 'whiteness' as a social construction was pioneered in other disciplines, including history, film and literary studies, where authors such as Richard Dyer (1988) argued that 'whiteness' was so taken for granted as to be almost invisible. Toni Morrison (1992) also noted the category's apparent invisibility in US literary studies, abhorring the critical blindness that sees 'whiteness' as mute, meaningless and unfathomable. Historians have since explored the role of 'white mythologies' in British colonial history (McClintock, 1995; Young, R., 1990) and documented how some US ethnic groups accepted the 'wages of whiteness' in order to distinguish themselves from African-Americans (Roediger, 1992).

Within geography, Alastair Bonnett (1992) has drawn attention to the neglect of anti-racist education in the 'white highlands' of Britain, away from the main centres of ethnic minority settlement, asserting the importance of distinguishing between specific forms of 'whiteness' and their implications for anti-racist education (Bonnett, 1997). Feminists have contributed to these debates through ETHNOGRAPHIC and historical research on the social construction of 'whiteness' (Frankenberg, 1993a and b), drawing attention to the mutual constitution of racialised and gendered identities (Hall, C., 1992; Ware, 1992). 'Whiteness' has come to be recognised as an historically specific social formation, shaped within a racialised problematic (Jackson, 1998; Kobayashi and Peake, 1994). PJ

Witch/crone Some feminists today identify as witches, or 'crones' (wise old women), developing, learning and practising spiritual and healing crafts (Starhawk, 1982). They proudly reclaim these terms drawn from pre-Christian earth-based and pagan traditions which stress the interconnectedness of humans and nature. CN

Woman A term used, often in inverted commas, to refer to the female subject; it is a deeply problematic term as it comes with theoretical and political connotations of UNIVERSALISM or ESSENTIALISM, despite Simone de Beauvoir's argument in her classic text *The Second Sex* (1949) that 'one is not born, but becomes a woman'. This becoming is variously explained through, for example, entry into the phallocentric symbolic order in which woman is, according to Lacan, defined as lack, or through the regulatory regime of COMPULSORY HETEROSEXUALITY, in which woman is defined by her attraction and opposition to man (Butler, 1990; Rich, 1980). According to some lesbian feminists, therefore, a lesbian is not a 'woman' as she escapes from this compulsory regulation (Wittig, 1992).

In contemporary feminist writing the term women is now more common as the specificity of the multiple ways of becoming a woman and doing gender are recognised. Indeed, feminists influenced by DECONSTRUCTION have argued that SUBJECTIVITY and IDENTITY are now so fluid and complex that a single subject position is impossible, thus seeming to challenge the whole basis of feminist politics. However, some feminists still celebrate specifically female ways of being or female attributes, for example the caring ethic of women (Gilligan, 1982), the relational construction of femaleness in Chodorow's (1978) OBJECT RELATIONS THEORY or female ways of talking and interacting (Spender, 1980; Tannen, 1994). In geographical discussions of methods, the question about a specifically female or feminist way of doing research has been discussed. LM

See also PSYCHOANALYSIS; SELF; SUBJECT/SUBJECTIVITY.

Woman question, the A term used by nineteenth-century feminists and socialists to refer to struggles for female emancipation; the great writer Mary

Wollstonecraft (1792) titled her famous plea for emancipation and equality the 'Vindication of the rights of woman'. LM

Women and Geography Study Group (WGSG) This group started from small beginnings in the late 1970s, first as an informal group of geographers who met regularly and then as a working party under the auspices of the INSTITUTE OF BRITISH GEOGRAPHERS (IBG), the then professional organisation for geographers working in higher education institutions. In 1980, the working party achieved more formal recognition and became one of the established study groups of the Institute. Anxiety about its aims resulted in its rather timid title – women and geography – neither GENDER nor FEMINISM were then acceptable terms to the council of the IBG, at least in the name of the group. However, in the twofold statement of its aims, bolder language was used. The group was established (1) to encourage the study of the geographical implications of gender differentiation and undertake geographical research from a feminist perspective and (2) to exchange ideas and information and to work to improve the position of women within the discipline. The last part of the second aim was omitted from the statement in the second book published by the WGSG (1997) but, as feminists have long argued, change in women's position is an important part of the definition of feminism. The WGSG published its first book in 1984 and both this one and the latter were not only published under the name of the group rather than the individuals involved but also the royalties were paid to the group to assist in conference and meeting attendance by its members. Groups with similar aims exist in other countries, for example the Canadian WGSG. In the USA there are two groups: one with a research focus, GENDER PERSPECTIVES ON WOMEN (GPOW), and the other with the same aim of improving women's position as academic geographers, the Commission on the Status of Women in Geography.

In Britain in the mid-1990s the Institute of British Geographers merged with the ROYAL GEOGRAPHICAL SOCIETY (RGS) and its members became Fellows of the RGS/IBG. This merger was in the face of opposition from RADICAL GEOGRAPHERS, including many members of the WGSG who regard the RGS as a PATRIARCHAL and COLONIAL institution with its history of involvement in IMPERIAL adventures. In addition, the RGS accepts corporate sponsorship from organisations such as Royal Dutch Shell whose activities in oil exploration have adverse environmental impacts. In an infamous case in Nigeria the Ongoni people mobilised to prevent Shell's activities and were savagely treated by the Nigerian Government. Some members of the WGSG have remained members of the merged RGS/IBG, hoping to achieve change from within, whereas others have resigned or not joined at all. Membership of the Group is open to non-RGS/IBG members for a small fee. The group publishes a newsletter, organises a conference in September each year and sponsors sessions at the annual RGS/IBG conference each January. It also organises occasional study weekends. LM

Women in Development (WID) A general term used to refer to an approach to development issues and programmes in Third World countries that both took women's issues and interests into account and involved women in these programmes. While WID was to some extent important in increasing women's independence and their EMPOWERMENT, it underestimated the structural inequalities between women and men and has now largely been replaced by the approach known as GENDER AND DEVELOPMENT (GAD).   LM
See also DEVELOPMENT; UNITED NATIONS; WORLD: FIRST, SECOND, THIRD AND FOURTH.

Women's movement/women's liberation movement See SECOND WAVE FEMINISM/SECOND WAVE WOMEN'S MOVEMENT; see also FEMINISM; SUFFRAGE/ SUFFRAGE CAMPAIGNS

Women of colour See COLOUR, WOMEN OF

Work/work force/employment The application of mental or physical effort to carrying out tasks that serve human needs. The term has been extensively contested and has different meanings in different discourses, as feminists have documented. Since the 1970s they have challenged notions of work based on the production or manufacture of goods and services for the market. This definition, narrowly based on renumerated work (or employment), renders the home invisible (see Olsen, 1978). Unpaid DOMESTIC LABOUR and childcare, for example, also constitute work and must be encompassed within a sound definition. As McDowell and Pringle (1992) suggest, this type of work is essential in producing socialised individuals and the current and future work force.

Feminists, therefore, have attacked notions of work which are in opposition to the HOME. But, as Pahl (1988) points out, it is only since the Industrial Revolution that work (or the PUBLIC sphere) and home (the PRIVATE sphere) have become spatially separated and the former tantamount to employment. Feminist historians analysing women's work in the pre-industrial West have shown that women participated in production and played a key role in, for example, COTTAGE INDUSTRY (see Tilly and Scott, 1978). Some suggest that the onset of industrialisation and the emergence of 'separate spheres' was unfavourable for women because 'work' became primary to masculine identity and 'home' to feminine identity (see Clark, 1992). Women's involvement in the work force, therefore, was considered secondary to their domestic responsibilities. The 'separate spheres' debate, however, can be criticised on a number of counts. For example, the growing number of women who perform HOMEWORK offers a challenge to the thesis (see Allen and Wolkowitz, 1987; Bradley, 1989). The argument is also ethnocentric. The debate ignores the fact that a sizeable number of women in industrialising developing countries work in the informal sector, and, although they may be paid, this type of work is often unrecognised in official employment statistics (see Momsen, 1991). The political and practical problems of how to count work to include the informal sector, as well as domestic and caring labour, in both 'the West' and developing nations are well documented, although few realistic solutions have been offered (see Malos, 1995).

Clearly, a highly charged debate has emerged about women's historical role in PRODUCTION and the impact of industrialisation on women. It is generally agreed, however, that since the end of World War II there has been a continuous rise in the number of women in employment across the globe, typically referred to as the feminisation of the work force (a notable exception to this argument is Hakim (1996)). It is also clear that women are confined to relatively low-paid jobs which are classified as unskilled or semi-skilled, and are increasingly entering FLEXIBLE or non-standard forms of employment, such as part-time work and job-sharing. Hakim (1979) and McDowell (1989), among others, have explored how women's employment is segregated, both horizontally, in a limited number of occupations, and vertically, so that women are employed at the bottom of the occupational ladder. It is the concentration of women in predominately stereotyped 'female' jobs, such as garment manufacture, and the probability that, even within relatively integrated economies, women and men tend to work in different occupations, that has led to the development of various theories that seek to explain the gender division of labour (for a summary of the main schools of thought see Dex, 1985; Walby, 1990). Feminist geographers, for example, have sought a spatial explanation to occupational SEGREGATION, with most studies focusing on the significance of local variations within cities, labour-market catchment areas and women's journey-to-work patterns (Hanson and Pratt, 1995).

The allocation of tasks on the basis of sex is not restricted to the public sphere. The domestic division of labour has also been widely documented. Since the early 1980s researchers have questioned whether it is becoming more egalitarian in light of the FEMINISATION OF THE WORK FORCE (Cooper and Lewis, 1993), the rise in male unemployment (Wheelock, 1990) and the technological revolution in household appliances (Cowan, 1983). Although men tend to do the more traditional male tasks, such as household repairs and gardening, the majority of studies on 'the West' conclude that the 'new man' is a myth and that women continue to do the bulk of housework and unpaid caring work. Analyses reveal that women in developing countries also bear the brunt of these tasks (Momsen, 1991).

It is ironic that only three decades ago it was assumed that women were excluded from work. Feminists reacted to this ANDROCENTRIC and inaccurate picture, and through broadening the standard definitions of work and through publishing a myriad of empirical and theoretical studies of women's work-force participation they have provided a more accurate analysis suggesting that the fundamental predicament women face is not one of exclusion, but that they have too much work (McDowell and Pringle, 1992). Evidently, the participation of women in paid and unpaid work is spatially uneven, and the explanations for women's position are diverse, so perhaps feminist geographers have a particular perspective to offer the debate.   RK

World: First, Second, Third, Fourth The terms First World, Second World and Third World reflect Cold War conceptualisations of space. Originally, the Second World was the communist bloc and the First World represented capitalist countries. In this scenario the Third World comprised countries pursuing a 'third way' (drawing on the notion of the third estate in the French revolution), following a trajectory which was neither capitalist nor communist. In the post-war period the Third World was associated with the non-aligned movement as decolonisation created NATION-STATES with a new voice on the world stage.

Currently, the Third World is synonymous with POVERTY and under-development. Consequently, attempts have been made to develop alternative conceptualisations of the world. Brandt (1980) describes a rich 'North' (the northern hemisphere plus Australasia) and a poor 'South', while World Systems analysis uses cores, peripheries and semi-peripheries (Taylor, 1995). Some non-government organisations use the term 'Two-Thirds World' as a play on 'Third World' thus emphasising that the majority of the world lives in poverty. As the WGSG has argued, however, even though the fall of the Berlin Wall means that the First World–Third World dichotomy has outlived capitalist–communist binaries,

> the continued use of these terms, both academically and in everyday language, testifies to the time, effort and resources which are deployed in maintaining and reinforcing the boundaries drawn between both sides of [this] dichotomy, and the implicit hierarchies which might exist within it (1997: 113–14).

The hierarchies which exist within Third World–First World are revealed by the recent emergence of a further classification, Fourth World. This term most commonly refers to the poorest Third World countries, thereby distinguishing between newly industrialising countries (NICs) and countries experiencing extreme poverty. The Fourth World, however, sometimes also refers to first nation and native peoples seeking sovereignty over their homelands (Griggs, 1992).

Since the work of Ester Boserup (1970), feminists have focused on another hierarchy within the 'Third World'. Boserup argued that women's roles are invisible in development planning and that processes of modernisation and INDUSTRIALISATION increase gender inequality. Boserup's work became the starting point for the WOMEN IN DEVELOPMENT (WID) and GENDER AND DEVELOPMENT (GAD) paradigms promoted under the UN Decade for Women 1976–85 (Rathgeber, 1990). These approaches not only aim to make women's roles more visible (Momsen and Townsend, 1987) but also emphasise the connections between development processes and changing gender relations (Moser, 1993a; Ostergaard, 1992).

Despite the growing emphasis on gender relations in development planning, Western feminists have been criticised for representing Third World women as passive victims, objectified and frozen in time and space (Mohanty, 1991a). Sarah Radcliffe (1994) argues that while some accounts by Western feminists exoticise and homogenise 'Third World women', other accounts, including those of geographers (feminist or otherwise), see Third World women as 'privileged loci of knowledge rather than active, experiencing subjects' (Radcliffe, 1994: 26). Radcliffe suggests that, by citing a few well-known Third World women academics, some geographers disclaim the right to speak about Third World women. Such an approach, she suggests, is dangerous because, in so doing, geographers abdicate the responsibility of engaging critically and personally with the global power relations which help classify the world into First World–Third World binaries in the first place.   NL

Writing

> In our common struggle and in our writing we reclaim our tongues. We wield a pen as a tool, a weapon, a means of survival, a magic wand that will attract power, that will draw self-love into our bodies (Anzaldúa, 1981: 163).

Retrieving women's written works, especially diaries, letters, memoirs and TRAVEL accounts, constituted part of 'stage one' of writing women's lives and experiences into the corpus of Anglo-American geography. Much, if not all, of geographic research was gender-blind through the 1970s, erroneously generalising theories and empirical findings from men's writings and experiences alone. In the early 1980s several feminist geographers called not only for making women visible in geographic research but also for questioning why PATRIARCHAL social relations globally and historically had severely restricted women's movements and imaginaries (WGSG, 1984). Feminist historical and cultural geographers in particular relied on primary written sources to read against the grain of traditional geography in reconstructing or reinterpreting places and landscapes from women's point of view, for instance in reading women's travel and exploration as heroic escape from confining social parameters (Middleton, 1982).

Throughout the 1980s VOICES of women of colour across disciplines initiated a critique of the category 'woman' and exposed the RACISM, classism and SEXISM apparent in both male-authored and female-authored texts (Norwood and Monk, 1987). And, while many feminists denounced 'high theory' as a practice aimed to protect white male privilege and prestige, many turned to DECONSTRUCTIVE analyses of writing. The deconstructive turn (after Saussure and Derrida) challenged the individualism and unified subjectivity of writers and offered a new way to think about writing as fragmentary, relational, lacking in essential meaning and based on problematical DUALISMS. From a POST-STRUCTURALIST perspective, all experiences and realities are fundamentally DISCURSIVE and linguistic. Feminist post-structuralists argue that it is not enough simply to retrieve women's writing, but instead one must work within a broadly conceived problematic of how reality is constructed through language, signs and discourses. After Foucault, these themes became inextricably linked to tensions between knowledge and POWER. Thus an overarching interest in post-structuralist feminism has been in how truth claims are produced. Feminist historian Joan Scott (1988a) argues that history is not a study of 'what happened', to be discovered and transmitted by the historian but is rather a set of practices for producing knowledge and power. Feminist geographers examine the contingent socio-geographic relations that produce multiple discursive subjectivities and multiple discursive meanings of space and place. The constitution of IDENTITY over SPACE and PLACE, interpretations of historical and cultural landscapes or forms, representations of gender, race, class and sexuality in colonial and imperial settings and the recovery of 'agency' and resistance of SUBALTERN subjects, have all become important in feminist geographic analyses of writing (Blunt, 1994; Blunt and Rose, 1994; Jones *et al.*, 1997; Pile and Thrift, 1995).

Some geographers have also called attention to the phallocentrism and masculinism of geographic writing, not only its content but also its form and structure. Drawing primarily on feminist critiques of Lacanian psychoanalytic models of gender identity formation (e.g. Cixous, 1981b), these geographers propose strategies that move away from the authoritative, dualistic and rationalist modes of discourse and categories of language. They argue for more open-ended, unbounded and non-linear writing, which often means binding together LANGUAGE and CORPOREALITY (Bondi, 1997; Pile and Thrift, 1995; Rose, 1993a).   KM

See also NARRATIVE; REPRESENTATION; TEXT.

Bibliography

Abramovitz, M. 1988: *Regulating the Lives of Women: Social Welfare Policy from Colonial Times to the Present.* Boston, MA: South End Press.

Acker, J. 1990: Hierarchies, jobs, bodies: a theory of gendered organisations *Gender and Society* 4, 139–58.

Adam, B. 1990: *Time and Social Theory.* Cambridge: Polity Press.

Adam, B. 1995: *Timewatch: The Social Analysis of Time.* Cambridge: Cambridge University Press.

Adams, P. and Cowie, E. (eds) 1990: *The Woman in Question.* Cambridge, MA: MIT Press.

Aglietta, M. 1979 [1976]: *A Theory of Capitalist Regulation: The US Experience.* London: New Left Books.

Agnew, J. 1987: *Place and Politics: The Geographical Mediation of State and Society.* London: Allen and Unwin.

Agnew, J. 1989: Sameness and difference: Hartshorne's *The Nature of Geography* and geography as areal variation. In Entrikin, J. and Brunn, S. (eds), *The Nature of Geography,* Occasional Publication, Association of American Geographers.

Agnew, J. and Corbridge, S. 1995: *Mastering Space: Hegemony, Territory and International Political Economy.* London: Routledge.

Aitken, L. and Griffin, G. 1996: *Gender Issues in Elder Abuse.* London: Sage.

Aitken, S. and Herman, T. 1997: Gender, power and crib geography: transitional spaces and potential places. *Gender, Place and Culture* 4 (1), 63–88.

Ajani, A. 1994: Opposing oppositional identities: fabulous boys and the refashioning of subculture theory. Unpublished MA thesis, Department of Anthropology, Stanford University, Stanford, CA.

Albinksi, N.B. 1988: *Women's Utopias in 19th and 20th Century Fiction.* London: Routledge.

Alcoff, L. 1988: Cultural feminism versus poststructuralism: the identity crisis in feminist theory *Signs: Journal of Women in Culture and Society* 13, 405–36.

Alcoff, L. 1990: Feminist politics and Foucault: the limits to a collaboration. In Dalley, A. and Scott, C. *Crises in Continental Philosophy.* Albany: SUNY Press.

Alcoff, L. 1996: Feminist theory and social science. In Duncan, N. (ed.), *Bodyspace.* London: Routledge.

Ali, Y. 1992: Muslim women and the politics of ethnicity and culture in Northern England. In Sahgal, G. and Yuval-Davis, N. (eds), *Refusing Holy Orders.* London: Virago.

Allen, J. 1990: Does feminism need a theory of the state? In Watson, S. (ed.), *Playing the State.* Sydney: Allen & Unwin.

Allen, J. 1994: Restructuring the world: review article. *Work, Economy & Society* 8 1), 113–26.

Allen, J. and Henry, N. 1997: Ulrich Beck's Risk Society at work: labour and employment in the contract services industries. *Transactions of the Institute of British Geographers* 22, 180–96.

Allen, J., Massey, D. and Cochrane, A. 1998: *Rethinking the Region*. London: Routledge.

Allen, P.G. 1984: *The Sacred Hoop: Recovering the Feminine in American Traditions*. Boston: Beacon Press.

Allen, S. and Wolkowitz, C. 1987: *Homeworking: Myths and Realities*. London: Macmillan.

Allen, T. and Thomas, A. (eds) 1992: *Poverty and Employment in the 1990s*. Oxford: Oxford University Press/Buckingham: Open University Press.

Althusser, L. 1971: Ideology and ideological state apparatuses. In *Lenin and Philosophy and Other Essays*, trans. B. Brewster. New York: Monthly Review Press.

Althusser, L. 1985 [1971]: Ideology and state apparatuses (notes towards an investigation). Reprinted in Beechy, V. and Donald, J. (eds), *Subjectivity and Social Relations*. Milton Keynes: Open University Press.

Alvarez, S. and Escobar, A. (eds) 1992: *The Making of Social Movements in Latin America: Identity, Strategy and Democracy*. Boulder, CO: Westview Press.

Amin, A. (ed) 1994: *Post-Fordism: A Reader*. Oxford: Basil Blackwell.

Amin, A. and Thrift, N. (eds) 1994: *Globalization, Institutions and Regional Development*. Oxford: Oxford University Press.

Andermahr, S., Lovell, T. and Walkowitz, C. 1997: *A Glossary of Feminist Theory*. London: Arnold.

Anderson, B. 1983: *Imagined Communities: Reflections on the Origin and Spread of Nationalism*. London: Verso.

Anderson, B. 1991: *Imagined Communities: Reflections on the Origin and Spread of Nationalism*, revised edition. London: Verso.

Anderson, H. 1992: *Utopian Feminism: Women's Movements in Fin-de-Siecle Vienna*. New Haven and London: Yale University Press.

Anderson, K. 1991: *Vancouver's Chinatown: Racial Discourse in Canada, 1875–1980*. Montreal: McGill–Queen's University Press.

Anderson, K. and Gale, F. (eds) 1992: *Inventing Places: Studies in Cultural Geography*. Melbourne: Longman Cheshire.

Andrusz, G. 1984: *Housing and Urban Development in the USSR*. London: Macmillan.

Anthias, F. and Yuval-Davis, N. 1983: Contextualising feminism. *Feminist Review* 14, 62–75.

Anthias F. and Yuval-Davis N. 1992*: Racialized Boundaries: Race, Nation, Gender, Colour and Class and the Anti-racist Struggle*. London: Routledge.

Anzaldúa, G. 1987. *Borderlands/La Frontera: The New Mestiza*. San Francisco, CA: Spinsters/Aunt Lute.

Anzaldúa, G. 1997: To live in the borderlands means you... . In McDowell, L.

and Sharp, J.P. (eds), *Space, Gender, Knowledge; Feminist Reading*. London: Arnold.

Appadurai, A. 1990b: Disjuncture and difference in the global cultural economy *Theory, Culture and Society* 295–310.

Appadurai, A. 1996: *Modernity at Large: Cultural Dimensions of Globalization*. Minneapolis: University of Minnesota Press.

Ardner, S. (ed.) 1975: *Perceiving Women*. New York: Halsted Press.

Arens, W. 1981: Professional football: an American symbol and ritual. In Arens, W. and Montague, S.P. (eds), *The American Dimension: Cultural Myths and Social Realities*, 2nd edn. Sherman Oaks, CA: Alfred Publishing.

Armstrong J. 1982: *Nations before Nationalism*. Chapel Hill, NC: University of North Carolina Press.

Ashcroft, B., Griffiths, G. and Tiffin, H. 1989: *The Empire Writes Back*. London: Routledge.

Ashcroft, B., Griffiths, G. and Tiffin, H. (eds) 1995: *The Postcolonial Studies Reader*. London: Routledge.

Attwood, L. 1990: *The New Soviet Man and Woman*. Basingstoke, Hants: Macmillan Education.

Atwood, M. 1986: *The Handmaid's Tale*. London: Cape.

Auerbach, E. 1953: *Mimesis: The Representation of Reality in Western Literature*. Princeton, NJ: Princeton University Press.

Baca Zinn, M. and Thorton Dill, B. (eds) 1994: *Women of Color in U.S. Society*. Philadelphia, PA: Temple University Press.

Baer, W. 1986: Housing in an internationalizing region: housing stock dynamics in Southern California and the dilemmas of fair share. *Environment and Planning D: Society and Space* 4, 337–50.

Bagguley, P., Mark-Lawson, J., Shapiro, J. *et al.* 1990: *Restructuring: Place, Class and Gender*. London: Sage.

Baishya, P. 1989: *Small and Cottage Industries: A Study in Assam*. Delhi: Manas Publications.

Bakhtin, M.M. 1981: *The Dialogic Imagination: Four Essays,* trans. C. Emerson and M. Holquist. Austin: University of Texas Press.

Bakshi, P., Goodwin, M., Painter, J. and Southern, A. 1995: Gender, race, and class in the local welfare state: moving beyond regulation theory in analysing the transition from Fordism. *Environment and Planning A* 27, 1539–54.

Balchin, N. 1972: Graphicacy. *Geography* 57, 185–92.

Bammer, A. 1991: *Partial Visions: Feminism and Utopianism in the 1970s*. London: Routledge.

Bammer, A. (ed.) 1994: *Displacements: Cultural Identities in Question*. Bloomington, IN: Indiana University Press.

Banks, O. 1981: *The Faces of Feminism*. London: Martin Robertson.

Barff, R. and Austen, J. 1993: 'It's gotta be da shoes': domestic manufacturing, international subcontracting, and the production of athletic footwear. *Environment and Planning A* 25, 1103–14.

Barlow, G.W. 1980: The development of sociobiology: a biologist's perspective.

In Barlow. G.W. and Silverberg J. (eds), *Sociobiology: Beyond Nature/Nurture? Reports, Definitions and Debate*. Boulder, CO: Westview Press.

Barnes, T. and Duncan, J. (eds) 1992: *Writing Worlds: Discourse, Text and Metaphor in the Representation of Landscape*. London: Routledge.

Barnes, T. and Gregory, D. (eds) 1997: *Reading Human Geography: The Politics and Poetics of Inquiry*. London: Arnold.

Barnett, C. 1997: Book reviews: Thirdspace: journeys to Los Angeles and other real and imagined places. *Transactions of the Institute of British Geographers*. 22, 529–40.

Barrett, M. 1988: *Women's Oppression Today: The Marxist Feminist Encounter*. London: Verso.

Barrett, M. 1989: Some different meanings of the concept of 'difference': feminist theory and the concept of ideology. In Meese, E. and Parker, A. (eds), *The Difference Within: Feminism and Critical Theory*. Amsterdam: John Benjamin.

Barrett, M. 1991: *The Politics of Truth: From Marx to Foucault*. Cambridge: Polity Press.

Barrett, M. and McIntosh, M. 1982: *The Anti-social Family*. London: Verso.

Barrett, M. and Phillips, A. (eds) 1992: *Destabilising Theory: Contemporary Feminist Debates*. Cambridge: Polity Press.

Barthes, R. 1975: *The Pleasure of the Text*, trans. R. Miller. London: Jonathan Cape.

Barthes, R. 1977: The Death of the Author. In *Image–Music–Text*, trans. S. Heath. New York: Hill and Wang.

Bartkowski, F. 1989: *Feminist Utopias*. Lincoln, NE: University of Nebraska Press.

Bartlett, K. and Kennedy, R. (eds) 1991: *Feminist Legal Theory: Readings in Law and Gender*. Boulder, CO: Westview Press.

Battersby, C. 1994: *Gender and Genius: Towards a Feminist Aesthetics*. London: Women's Press.

Baudelaire, C. 1964 [1863]: The painter of modern life. In Baudelaire, C., *The Painter of Modern Life and Other Essays*. London: Phaidon.

Baudrillard, J. 1983: *Simulations*. New York: Semiotext(e).

Beale, J. 1987: *Women in Ireland: Voices of Change*. London: Macmillan.

Beauregard, R. 1986: The chaos and complexity of gentrification. In Smith, N. and Williams, P. (eds), *Gentrification of the City*. Boston, MA: Allen and Unwin.

Beaverstock, J. 1994: Re-thinking skilled international labour migration: World Cities and banking organisations. *Geoforum* 25, 325–38.

Beck, U. 1994 [1986]: *Risk Society: Towards a New Modernity*. London: Sage.

Beck, U., Giddens, A. and Lash, A. (eds) 1994: *Reflexive Modernization*. Cambridge: Polity Press.

Beechey, V. 1987: *Unequal Work*: London: Verso.

Beechey, V. and Perkins, T. 1987: *A Matter of Hours: Women, Part-time Work and the Labour Market*. Cambridge: Polity Press.

Beetham, M. 1996: *A Magazine of Her Own? Domesticity and Desire in the Woman's Magazine, 1800–1914.* London: Routledge.

Behar, R. and Gordon, D. (eds) 1995: *Women Writing Culture.* Berkeley, CA: University of California Press.

Bell, D. 1973: *The Coming of Post-industrial Society: A Venture in Social Forecasting.* New York: Basic Books.

Bell, D. and Valentine, G. (eds) 1995: *Mapping Desire: Geographies of Sexualities.* London: Routledge.

Bell, D. and Valentine, G. 1997: *Consuming Geographies: You are Where You Eat.* London: Routledge.

Bell, D., Binnie, J., Cream, J. and Valentine, G. 1994: All hyped up and no place to go. *Gender, Place and Culture* 1, 31–47.

Bell, D., Caplan, P. and Wazir, J. 1993: Gendered Fields: Women, Men and Ethnography. London: Routledge.

Bell, M. and McEwan, C. 1996: The admission of women Fellows to the Royal Geographical Society, 1892–1914; the controversy and the outcome. *The Geographical Journal* 162, 295–312.

Bell, M., Butlin, R.A. and Heffernan, M.J. (eds) 1995: *Geography and Imperialism, 1820–1940.* Manchester: Manchester University Press.

Bell, S. 1994: *Reading, Writing and Rewriting the Prostitute Body.* Bloomington, IN: Indiana University Press.

Bell, V. 1993: *Interrogating Incest: Feminism, Foucault and the Law.* London: Routledge.

Bendix, R. 1997: *In Search of Authenticity: The Formation of Folklore Studies.* Madison, WI: University of Wisconsin Press.

Benedict, R. 1934: *Patterns of Culture.* Boston, MA: Houghton Mifflin.

Benería, L. and Sen, G. 1982: Class and gender inequalities and women's role in economic development. *Feminist Studies* 3, 203–25.

Benhabib, S. 1986: *Critique, Norm, and Utopia: A Study of the Foundations of Critical Theory.* New York: Columbia University Press.

Benhabib, S. 1987: The generalised and concrete other. In Kittay, E. and Meyers, D. (eds), *Women and Moral Theory.* Savage MD: Rowman and Littlefield.

Benhabib, S. 1995: *Feminist Contentions: Philosophical Exchanges.* London: Routledge.

Benjamin, D. (ed.) 1995: *The Home: Words, Interpretations, Meanings and Environments.* Aldershot, Hants: Avebury.

Benjamin, J. 1988: *The Bonds of Love.* London: Virago.

Benstock, S. 1986: *Women of the Left Bank: Paris, 1900–1940.* Austin: University of Texas Press.

Benton, T. 1989: Marxism and Natural Limits. *New Left Review* 178, 51–86.

Benton, T. 1991: Biology and social science: why the return of the repressed should be given a (cautious) welcome. *Sociology* 25, 1–29.

Benton, T. 1992: Why the welcome needs to be cautious: a reply to Keith Sharp. *Sociology* 26, 225–32.

Benton, T. 1993: *Natural Relations: Ecology, Animal Rights and Social Justice.* London: Verso.

Berg, L. 1998: Reading (post)colonial history: masculinity, 'race', and rebellious natives in the Waikato, New Zealand, 1863. *Historical Geography* 26, forthcoming.

Berg L. and Kearns R. 1996: Naming as norming: 'race', gender, and the identity politics of naming places in Aotearoa/New Zealand. *Environment and Planning D: Society and Space* 14, 99–122.

Berg, M., Hudson, P. and Sonenscher, M. (eds) 1983: *Manufacture in Town and Country Before the Factory.* Cambridge: Cambridge University Press.

Berger, J. 1972: *Ways of Seeing.* London: British Broadcasting Corporation/ Harmondsworth, Middx: Penguin Books.

Berger, P. and Luckman, T. 1967: *The Social Construction of Reality.* London: Allen Lane.

Berman, M. 1983: *All That is Solid Melts into Air: The Experience of Modernity.* London: Verso.

Berry, B. and Marble, D.F. (eds) 1965: *Spatial Analysis: A Reader in Statistical Geography.* Englewood Cliffs, NJ: Prentice-Hall.

Berry, C.J. 1986: *Human Nature: Ideas in Political Theory.* London: Macmillan.

Berthoud, R. 1976: *The Disadvantages of Inequality: A Study of Social Deprivation.* London: MacDonald and Jane's.

Beynon, H. 1984: *Working for Ford.* Harmondsworth, Middx: Penguin Books.

Bhabha, H. 1983: The Other question: Homi K. Bhabha reconsiders the stereotype and colonial discourse. *Screen* 24, 18–36.

Bhabha, H. 1984a: Representation and the colonial text: a critical exploration of some forms of mimeticism. In Gloversmith, F. (ed.), *The Theory of Reading.* Brighton, E. Sussex: Harvester.

Bhabha, H. 1984b: Of mimicry and man: the ambivalence of colonial discourse. *October* 28, 125–33.

Bhabha, H. 1986: Foreword: remembering Fanon – self, psyche and the colonial condition. In Fanon, F., *Black Skin, White Masks.* London: Pluto Press.

Bhabha, H. 1990: The third space: interview with Homi Bhabha. In Rutherford, J. (ed.) *Identity: Community, Culture and Difference.* London, Lawrence and Wishart.

Bhabha, H. 1991: The postcolonial critic. *Arena* 96, 47–63.

Bhabha, H. 1994: *The Location of Culture.* London: Routledge.

Bhaskar, R. 1975: *A Realist Theory of Science.* Leeds: Leeds Books.

Bhaskar, R. 1979: *The Possibility of Naturalism: A Philosophical Critique of the Contemporary Human Sciences.* Brighton, E. Sussex: Harvester Press.

Bhaskar, R. 1986: *Scientific Realism and Human Emancipation.* London: Verso.

Biehl, J. 1991: *Rethinking Ecofeminist Politics.* Boston, MA: South End Press.

Biggs, S. 1993: *Understanding Ageing: Images, Attitudes and Professional Practice.* Buckingham: Open University Press.

Birke, L. 1986: *Women, Feminism and Biology: the Feminist Challenge.* Brighton, E. Sussex: Wheatsheaf.

Birke, L. and Silvertown, J. 1984: *More than the Parts: Biology and Politics*. London: Pluto Press.

Biyani, C. 1995: The needs of refugee women: a human rights perspective. *Gender and Development* 3, 29–35.

Blackaby, F. (ed.) 1979: *Deindustrialization*. London: William Heinemann.

Blasius, M. 1994: *Gay and Lesbian Politics: Sexuality and the Emergence of a New Ethic*. Philadelphia, PA: Temple University Press.

Blaut, J. 1987: *The National Question: Decolonising the Theory of Nationalism*. London: Zed Books.

Blea, I.I. 1992: *La Chicana and the Intersection of Race, Class, and Gender*. New York: Praeger.

Blomley, N. 1994: *Law, Space and the Geographies of Power*. London: Guildford Press.

Bloom, A. 1987: *The Closing of the American Mind*. Harmondsworth, Middx: Penguin Books.

Bluestone, B. and Harrison, B. 1982: *The De-industrialization of America*. New York: Basic Books.

Blum, V. and Nast, H. 1996: Where's the difference? The heterosexualisation of alterity in Henri Lefebvre and Jacques Lacan. *Environment and Planning D: Society and Space* 14, 559–80.

Blumer, H. 1969: *Symbolic Interactionism*. Englewood Cliffs, NJ: Prentice-Hall.

Blunt, A. 1994: *Travel, Gender, and Imperialism: Mary Kingsley and West Africa*. New York: Guildford Press.

Blunt, A. and Rose, G. (eds) 1994: *Writing, Women and Space: Colonial and Postcolonial Geographies*. New York and London: Guildford Press.

Bly, R. 1990: *Iron John: A Book About Men*. Reading, MA: Addison-Wesley.

Boas, F. 1955: *Race, Language and Culture*. New York: Macmillan.

Bock, G. and James, S. (eds) 1992: *Beyond Equality and Difference: Citizenship, Feminist Politics and Female Subjectivity*. London: Routledge.

Boddy, J. 1995: Managing tradition: 'superstition' and the making of national identity among Sudanese women refugees. In James, W. (ed.), *The Pursuit of Certainty: Religious and Cultural Formulations*. London and New York: Routledge.

Bondi, L. 1990a: Progress in geography and gender: feminism and difference. *Progress in Human Geography* 14, 438–45.

Bondi, L. 1990b: Feminism, postmodernism and geography: space for women? *Antipode* 22, 156–68.

Bondi, L. 1991: Gender divisions and gentrification: a critique. *Transactions of the Institute of British Geographers* 16, 190–8.

Bondi, L. 1992a: Gender and dichotomy. *Progress in Human Geography* 16, 98–104.

Bondi, L. 1992b: Gender symbols and urban landscapes. *Progress in Human Geography* 16, 157–70.

Bondi, L. 1997: In whose words? On gender identities, knowledge, and writing practices. *Transactions of the Institute of British Geographers* 22, 245–58.

Bondi, L. and Domosh, M. 1992: Other figures in other places: on feminism, postmodernism, and geography. *Environment and Planning D: Society and Space* 10, 199–215.

Bonner, F *et al.* (eds) 1992: *Imagining Women: Cultural Representations and Gender*. Cambridge: Polity Press.

Bonnett, A. 1992: Anti-racism in 'white' areas: the example of Tyneside. *Antipode* 24, 1–15.

Bonnett, A. 1996: The new primitives: identity, landscape and cultural appropriation in the mythopoetic men's movement. *Antipode* 28, 273–29.

Bonnett, A. 1997: Geography, 'race' and Whiteness: invisible traditions and current challenges. *Area* 29, 193–9.

Bookchin, M. 1989: *Remaking Society*. New York: Black Rose Books.

Booth, C., Darke, J. and Yeandle, S. (eds) 1996: *Changing Places: Women's Lives in the City*. London: Paul Chapman.

Bordo, S. 1993: *Unbearable Weight*. Berkeley and Los Angeles, CA: University of California Press.

Boserup, E. 1970: *Women's Role in Economic Development*. New York: St Martin's Press.

Boston, S. 1980: *Women Workers and the Trade Unions*. London: Lawrence and Wishart.

Bottomley, A. (ed.) 1996: *Feminist Perspectives on the Foundational Subjects of Law*. London: Cavendish.

Bourdieu, P. 1977: *Outline of a Theory of Practice*. Cambridge: Cambridge University Press.

Bourdieu, P. 1984: *Distinction: A Social Critique of the Judgement of Taste*. Cambridge, MA: Harvard University Press.

Bourdieu, P. and Passeron, J. 1977: *Reproduction in Education, Society and Culture*. London: Sage.

Bowlby, J. 1979: *The Making and Breaking of Affectional Bonds*. London: Tavistock.

Bowlby, R. 1985: *Just Looking: Consumer Culture in Dreiser, Gissing and Zola*. New York: Methuen.

Bowlby, S. 1990: Women, work and the family: constraints and control. *Geography* 75, 17–26

Boyer, M.C. 1994: *The City of Collective Memory: Its Historical Imagery and Architectural Entertainments*. Cambridge, MA: MIT Press.

Boyle, P.J. and Halfacree, K.H. (eds) 1998: *Migration and Gender in the Developed World*. London: Routledge.

Boys, J. 1984: Is there a feminist analysis of architecture? *Built Environment* 10, 25–34.

Boys, J. 1998: Beyond maps and metaphors? Re-thinking the relationships between architecture and gender. In Ainley, R. (ed.), *New Frontiers of Space, Bodies and Gender*. London: Routledge.

Bozzoli, B. and Delius, P. 1990: Radical history and South African society. *Radical History Review* 46/47, 13–45.

Bradford, M.G., Robson, B.T. and Tye, R. 1995: Constructing an urban deprivation index: a way of meeting the need for flexibility. *Environment and Planning A* 27, 519–33.

Bradley, H. 1989: *Men's Work, Women's Work*. Cambridge: Polity Press.

Brah, A. 1992: Difference, diversity and differentiation. In James, D. and Rattansi, A. (eds), *Race, Culture and Difference*. London: Sage.

Brah, A. 1996: *Cartographies of Diaspora: Contesting Identities*. London: Routledge.

Braidotti, R. 1994a: Body images and the pornography of representation. In Lennon, K. and Whitford, M. (eds), *Knowing the Difference: Feminist Perspectives in Epistemology*. London: Routledge.

Braidotti, R. 1994b: *Nomadic Subjects*. New York: Columbia University Press.

Braidotti, R. and Butler, J. 1995: Feminism by any other name. *Differences* 6, 27–61.

Brandt, W. 1980: North–south, a programme for survival: report of the Independent Commission on International Development Issues. London: Pan Books.

Braverman, H. 1974: *Labour and Monopoly Capital: The Degradation of Work in the Twentieth Century*. New York: Monthly Review Press.

Brettell, C. and Sargeant, C. 1992: Gender in Cross-cultural Perspective. Englewood Cliffs, NJ: Prentice-Hall.

Bridger, S., Kay, R. and Pinnick, K. 1996: *No More Heroines? Russia, Women and the Market*. London: Routledge.

Brock, R.N. and Thistlethwaite, S.B. 1996: *Casting Stones: Prostitution and Liberation in Asia and the United States*. Minneapolis, MN: Augsburg Fortress Press.

Brod, H. (ed.) 1987: *The Making of Masculinities: The New Men's Studies*. London: Allen and Unwin.

Brod, H. and Kaufman, M. (eds) 1994: *Theorizing Masculinities*. Thousand Oaks, CA: Sage.

Broe, M.L. and Ingram, A. 1989: *Women's Writing in Exile*. Chapel Hill, NC: University of North Carolina Press.

Brown, M. 1995a: Sex, scale and the 'New urban politics'. In Bell, D. and Valentine, G. (eds), *Mapping Desire: Geographies of Sexualities*. London: Routledge.

Brown, M. 1995b: Ironies of distance: an ongoing critique of the geographies of AIDS. *Environment and Planning D: Society and Space* 13, 159–83.

Brown, M. 1997: *Replacing Citizenship: AIDS Activism and Radical Democracy*. New York: Guildford Press.

Brown, M. and Madge, N. 1982: *Despite the Welfare State: A Report on the SSRC/DHSS Programme of Research into Transmitted Deprivation*. London: Heinemann Educational.

Brown, W. 1995: *States of Injury: Power and Freedom in Late Modernity*. Princeton, NJ: Princeton University Press.

Brownmiller, S. 1975: *Against Our Will: Men, Women and Rape*. London: Martin, Secker and Warburg/Buckingham: Open University Press.

Bruner, E. 1994: Abraham Lincoln as authentic reproduction: a critique of postmodernism. *American Anthropologist* 96, 397–415.

Bruno, G. 1993: *Streetwalking on a Ruined Map: Cultural Theory and the City Films of Elvira Notari*. Princeton, NJ: Princeton University Press.

Bruno, M. 1996: Employment strategies and the formation of new identities in the service sector in Moscow. In Pilkington, H. (ed.), *Gender, Generation and Identity in Contemporary Russia*. London: Routledge.

Buckley, M. (ed.) 1997: *Post-Soviet Women: From the Baltic to Central Asia*. Cambridge: Cambridge University Press.

Buck-Morss, S. 1989: *The Dialetics of Seeing: Walter Benjamin and the Arcades Project*. Cambridge, MA: MIT Press.

Bulmer, M. 1984: Local inequality: sociability, isolation and loneliness as factors in the differential provision of neighbourhood care. Paper presented to the Social Administration Association Conference, University of Kent.

Burchell, G., Gordon, G. and Miller, P. (eds) 1991: *The Foucault Effect: Studies in Governability*. London and New York: Harvester Wheatsheaf.

Burgess, A. 1997: *Fatherhood Reclaimed: The Making of a Modern Father*. London: Vermillion.

Burgin, V. (ed.) 1986: *The End of Art Theory: Criticism and Postmodernity*. Houndsmills: Macmillan.

Burke, P. 1997: History as social memory. In Burke, P. *Varieties of Cultural History*. Cambridge: Polity Press.

Burrows, R. 1997: Cyberpunk as social theory: William Gibson and the sociological imagination. In Westwood, S. and Williams, J. (eds), *Imagining Cities: Scripts, Signs, Memory*. London: Routledge.

Burrows, R., Gilbert, N. and Pollert, A. 1991: Introduction: Fordism, post-Fordism and economic flexibility. In Gilbert, N., Pollert, A. and Burrows, R. (eds), *Fordism and Flexibility: Divisions and Change*. London: Macmillan.

Butler, J. 1990: *Gender Trouble: Feminism and the Subversion of Identity*. New York: Routledge.

Butler, J. 1993: *Bodies that Matter: On the Discursive Limits of 'Sex'*. New York: Routledge.

Butler, J. 1995: Contingent Foundations. In Benhabib, S., Butler, J., Cornell, D. and Fraser, N., *Feminist Contentions: A Philosophical Exchange*. New York: Routledge.

Butler, J. and Scott, J. (eds) 1992: *Feminists Theorize the Political*. London: Routledge.

Butler, R. and Pearce, D. (eds) 1995: *Change in Tourism: People, Places, and Processes*. London: Routledge.

Butler, T. and Hamnett, C. 1994: Gentrification, class, and gender: some comments on Warde's 'Gentrification as consumption'. *Environment and Planning D: Society and Space* 12, 477–93.

Buttimer, A. 1974: Values in geography. *Resource Paper 24*. Association of American Geographers, Washington, DC.

Buttimer, A. 1984: Musing on Helicon: root metaphors and geography. *Geoscience and Man* 24, 55–62.

Camilleri, J. and Falk, J. 1992: *The End of Sovereignty? The Politics of a Shrinking and Fragmenting World*. Aldershot, Hants: Edward Elgar.

Campbell, B. 1984: *Wigan Pier Revisited*. London, Virago.

Campbell, B. 1987: *The Iron Ladies: Why do Women Vote Tory?* London: Virago.

Campbell, B. 1993: *Goliath: Britain's Dangerous Places*. London: Methuen.

Campbell, D. 1992: *Writing Security: United States Foreign Policy and the Politics of Identity*. Minneapolis, MN: University of Minnesota Press.

Caplan, P. and Bujra, J. (eds) 1979: *Women United, Women Divided: Comparative Studies of Ten Contemporary Cultures*. Bloomington, IN: Indiana University Press.

Carby, H. 1981: White women listen! Black feminism and the boundaries of sisterhood. In Centre for Contemporary Cultural Studies, *The Empire Strikes Back: Race and Racism in 1970s Britain*. London: Hutchinson.

Carlstein, T., Parkes, D. and Thrift, N. 1978: *Timing Space and Spacing Time*. London: Arnold.

Carney, J. and Watts, M. 1990: Manufacturing dissent: work, gender and the politics of meaning in a peasant society. *Africa* 60, 207–40.

Carroll, W.K. (ed.) 1992: *Organizing Dissent: Contemporary Social Movements in Theory and Practice*. Toronto: Garamond Press.

Carter, E., Donald, J. and Squires, J. (eds) 1993: *Space and Place: Theories of Identity and Location*. London: Lawrence and Wishart.

Castells, M. 1978: *City, Class and Power*. London: Macmillan.

Castells, M. 1983: *The City and the Grassroots. A Cross-cultural Theory of Urban Social Movements*. Berkeley, CA: University of California Press.

Castells, M. 1996: *The Rise of the Network Society*. Oxford: Basil Blackwell.

Castells, M. 1997: *Power and Identity*. Oxford: Basil Blackwell.

Castilo, A. 1994: *Massacre of the Dreamers: Essays on Xicanisma*. Albuquerque, NM: University of New Mexico Press.

Chambers, I. 1997: Maps, movies, music and memory. In Clarke, D.B. (ed.), *The Cinematic City*. London: Routledge.

Chant S. (ed.) 1992: *Gender and Migration in Developing Countries*. London: Belhaven Press.

Chant, S. 1997: *Women-headed Households: Diversity and Dynamics in the Developing World*. Basingstoke, Hants: Macmillan.

Chant, S. and McIlwaine, C. 1995: Gender and export manufacturing in the Philippines: continuity or change in female employment? The case of the Mactan Export Processing Zone. *Gender, Place and Culture* 2(2), 149–78.

Chant, S. and Radcliffe, S. 1992: Migration and development: the importance of gender. In Chant, S. (ed.), *Gender and Migration in Developing Countries*. London: Belhaven Press.

Chapkis, W. 1997: *Live Sex Acts: Women Performing Erotic Labor*. New York: Routledge.

Chaudhuri, N. and Strobel, E. (eds) 1992: *Western Women and Imperialism: Complicity and Resistance*. Bloomington, IN: Indiana University Press.

Chodorow, N. 1978: *The Reproduction of Mothering: Psychoanalysis and the Sociology of Gender*. Berkeley, CA, University of California Press.

Chodorow, N. 1979: Gender, relation and difference in psychoanalytic perspective. Reprinted in Chodorow, N. 1989: *Feminism and Psychoanalytic Theory*. Cambridge: Polity Press.

Chodorow, N. 1994: *Femininities, Masculinities, Sexualities*. London: Free Association Books.

Chomsky, N. 1996: *Power and Prospects: Reflections on Human Nature and the Social Order*. London: Pluto.

Chouinard, V. 1996: Gender and class identities in process and in place: the local state as a site of gender and class formation. *Environment and Planning A* 28, 1485–506.

Chouinard, V. and Grant, A. 1995: On not being anywhere near 'The Project': revolutionary ways of putting ourselves in the picture. *Antipode* 27, 137–66.

Chouinard, V., Fincher, R. and Webber, R. 1984: Empirical research in human geography. *Progress in Human Geography* 8, 347–80.

Chowdhry, P. 1994: *The Veiled Women; Shifting Gender Equations in Rural Haryana 1880–1990*. Delhi: Oxford University Press.

Christ, C. and Plaskow, J. (eds) 1979: *Womanspirit Rising: A Feminist Reader in Religion*. San Francisco, CA: Harper and Row.

Cixous, H. 1981a: Sorties. In Marks, E. and de Courtivron, I. (eds), *New French Feminisms*. Brighton, E. Sussex: Harvester.

Cixous, H. 1981b: Castration or decapitation? *Signs* 7, 41–55.

Cixous, H. 1986 [1975]: *The Newly Born Woman*, with C. Clement, trans. B. Wing. Manchester: Manchester University Press.

Cixous, H. 1990: The two countries of writing. In Flower MacCannell, J. (ed.), *The Other Perspective in Gender and Culture: Rewriting Women and the Symbolic*. New York: Columbia University Press.

Clark, A. 1992: *Working Life of Women in the Seventeenth Century*. London: Routledge.

Clark, D. 1993: Commodity lesbianism. In Abelove, H., Barale, M.A. and Halperin, D.M. (eds), *The Lesbian and Gay Studies Reader*. London: Routledge.

Clark, G., Dear, M. 1984: *State Apparatus: Structures and Language of Legitimacy*. Boston, MA: Allen and Unwin.

Cliff, D. 1993: 'Under the wife's feet': renegotiating gender divisions in early retirement. *The Sociological Review* 41, 30–53.

Cliff, M. 1980: *Claiming an Identity They Taught Me to Despise*. Waterburn, MA: Persephone Press.

Clifford, J. 1988: *The Predicament of Culture: Twentieth-century Ethnography, Literature, and Art*. Cambridge, MA, and London: Harvard University Press.

Clifford, J. 1992: Travelling cultures: In Grossberg, L. *et al.* (eds), *Cultural Studies*. London: Routledge.

Clifford, J. 1994: Diasporas. *Cultural Anthropology* 9, 302–38.

Clifford, J. 1997: *Routes: Travel and Translation in the Late Twentieth Century.* Cambridge MA: Harvard University Press.

Clifford, J. and Marcus, G. (eds) 1986: *Writing Culture: The Poetics and Politics of Ethnography.* Berkeley and Los Angeles, CA: University of California Press.

Cloke, P. 1990: Rural geography and political economy. In Peet, R. and Thrift, N. (eds), *New Models in Geography, Volume 1.* London: Unwin Hyman.

Cloke, P. and Little, J. 1990: *The Rural State? Limits to Planning in Rural Society.* Oxford: Clarendon Press.

Cloke, P. and Little, J. 1997: *Contested Countryside Cultures: Otherness, Marginalisation and Rurality.* London: Routledge.

Cloke, P. *et al.* 1994: *Writing the Rural: Five Rural Cultures.* London: Paul Chapman.

Cloke, P., Philo, C. and Sadler, D. 1991: *Approaching Human Geography: An Introduction to Contemporary Theoretical Debates.* London: Paul Chapman.

Cock, J. 1993: *Women and War in South Africa.* Cleveland, OH: The Pilgrim Press.

Cockburn, C. 1977: *The Local State: Management of Cities and People.* London: Pluto Press.

Cockburn, C. 1983: *Brothers: Male Dominance and Technological Change.* London: Pluto Press.

Cockburn, C. 1991: *In the Way of Women: Men's Resistance to Sex Equality in Organisations.* Basingstoke, Hants: Macmillan.

Cohen, P. 1988: The perversions of inheritance: studies in the making of multi-racist Britain. In Cohen, P. and Bains, H., *Multi-racist Britain.* London: Macmillan.

Cohen, P. 1993: Home rules: some reflections on racism and nationalism in everyday life. The New Ethnicities Unit, University of East London, London.

Cohen, R. 1997: *Global Diasporas: An Introduction.* London: UCL Press.

Colborn, T., Dumanoski, D. and Myers, J.P. 1997: *Our Stolen Future.* London: Abacus.

Collard, A. and Contrucci, J. 1988: *The Rape of the Wild.* London: Women's Press.

Collier, J.F. and Yanagisako, S. 1989: Theory in anthropology since feminist practice. *Critical Anthropology* 9, 27–37.

Collier, R. 1998: *Masculinities, Crime and Criminology: Men, Heterosexuality and the Criminal(ised) Other.* London: Sage.

Collins, P. 1997: Comment on Hekman's truth and method: feminist standpoint theory revisited: where's the power? *Signs: Journal of Women in Culture and Society* 22, 375–81.

Collins, P. 1989: The social construction of Black feminist thought. *Signs: Journal of Women in Culture and Society* 14, 745–73.

Collins, P.H. 1992: *Black Feminist Thought: Knowledge, Consciousness, and the Politics of Empowerment*. London: Routledge.

Comaroff, J. and Comaroff, J. 1991: *Of Revelation and Revolution: Christianity, Colonialism and Consciousness in South Africa*. Chicago, IL: Chicago University Press.

Comaroff, J. and Comaroff, J. (eds) 1993: *Modernity and its Malcontents: Ritual and Power in Postcolonial Africa*. Chicago: Chicago University Press.

Connell, R.W. 1987: *Gender and Power: Society, the Person and Sexual Politics*. Oxford: Polity Press.

Connell, R.W. 1995: *Masculinities*. Berkeley, CA: University of California Press.

Connolly, C. 1991: Washing our linen: one year of Women against Fundamentalism. *Feminist Review* 37, 68–77.

Cook, I., Crang, P. and Thorpe, M. 1999: Eating into Britishness: multicultural imaginaries and the identity politics of food. In Roseneil, S. and Seymour, J. (eds), *Practising Identities: Power and Resistance*. London: Macmillan.

Cooke, P. 1983: Labour market discontinuity and spatial development. *Progress in Human Geography* 7, 543–65.

Cooke, P. (ed.) 1989: *Localities: The Changing Face of Urban Britain*. London: Unwin Hyman.

Coombes, M., Raybould, S. and Wong, C. 1995: *Towards an Index of Deprivation: A Review of Alternative Approaches*. London: HMSO.

Cooper, C. and Lewis, S. 1993: *The Workplace Revolution: Managing Today's Dual-career Families*. London: Kogan Page.

Cooper, D.E. and Palmer, J.A. 1992: *The Environment in Question: Ethics and Global Issues*. London and New York: Routledge.

Coote, A. and Campbell, B. 1982: *Sweet Freedom: The Struggle for Women's Liberation*. London: Pan Books.

Corbridge, S., Martin, R. and Thrift, N. (eds) 1994: *Money, Power and Space*. Oxford: Basil Blackwell.

Corfield, P. 1995: *Power and the Professions in Britain 1700–1850*. London and New York: Routledge.

Cornell, D. 1991: *Beyond Accommodation*. London and New York: Routledge.

Corrin, C. (ed.) 1992: *Superwomen and the Double Burden*. London: Scarlet Press.

Cosgrove, D. 1985a: Prospect, perspective and the evolution of the landscape idea. *Transactions of the Institute of British Geographers* 10, 45–62.

Cosgrove, D. 1985b: *Social Formation and Symbolic Landscape*. London: Croom Helm.

Cosgrove, D. 1988 [1984]: *Social Formation and Symbolic Landscape*, 2nd edn. Madison, WI: University of Wisconsin Press.

Cosgrove, D. and Daniels, S. (eds) 1988: *The Iconography of Landscape: Essays on the Symbolic Representation, Design and Use of Past Environments*. Cambridge: Cambridge University Press.

Cosgrove, D. and Domosh, M. 1993: Author and authority: writing the new cultural geography. In Duncan, J. and Ley, D. (eds), *Place/Culture/Representation*. London and New York: Routledge.

Cosgrove, D. and Jackson, P. 1987: New directions in cultural geography. *Area* 19, 95–101.

Cotterill, P. 1992 'But for freedom, you see, not to be a babyminder': women's attitudes towards grandmother care. *Sociology* 26, 603–18.

Cowan, R. 1983: *More Work for Mother: The Ironies of Household Technology from the Open Hearth to the Microwave*. New York: Basic Books.

Coward, R. 1984: *Female Desire*. London: Paladin.

Coward, R. 1998: Mad about the girls. *Guardian* 23 March, 13.

Cox, K. (ed.) 1997: *Spaces of Globalization: Reasserting the Power of the Local*. London: Guildford.

Crang, M. 1994: On the heritage trail: maps of and journeys to olde Englande. *Environment and Planning D: Society and Space* 12, 341–55.

Cream, J. 1993a: Unpublished PhD dissertation; available in the Senate House Library, University of London.

Cream, J. 1993b: Child sexual abuse and the symbolic geographies of Cleveland. *Environment and Planning D* 11, 231–46.

Cream, J. 1995a: Women on trial: a private pillory? In Pile, S. and Thrift, N. (eds), *Mapping the Subject: Geographies of Cultural Transformation*. London: Routledge.

Cream, J. 1995b: Re-solving riddles: the sexed body. In Bell, D. and Valentine G. (eds), *Mapping Desire*. London: Routledge.

Cresswell, T. 1994: Putting women in their place: the carnival at Greenham Common. *Antipode* 26, 35–58.

Crompton, R. 1993: *Class and Stratification: An Introduction to Current Debates*. Cambridge: Polity Press.

Crompton, R. and Jones, G. 1984: *White-collar Proletariat: Deskilling and Gender in Clerical Work*. London: Macmillan.

Crompton, R. and Mann, M. 1986: *Gender and Stratification*. Cambridge: Polity Press.

Cuneo, C.J. 1990: *Pay Equity: The Labour–Feminist Challenge*. Toronto: Oxford University Press.

Curry, M. 1995: GIS and the inevitability of ethical inconsistency. In Pickles, J. (ed.), *Ground Truth*. New York: Guildford Press.

Curtis, S. and Taket, A. 1996: *Health and Societies: Changing Perspectives*. London: Arnold.

Dahl, R. 1961: *Who Governs?* New Haven, CT: Yale University Press.

Dalby, S. 1994: Gender and critical geopolitics: reading security discourse in the new world order. *Environment and Planning D: Society and Space* 12, 595–612.

Dale, J. and Foster, P. 1986: *Feminists and State Welfare*. London: Routledge.

Dalla Costa, M. and James, S. 1972: *The Power of Women and the Subversion of the Community*. Bristol: Falling Wall Press.

Daly, G. 1996: *Homeless: Policies, Strategies and Lives on the Street*. London and New York: Routledge.

Daly, M. 1968: *The Church and the Second Sex*. New York: Harper and Row.

Daly, M. 1973: *Beyond God the Father: Towards a Philosophy of Women's Liberation*. Boston, MA: Beacon Press.

Daly, M. 1978: *Gyn/Ecology: The Metaethics of Radical Feminism*. Boston, MA: Beacon Press.

Daly, M. 1987: *Webster's First New Intergalactic Wickedary of the English Language*. Boston, MA: Beacon Press.

Daniels, J. 1996: *White Lies, Race, Class, Gender, and Sexuality in White Supremacist Discourse*. New York: Routledge.

Daniels, S. 1989: Marxism, culture, and the duplicity of landscape. In Peet, R. and Thrift, N. (eds), *New Models in Geography, Volume 2*. London: Unwin Hyman.

Daniels, S.J. 1993: *Fields of Vision*. Cambridge: Polity Press.

Davidoff, L. 1995: *Worlds Between: Historical Perspectives on Gender and Class*. Cambridge: Polity Press.

Davidoff, L. and Hall, C. 1987: *Family Fortunes: Men and Women of the English Middle Class, 1780–1850*. London: Hutchinson.

Davidoff L., L'Esperance J. and Newby, H. 1976: Landscape with figures: home and community in English society. In Mitchell, J. and Oakley, A. (eds), *The Rights and Wrongs of Women*. Harmondsworth, Middx: Penguin Books.

Davies, K. 1990: *Women and Time: The Weaving of the Strands of Everyday Life*. Aldershot, Hants: Gower, Avebury.

Davis, A.Y. 1982: *Women, Race and Class*. London: Women's Press.

Davis, F. 1979: *Yearning for Yesterday: A Sociology of Nostalgia*. New York: The Free Press.

Davis, M. 1990: *City of Quartz: Excavating the Future in Los Angeles*. London: Verso.

Dawkins, R. 1978: *The Selfish Gene*. St Albans: Granada.

Dawkins, R. 1986: *The Blind Watchmaker*. Oxford: Oxford University Press.

Dawson, G. 1994: *Soldier Heroes: British Adventure, Empire and the Imaginings of Masculinities*. London: Routledge.

Dear, M. 1986: Postmodernism and planning. *Environment and Planning D: Society and Space* 4, 367–84.

Dear, M. and Wolch, J. 1987: *Landscapes of Despair: From De-institutionalization to Homelessness*. Princeton, NJ: Princeton University Press.

d'Eaubonne, F. 1994: The time for ecofeminism. In Merchant, C. (ed.), *Ecology*. Atlantic Highlands, NJ: Humanities Press.

de Beauvoir, S. 1972 [1949]: *The Second Sex*. Harmondsworth, Middx: Penguin Books.

de Lauretis, T. 1986: *Feminist Studies/Critical Studies*. Bloomington, IN: Indiana University Press.

de Lauretis, T. 1987: *Technologies of Gender: Essays on Theory, Film and Fiction*. London: Macmillan/Bloomington, IN: Indiana University Press.

Deleuze, G. and Guattari, F. 1983: *Anti-Oedipus*. Minneapolis, MN: University of Minnesota Press.

Deleuze, G. and Guattari, F. 1986: What is a minor literature? In Kafka, F., trans. Polan, D. *Towards a Minor Literature*. Minneapolis, MN: University of Minnesota Press.

Deleuze, G. and Guattari, F. 1987: *A Thousand Plateaus. Minneapolis*, MN: University of Minnesota Press.

Delphy, C. 1984: *Close to Home: A Materialist Analysis of Women's Oppression*. London: Hutchinson.

Delphy, C. and Leonard, D. 1992: *Familiar Exploitation: A New Analysis of Marriage in Contemporary Western Societies*. Cambridge: Polity Press.

Demeritt, D. 1996: Social theory and the reconstruction of science and geography.*Transactions of the Institute of British Geographers* 21, 484–503.

Department of Health 1990*: Caring For People: Community Care into the Next Decade and Beyond. Policy Guidance*. London: HMSO.

Derrida, J. 1973 [1967]: *'Speech and Phenomena' and Other Essays on Husserl's Theory of Signs*, trans. D.B. Allison. Evanston, IL: Northwestern University Press.

Derrida, J. [1967] 1974: *Of Grammatology*. Baltimore, MD: Johns Hopkins University Press.

Derrida, J. 1981 [1972]: *Dissemination*, trans. Johnson, B. London: Athlone Press.

Derrida, J. 1997: *Deconstruction in a Nutshell: A Conversation with Jacques Derrida*, edited with a commentary by J.D. Caputo. New York: Fordham University Press.

Deutsche, R. 1991: Boys town. *Environment and Planning D: Society and Space* 9, 5–30.

Deutsche, R. 1997: *Evictions: Art and Spatial Politics*. Cambridge, MA: MIT Press.

DeVault, M. 1996: Talking back to sociology: distinctive contributions of feminist methodology. *Annual Review of Sociology* 22, 29–50.

Dex, S. 1985: *The Sexual Division of Work*. Hemel Hempstead, Herts: Harvester Wheatsheaf.

Dex, S. 1988: Gender and the labour market. In Gaillie, D. (ed.), *Employment in Britain*. Oxford: Basil Blackwell.

Diamond, I. and Orenstein, G.F. (eds) 1900: *Reweaving the World*. San Francisco: Sierra Club Books.

Dicken, P. 1992: *Global Shift: The Internationalisation of Economic Activity*, 2nd edn. London: Paul Chapman.

Dicken, P. 1994: Global–local tensions: firms and states in the global space-economy. *Economic Geography* 70, 101–28.

Dickens, P. 1988: *One nation? Social Change and the Politics of Locality*. London: Pluto Press.

di Leonardo, M. 1991: *Gender at the Crossroads of Knowledge: Feminist Anthropology in the Postmodern Era*. Berkeley and Los Angeles, CA: University of California Press.

Dinnerstein, D. 1976: *The Mermaid and the Minotaur*. New York: Harper and Row.

Dirks. N. (ed.) 1992: *Colonialism and Culture*. Ann Arbor, MI: University of Michigan Press.

Dirlik, A. 1994: The postcolonial aura: Third World criticism in the age of global capitalism. *Critical Inquiry* 20, 328–56.

Di Stefano, C. 1990: Dilemmas of difference: feminism, modernity and postmodernity. In Nicholson, L. (ed.), *Feminism/Postmodernism*. London: Routledge.

Doane, M.-A. 1982: Film and the masquerade: theorising the female spectator. *Screen* 23, 74–87.

Dobash, R. and Dobash, R. 1992: *Women, Violence and Social Change*. London: Macmillan.

Dobson, A. 1995: *Green Political Thought*. London: Routledge.

Domosh, M. 1991: Toward a feminist historiography of geography. *Transactions of the Institute of British Geographers* 16, 95–104.

Domosh, M. 1996: *Invented Cities: The Creation of Landscape in Nineteenth-century New York and Boston*. New Haven, CT, and London: Yale University Press.

Domosh, M. 1997: Geography and gender: the personal and the political. *Progress in Human Geography* 21, 81–8.

Donald, J. 1992: Metropolis: the city as text. In Bocock, R. and Thompson, K. (eds), *Social and Cultural Forms of Modernity*. Cambridge: Polity Press (in association with the Open University).

Dorn, M. and Laws, G. 1994: Social theory, body politics and medical geography: extending Kearns's invitation. *The Professional Geographer* 46, 106–10.

Dowler, L. 1988: And they think I'm just a nice old lady: Women and war in Belfast, Northern Ireland. *Gender, Place and Culture* 5, 159–76.

Doyal, L. 1994: H.I.V. and Aids: putting women on the global agenda. In Doyal, L., Naidoo, J. and Wilton, T. *Aids: Setting a Feminist Agenda*. London: Taylor and Freeman.

Doyal, L. and Gough, I. 1991: *A Theory of Human Need*. London: Macmillan.

Driver, F. 1992: Geography's empire: histories of geographical knowledge. *Environment and Planning D: Society and Space* 10, 23–40.

Driver, F. 1994: Bodies in space: Foucault's account of disciplinary power. In Jones, C. and Porter, R. (eds), *Re-assessing Foucault*. London: Routledge.

Driver, F. and Gilbert, D. 1998: Heart of empire? Landscape, space, and performance in imperial London. *Environment and Planning D: Society and Space* 16, 11–28.

Driver, F., Matlass, D., Rose, G., Barnett, C. and Livingstone, D. 1995: Geographical traditions: rethinking the history of geography. *Transactions of the Institute of British Geographers* 20, 403–22.

DuBois, E.C. and Ruiz, V.L. (eds) 1990: *Unequal Sisters: A Multicultural Reader in U.S. Women's History*. New York: Routledge.

DuCille, A. 1997: *Skin Trade*. Cambridge, MA: Harvard University Press.

Duncan, J.S. 1980: The 'superorganic' in American cultural geography. *Annals of the Association of American Geographers* 70, 181–98.

Duncan, J.S. 1990: *The City as Text: The Politics of Landscape Interpretation in the Kandyan Kingdom.* Cambridge: Cambridge University Press.

Duncan, J.S. 1994: Landscape. In Johnston, R.J., Gregory, D. and Smith, D. (eds*), The Dictionary of Human Geography*, 3rd edn. Oxford and Cambridge: Basil Blackwell.

Duncan, J.S. and Duncan, N. 1988: (Re)reading the landscape. *Environment and Planning D: Society and Space* 6, 117–26.

Duncan, J.S. and Duncan, N. 1997: Deep suburban irony: the perils of democracy in Westchester County. In R. Silverstone, R. (ed.), *Visions of Suburbia.* London: Routledge.

Duncan, J.S. and Ley, D. (eds) 1993: *Place/Culture/Representation.* London and New York: Routledge.

Duncan, N. (ed.) 1996a: *Body Space: Destabilizing Geographies of Gender and Sexuality.* London and New York: Routledge.

Duncan, N. 1996b: Renegotiating gender and sexuality in public and private spaces. In Duncan, N.G. (ed.), *Body Space: Destabilizing Geographies of Gender and Sexuality.* New York, Routledge.

Duncan, N. 1996c: Postmodernism in human geography. In Earle, C., Mathewson, K. and Kenzer, M. (eds), *Concepts in Human Geography.* London: Rowman and Littlefield.

Duncan N. and Sharp J.P. 1993: Confronting representation(s). *Environment and Planning D: Society and Space* 11, 473–86.

Duncan, S. 1989: What is locality? In Peet, R. and Thrift, N. (eds), *New Models in Geography Volume 2.* London: Unwin Hyman.

Duncan, S. and Goodwin, M. 1988: *The Local State and Uneven Development: Behind the Local Government Crisis.* Cambridge: Polity Press.

Duncan, S. and Savage, M. 1989: Space, scale and locality. *Antipode* 21, 179–206.

Dunleavy, P. and O'Leary, B. 1987: *Theories of the State: The Politics of Liberal Democracy.* London: Macmillan.

Dworkin, A. 1976: *Our Blood: Prophecies and Discourses on Sexual Politics.* London: Harper and Row.

Dworkin, A. 1981: *Pornography: Men Possessing Women.* London: Women's Press.

Dyck, I. 1989: Integrating home and wage workplace: women's daily lives in a Canadian suburb. *Canadian Geographer* 33, 329–341.

Dyck, I. 1990: Space, time, and renegotiating motherhood: an exploration of the domestic workplace. *Environment and Planning D: Society and Space* 8, 459–83.

Dyck, I. 1995a: Hidden geographies: the changing lifeworlds of women with disabilities. *Social Science and Medicine* 40, 307–20.

Dyck, I. 1995b: Putting chronic illness 'in place'. Women immigrants' accounts of their health care. *Geoforum* 26, 247–60.

Dyck, I. 1996: Women with disabilities and everyday geographies: home space and the contested body. In Kearns, R.A. and Gesler, W.M. (eds), *Putting Health into Place.* Syracuse, NY: Syracuse University Press.

Dyck, I. and Kearns, R. 1995: Transforming the relations of research: towards culturally safe geographies of health and healing. *Health and Place* 1, 137–47.

Dyer, R. 1988: White. *Screen* 29, 44–64.

Eagleton, T. 1983: *Literary Theory: An introduction*. Minneapolis, MN: University of Minnesota Press.

Eagleton, T. 1991: *Ideology: An Introduction*. London: Verso.

Eco, U. 1986: *Travels in Hyperreality: Essays*. San Diego, CA: Harcourt Brace Jovanovich.

Edwards, T. 1992: The Aids dialectics: awareness, identity death and sexual politics. In Plummer, K. (ed.), *Modern Homosexualities: Fragments of Lesbian and Gay Experience*. New York: Routledge.

Einhorn, B. 1991: Where have all the women gone? Women and the women's movement in East Central Europe. *Feminist Review* 39, 16–36.

Einhorn, B. 1993: *Cinderella Goes to Market*. London: Verso.

Elias, N. 1992: *Time*. Oxford: Oxford University Press.

Elliott, B. and Wallace, J.-A. 1994: *Women Artists and Writers: Modernist (Im)positionings*. London: Routledge.

Elshtain, J.B. 1981: *Public Man, Private Woman*. Oxford: Martin Robertson.

Elshtain, J. 1987: *Women and War*. New York: Basic Books.

Elshtain, J. and Tobias, S. 1990: *Women, Militarism, and War*. Maryland, MD: Rowman and Littlefield.

Elson, D. 1991: Structural adjustment: its effect on women. In Wallace, T. and March, C. (eds), *Changing Perceptions: Writings on Gender and Development*. Oxford: Oxfam.

Elson, D. (ed.) 1995: *Male Bias in the Development Process*, 2nd edn. Manchester: Manchester University Press.

Elson, D. and Pearson, R. 1981: Nimble fingers make cheap workers: an analysis of women's employment in Third World export manufacturing. *Feminist Review* 7, 87–107.

Elson, D. and Pearson, R. 1984: The subordination of women and the internationalisation of factory production. In Young, K., Wolkowitz, C. and McCullagh, R. (eds), *Of Marriage and the Market. Women's Subordination Internationally and its Lessons*. London: Routledge.

Elster, J. 1984: *Ulysses and the Siren*. Cambridge: Cambridge University Press.

Engels, F. 1985 [1884]: *The Origin of the Family, Private Property and the State*. Introduction by M. Barrett. Harmondsworth, Middx: Penguin Books.

England, K. 1994: Getting personal: reflexivity, positionality and feminist research. *The Professional Geographer* 46(1), 80–9.

England, K. (ed.) 1997 *Who Will Mind the Baby?* London: Routledge.

Enloe, C. 1983: *Does Khaki Become You? The Militarisation of Women's Lives*. Boston, MA: South End Press.

Enloe, C. 1989: *Bananas, Beaches & Bases: Making Feminist Sense of International Politics*. Berkeley, CA: University of California Press.

Enloe, C. 1993: *The Morning After: Sexual Politics at the End of the Cold War*.

Berkeley, CA: University of California Press.

Entrikin, N. 1991: *The Betweenness of Place: Towards a Geography of Modernity*. London: Macmillan.

Environment and Planning A 1991: Special issue: new perspectives on the locality debate. *Environment and Planning A* 23, 155–308.

Escobar, A. 1995: *Encountering Development: The Making and Unmaking of the Third World*. Princeton, NJ: Princeton University Press.

Escott, K. and Whitfield, D. 1995: The gender impact of CCT in local government Equal Opportunities Commission (EOC) Research Discussion Series 12. Manchester: EOC.

Etzioni, A. 1993: *The Spirit of Community: Rights, Responsibilities and the Communitarian Agenda*. New York: Crown.

Evans, J. 1995: *Feminist Theory Today: An Introduction to Second Wave Feminism*. London: Sage.

Evans, P. and Wekerle, G. (eds) 1997: *Women and the Canadian Welfare State*. Toronto: University of Toronto Press.

Evans-Pritchard, E. 1937: *Witchcraft, Oracles and Magic among the Azande of the Anglo-American Sudan*. Oxford: Clarendon Press.

Eyles, J. and Perri, E. 1993: Life history as a method: an Italian–Canadian family in an industrial city. *The Canadian Geographer* 37, 104–19.

Fabian, J. 1983: *Time and the Other: How Anthropology Makes Its Object*. New York: Columbia University Press.

Fainstein, S., Gordon, I. and Harlow, M. (eds) 1992: Divided Cities. Oxford: Basil Blackwell.

Fairbairn, R. 1952: *Psycho-analytic Studies of the Personality*. London: Routledge and Kegan Paul. Also published as *An Object Relations Theory of the Personality*. New York: Basic Books.

Falconer Al-Hindi, K. and Staddon, C. 1997: The hidden histories and geographies of neotraditional town planning: the case of Seaside, Florida. *Environment and Planning D: Society and Space* 15, 349–72.

Falk, P. and Campbell, C. (eds) 1997: *The Shopping Experience*. London: Sage.

Faludi, S. 1992: *Backlash: The Undeclared War Against American Women*. London: Virago.

Fawcett, B., Featherstone, B., Hearn, J. and Toft, C. 1996: *Violence and Gender Relations*. London: Sage.

Featherstone, M. 1993: Global and Local Cultures. In Bird, J. Curtis, B., Putnam, T. *et al.* (eds*), Mapping the Futures: Local Culture, Global Change*. London: Routledge.

Featherstone, M. and Burrows, R. (eds) 1995: *Cyberspace/Cyberbodies/Cyberpunk: Cultures of Technological Embodiment*. London: Sage.

Featherstone, M. and Lash, S. 1996: Globalization, Modernity and the Spatialization of Social Theory: An Introduction. In Featherstone, M., Lash, S. and Robertson, R., *Global Modernities*. London: Sage.

Feinberg, L. 1992: Transgender liberation: a movement whose time has come. *World View Forum*.

Felski, R. 1995: *The Gender of Modernity*. Cambridge, MA: Harvard University Press.

Feminist Review 1987: *Sexuality. A Reader*. London: Virago.

Fentress, J. and Wickham, S. 1992: *Social Memory*. Oxford: Basil Blackwell.

Ferguson, K. 1993: *The Man Question: Visions of Subjectivity in Feminist Theory*. Berkeley, CA: University of California Press.

Fincher, R. 1991: Caring for workers' dependents: gender, class and local state practice in Melbourne. *Political Geography Quarterly* 10, 356–81.

Fincher, R. 1993: Women, the state and the life course. In Katz, C. and Monk, J. (eds), *Full Circles: Geographies of Women over the Lifetime*. London: Routledge.

Fincher, R. and McQuillen, J. 1989: Women in urban social movements. *Urban Geography* 10, 604–14.

Firestone, S. 1972: *The Dialectic of Sex*. London: Paladin.

Fish, S. 1980: *Is there a Text in this Class? The Authority of Interpretive Communities*. Cambridge, MA: Harvard University Press.

Fitzpatrick, D. 1986: A share of the honeycomb? Education, emigration and Irish women. *Continuity and Change* 1, 217–34.

Fitzsimmons, M. 1989: The matter of nature. *Antipode* 21, 106–20.

Flax, J. 1987: Postmodernism and gender relations in feminist theory. *Signs* 12, 621–43.

Flax, J. 1990: *Thinking Fragments: Psychoanalysis, Feminism and Postmodernism in the Contemporary West*. Berkeley and Los Angeles, CA: University of California Press.

Flax, J. 1993: *Disputed Subjects: Essays on Psychoanalysis, Politics and Philosophy*. London: Routledge.

Foord, J. and Gregson, N. 1986: Patriarchy: towards a reconceptualisation. *Antipode* 18, 186–211.

Forbes, G. 1990 Caged tigers: 'first wave' feminists in India. *Women Studies International Forum* 5, 525–36.

Forrest, B. 1995: West Hollywood as a symbol: the significance of place in the construction of a gay identity. *Environment and Planning D: Society and Space* 13, 133–57.

Foucault, M. 1967: *Madness and Civilisation: A History of Insanity in an Age of Reason*. London: Tavistock Publications.

Foucault, M. 1970: *The Order of Things: An Archaeology of the Human Sciences*. London: Tavistock Publications.

Foucault, M. 1973: *The Birth of the Clinic*. London: Tavistock Publications.

Foucault, M. 1977: *Discipline and Punish: The Birth of the Prison*. London: Allen Lane.

Foucault, M. 1978: *The History of Sexuality Volume I: An Introduction*. London: Allen Lane.

Foucault, M. 1980: *Power/Knowledge: Selected Interviews and Other Writings, 1972–1977*. New York: Pantheon Books.

Fox-Genovese, E. 1986: The claims of a common culture: gender, race, class and the canon. *Salmagundi* 72, 119–32.

Fox Keller, E. and Longino, H. 1996: *Feminism and Science*. Oxford: Oxford University Press.

Franck, K.A. 1995: Questioning the American dream: recent housing innovations in the United States. In Gilroy, R. and Woods, R. (eds), *Housing Women*. London: Routledge.

Frankenberg, R. 1993a: *White Women: Race Matters*. London: Routledge/ Minneapolis, MN: University of Minnesota Press.

Frankenberg, R. 1993b: Growing up white: feminism, racism and the social geography of childhood. *Feminist Review* 45, 51–84.

Fraser, N. 1989: *Unruly Practices: Power, Discourse and Gender in Contemporary Social Theory*. Minneapolis, MN: University of Minnesota Press.

Fraser, N. 1997: *Justice Interruptus*. London and New York: Routledge.

Fraser, N. and Lacey, N. 1993: *The Politics of Community: A Feminist Critique of the Liberal-Communitarian Debate*. London and New York: Harvester Wheatsheaf.

Freud, S. 1916–17: *Introductory Lectures on Psychoanalysis. The Pelican Freud Library, Volume 1*. Harmondsworth, Middx: Penguin.

Freud, S. 1917: *Mourning and Melancholia. The Pelican Freud Library, Volume 11*. Harmondsworth, Middx: Penguin.

Freud, S. 1931: *Female Sexuality. The Pelican Freud Library, Volume 7*. Harmondsworth, Middx: Penguin.

Freud, S. 1933: *New Introductory Lectures on Psychoanalysis. The Pelican Freud Library, Volume 2*. Harmondsworth, Middx: Penguin.

Friedberg, A. 1993: *Window Shopping: Cinema and the Postmodern*. Berkeley: University of California Press.

Friedman, B. 1965: *The Feminine Mystique*. London: Pelican.

Fröbel, F., Heinrichs, J. and Kreye, O. 1980: *The New International Division of Labour*. Cambridge: Cambridge University Press.

Frye, M. 1983: Some reflections on separatism and power. In *The Politics of Reality*. Trumansburg, NY: The Crossing Press.

Funk, N. and Mueller, M. (eds) 1993: *Gender Politics and Post-communism*. London: Routledge.

Fuss, D. 1989: *Essentially Speaking: Feminism, Nature and Difference*. London: Routledge.

Fuss, D. 1991: *Insider–Outsider: Lesbian Theories*. London: Routledge.

Fussell, P. 1980: *Abroad: British Literary Traveling Between the Wars*. New York: Oxford University Press.

Fyfe, A. 1989: *Child Labour*. Cambridge: Polity Press.

Gable, E. and Handler, R. 1996: After authenticity at an American heritage site. *American Anthropologist* 98, 568–78.

Gadamer, H.-G. 1975: *Truth and Method*. New York: Seabury Press.

Gagnon, J.H. and Simon, W.S. 1974: *Sexual Conduct: The Social Sources of Human Sexuality*. Chicago, IL: Aldine.

Gailey, C. 1987: *Kinship to Kingship: Gender Hierarchy and State Formation in the Tongan Islands*. Austin, TX: University of Texas Press.

Gaines, J. 1988: White privilege and looking relations: race and gender in feminist film theory. *Screen* 29, 12–27.

Gallop, J. 1982: *The Daughter's Seduction: Feminism and Psychoanalysis*. Ithaca, NY: Cornell University Press.

Gallop, J. 1985: *Reading Lacan*. Ithaca, NY: Cornell University Press.

Ganguly, K. 1992: Migrant identities: personal memory and the construction of selfhood. *Cultural Studies* 6, 27–50.

Gardiner, J. 1997: *Gender, Care and Economics*. Basingstoke, Hants: Macmillan Education.

Gardner, C.B. 1995: *Passing By: Gender and Public Harassment in Public Space*. Berkeley, CA: University of California Press.

Garrison, W.L. and Marble, D.F. 1957: The spatial structure of agricultural activities. *Annals of the Association of American Geographers* 47, 137–44.

Gatens, M. 1996: *Imaginary Bodies: Ethics, Power and Corporeality*. London: Routledge.

Gearheart, S.M. 1978: *The Wanderground: Stories of the Hill Women*. Watertown, MA: Persephone Press.

Geertz, C. 1973: *The Interpretation of Cultures*. New York: Basic Books.

Gerth, H.H. and Mills, C.W. [1948] 1991: *From Max Weber: Essays in Sociology*. London: Routledge.

Gesler, W. 1992: Therapeutic landscapes: medical issues in the light of the new cultural geography. *Social Science and Medicine* 34, 735–46.

Getis, A. 1963: The determination of the location of retail activities with the use of a map transformation. *Economic Geography* 39, 1–22.

Gibson, K. 1991: Company towns and class processes. *Environment and Planning D: Society and Space* 9, 285–308.

Gibson, W. 1984: *Necromancer*. London: Gollancz.

Gibson, W. 1991: Academy leader. In M. Benedikt (ed.) *Cyberspace: First Steps*. Cambridge, MA: MIT Press.

Gibson-Graham, J.K. 1994: 'Stuffed if I know': reflections on post-modern feminist social research. *Gender, Place and Culture: A Journal of Feminist Geography* 1, 205–24.

Gibson-Graham, J.K. 1996: *The End of Capitalism (As We Knew It): A Feminist Critique of Political Economy*. Oxford: Basil Blackwell.

Giddens, A. 1979: *Central Problems in Social Theory*. London: Macmillan.

Giddens, A. 1981: *A Contemporary Critique of Historical Materialism*. London: Macmillan.

Giddens, A. 1984: *The Constitution of Society*. Cambridge: Polity Press.

Giddens, A. 1989: *The Nation State and Violence*. Cambridge: Polity Press.

Giddens, A. 1990: *The Consequences of Modernity*. Stanford, CA: Stanford University Press.

Giddens, A. 1991: *Modernity and Self-identity*. Cambridge: Polity Press.

Giddens, A. 1993: *Sociology*. Cambridge: Polity Press.

Giddens, A. 1995: Living in a post-traditional society. In Beck, U., Giddens, A. and Lash, S. (eds), *Reflexive Modernization: Politics, Tradition and Aesthet-*

ics in the Modern Social Order. Cambridge, Polity Press.

Gier, J. and Walton, J. 1987: Some problems with reconceptualising patriarchy. *Antipode* 19, 54–8.

Gilbert, M. 1994: The politics of location: doing feminist research at 'home'. *Professional Geographer* 46, 90–6.

Gilbert M. 1997: Identity, space and politics: a critique of the poverty debates. In Jones J.P., Nast, H. and Roberts, S. (eds), *Thresholds in Feminist Geography: Difference, Methodology, Representation.* Lanham, MD: Rowman and Littlefield.

Gilbert, S. and Gubar, S. 1979: *The Madwoman in the Attic: The Woman Writer and the Nineteenth-century Literary Imagination.* New Haven, CT: Yale University Press.

Gilbert, S. and Gubar, S. 1988: *No Man's Land: The Place of the Woman Writer in the Twentieth Century.* Volume 1: *The War of the Words.* New Haven: Yale University Press.

Gilbert, S. and Gubar, S. 1988: *No Man's Land: The Place of the Woman Writer in the Twentieth Century.* Volume 2: *Sexchanges.* New Haven: Yale University Press.

Gilligan, C. 1982: *In a Different Voice.* Cambridge, MA: Harvard University Press.

Gilligan, C. 1993: *In a Different Voice,* new edn. Cambridge, MA: Harvard University Press.

Gilroy, P. 1993: *The Black Atlantic.* London: Verso.

Gilroy, R. 1994: Women and owner occupation in Britain: first the prince and then the palace? In Gilroy, R. and Woods, R. (eds), *Housing Women.* London: Routledge.

Ginsburg, F.E. and Rapp, R. (eds) 1995: *Conceiving the New World Order: The Global Stratification of Reproduction.* Berkeley and Los Angeles, CA: University of California Press.

Gleber, A. 1997: Female flanerie and *Symphony of the City.* In von Ankum, K. (ed.) *Women in the Metropolis: Gender and Modernity in Weimar Culture.* Berkeley: University of California Press.

Gleeson, B.J. 1996: A geography for disabled people? *Transactions of the Institute of British Geographers: New Series* 21, 387–96.

Glendinning, V. and Millar, J. (eds) 1992: *Women and Poverty in Britain,* 2nd edn. London: Macmillan.

Glucksmann, M. 1990: *Women Assemble: Women Workers and the New Industries in Inter-war Britain.* London: Routledge.

Godlewska, A. and Smith, N. 1994: *Geography and Empire.* Oxford: Basil Blackwell.

Goffman, E. 1969: *The Presentation of Self in Everyday Life.* Harmondsworth, Middx: Penguin Books.

Gold, J. and Burgess, J. (eds) 1985: *Geography, the Media and Popular Culture.* London: Croom Helm.

Gold, J. and Ward, S. (eds) 1994: *Place Promotion.* Chichester, Sussex: John Wiley.

Golding, P. (ed.) 1986: *Excluding the Poor*. London: CPAG.

Gomez, L. 1997: An Introduction to Object Relations. London: Free Association Books.

Goodale, J. 1971: *Tiwi Wives: A Study of Women of Melville Island, Northern Australia*. Seattle, WA: University of Washington Press.

Goodey, J. 1997: Boys don't cry: masculinities, fear of crime and fearlessness. *British Journal of Criminology* 37, 401–18.

Gordon, L. (ed.) 1990: *Women, the State, and Welfare*. Madison, WI: University of Wisconsin Press.

Gorz, A. 1989: *A Critique of Economic Reason*. London: Verso.

Gottdiener, A. and Lagopoulos, A. (eds) 1986: *The City and the Sign*. New York: Columbia University Press.

Gould, S. 1977: *Ever Since Darwin*. New York: W.W. Norton.

Gould, S. 1981: *The Mismeasure of Man*. New York: Norton.

Graff, G. 1992: *Beyond the Culture Wars*. New York: W.W. Norton

Gramsci, A. 1971: Selections from Prison Notebooks, edited and translated by Hoare, Q. and Nowell Smith, G. London: Lawrence and Wishart.

Granovetter, M. 1985: Economic action and social structure: the problem of embeddedness. *American Journal of Sociology* 91, 481–510.

Granovetter, M. and Swedberg, R. (eds) 1992: *The Sociology of Economic Life*. Oxford: Westview Press.

Greed, C. 1994: *Women and Planning: Creating Gendered Realities*. London: Routledge.

Green, A. 1997: Income and wealth. In Pacione, M. (ed.), *Britain's Cities: Geographies of Division in Urban Britain*. London: Routledge, 179–202.

Greenberg, D. 1988: *The Construction of Homosexuality*. Chicago, IL: University of Chicago Press.

Greenberg, J.R. and Mitchell, S. 1983: *Object Relations in Psychoanalytic Theory*. Cambridge, MA: Harvard University Press.

Gregory, D. 1978: *Ideology, Science and Human Geography*. London: Hutchinson.

Gregory, D. 1981: Human agency and human geography. *Transactions of the Institute of British Geographers* 6, 1–18.

Gregory, D. 1989: Areal differentiation and postmodern human heography. In Gregory, D. and Walford, R. (eds), *New Horizons in Human Geography*. London: Macmillan, 67–96.

Gregory, D. 1993: Modernity. In Johnston, R., Gregory, D. and Smith, D.M. (eds), *The Dictionary of Human Geography*, 3rd edn. Oxford, Basil Blackwell, 388–92.

Gregory, D. 1994: *Geographical Imaginations*. Oxford: Basil Blackwell.

Gregson, N. 1989: On the (ir)relevance of structuration theory to empirical research. In Held, D. and Thompson, J.B. (eds), *Social Theory of the Modern Societies: Anthony Giddens and his Critics*. Cambridge: Cambridge University Press.

Gregson, N. and Crewe, L. 1994: Beyond the high street and the mall: boot fairs and the new geographies of consumption in the 1990s. *Area* 26, 261–7.

Gregson, N. and Foord, J. 1987: Patriarchy: comments on critics. *Antipode* 19, 371–5.

Gregson, N., and Lowe, M. 1993: Renegotiating the domestic division of labour? A study of dual-career households in the north east and south east of England. *The Sociological Review* 41, 475–505.

Gregson, N. and Lowe, M. 1994: *Servicing the Middle Classes: Class, Gender and Waged Domestic Labour in Contemporary Britain*. London: Routledge.

Gregson, N. and Lowe, M. 1995: 'Home'-making: on the spatiality of daily social reproduction in contemporary middle-class Britain. *Transactions of the Institute of British Geographers: New Series* 20, 224–35.

Grewal, I. and Caplan, C. (eds) 1994: *Scattered Hegemonies: Postmodernity and Transnational Feminist Practices*. Minneapolis, MN: University of Minnesota Press.

Griffin, S. 1978: *Woman and Nature: The Roaring Inside Her*. New York: Harper and Row.

Griffiths, M. and Whitford, M. (eds) 1988: *Feminist Perspectives in Philosophy*. London: Macmillan.

Griggs, R. 1992: The meaning of 'nation' and 'state' in the Fourth World. Occasional paper 18, Centre for World Indigenous Studies, Olympia WA.

Grossberg, L., Nelson, C. and Treichler, P. (eds) 1992: *Cultural Studies*. London: Routledge.

Grosz, E. 1987: Corporeal feminism. *Australian Feminist Studies* 5, 1–16.

Grosz, E. 1989: *Sexual Subversions*. Sydney: Allen and Unwin.

Grosz, E. 1990a: *Jacques Lacan. A Feminist Introduction*. London and New York: Routledge.

Grosz, E. 1990b: Inscriptions and body-maps: representations and the corporeal. In Threadgold, T. and Cranny-Francis, A. (eds), *Feminine, Masculine and Representation*. London: Allen and Unwin.

Grosz, E. 1994: *Volatile Bodies: Towards a Corporeal Feminism*. Indiana University Press.

Grosz, E. 1995: *Space, Time and Perversion: Essays on the Politics of Bodies*. London, Routledge.

Groth, P. and Bressi, T. (eds) 1997: *Understanding Ordinary Landscapes*. New Haven, CT: Yale University Press.

Guelke, L. 1978: Geography and logical positivism. In Herbert, D. and Johnston, R.J. (eds), *Geography and the Urban Environment*, vol. 2. New York: John Wiley.

Guha, R. (ed.) 1982: *Subaltern Studies 1*. Delhi: Oxford University Press.

Guha, R. and Spivak, G. (eds) 1988: *Selected Subaltern Studies*. New York: Oxford University Press.

Guntrip, H. 1961: *Personality Structure and Human Interaction: The Developing Synthesis of Psychodynamic Theory*. London: Hogarth.

Habermas, J. 1989: *The Structural Transformation of the Public Sphere*, trans. Burger, T. Cambridge, MA: MIT Press. Originally published in 1962: *Strukturwandel der Offentlicheit*. Berlin: Hermann Luchterhand Verlag.

Hadley, J. 1996: *Abortion: Between Freedom and Necessity*. London: Virago.

Hagan, L. (ed.) 1992: *Women Respond to the Men's Movement*. San Francisco, CA: Pandora.

Hagerstrand, T. 1973: The domain of human geography. In Chorley, R.J. (ed.), *Directions in Geography*. London: Methuen.

Hakim, C. 1979: *Occupational Segregation*. London: Department of Employment.

Hakim, C. 1996: *Key Issues in Women's Work: Female Heterogeneity and the Polarisation of Women's Employment*. London: Athlone.

Halbwachs, M. 1980 [1950]: *The Collective Memory*, trans. Ditter Jr, F.J. and Ditter, V.Y. New York: Harper Colophon Books.

Halford, S. 1988: Women's initiatives in local government ... where do they come from and where do they go? *Policy and Politics* 16, 251–9.

Halford, S., Savage, M. and Witz, A. 1997: *Gender, Careers and Organizations: Current Developments in Banking, Nursing and Local Government*. London: Macmillan.

Hall, C. 1992: *White, Male and Middle Class: Explorations in Feminism and History*. Cambridge: Polity Press.

Hall, C. 1996: Histories, empires and the post-colonial moment. In Chambers, I. and Curti, L. (eds), *The Post-colonial Question*. London: Routledge.

Hall, S. 1991: The local and the global. Globalisation and ethnicity. In King, A.D. (ed.), *Culture, Globalisation and the World System*. London: Macmillan.

Hall, S. 1992a: The question of cultural identity. In Hall, S., Held, D. and McGrew, T. (eds), *Modernity and its Futures*. Cambridge: Polity Press.

Hall, S. 1992b: The West and the rest: discourse and power. In Hall, S. and Gieben, B., *Formations of Modernity*. Cambridge: Polity Press.

Hall, S. 1993: Cultural identity and diaspora. In Williams, P. and Chrisman, L. (eds), *Colonial Discourse and Post-colonial Theory: A Reader*. Hemel Hempstead, Herts: Harvester Wheatsheaf.

Hall, S. 1995: New cultures for old. In Massey, D. and Jess, P. (eds), *A Place in the World*. Milton Keynes: Open University Press.

Hall, S. and Jefferson, T. (eds) 1976: *Resistance through Rituals: Youth Subcultures in Post-war Britain*. London: Routledge and Kegan Paul.

Hamnett, C., Sarre, P. and McDowell, L. 1989: *The Changing Social Structure*. London: Hodder and Stoughton.

Hampele, A. 1993: The organised women's movement in the collapse of the GDR: the Independent Women's Association (UFV). In Funk, N. and Mueller, M. (eds), *Gender Politics and Post-communism*. London: Routledge.

Handler, R. 1986: Authenticity. *Anthropology Today* 2, 2–4.

Handler, R. and Saxton, W. 1988: Dyssimulation: reflexivity, narrative, and the quest for authenticity in 'living'. *Cultural Anthropology* 3, 242–60.

Hannigan, J.A. 1995: *Environmental Sociology: a Social Constructionist Perspective.* London: Routledge.

Hansen, M. 1991: *Babel and Babylon: spectatorship in American silent films.* Cambridge, MA: Harvard University Press.

Hanson, S. and Pratt, G. 1995: *Gender, Work and Space.* London: Routledge.

Haraway, D. 1984: Teddy bear patriarchy: taxidermy in the Garden of Eden, New York City, 1908–1936. *Social Text* 11, 20–64.

Haraway, D. 1988: Situated knowledges: the science question in feminism and the privilege of partial perspective. *Feminist Studies* 14, 575–99.

Haraway, D. 1989: *Primate Visions: Gender, Race and Nature in the World of Modern Science.* New York: Routledge.

Haraway, D. 1991: *Simians, Cyborgs and Women: The Reinvention of Nature.* London: Free Association Books.

Haraway, D. 1997: *Modest_Witness@Second.Millennium.FemaleMan_Meets_Oncomouse: Feminism and Technology.* London: Routledge.

Harding, C. (ed.) 1992: *Wingspan: Inside the Men's Movement.* New York: St Martin's Press.

Harding, S. 1986: *The Science Question in Feminism.* Ithaca, NY: Cornell University Press.

Harding, S. 1987: *Feminism and Methodology, Social Science Issue.* Bloomington, IN: Indiana University Press.

Harding, S. 1990: Feminism, science and the anti-Enlightenment critiques. In Nicholson, L. (ed.), *Feminism/Postmodernism.* London: Routledge.

Harding, S. 1991: *Whose Science? Whose Knowledge? Thinking from Women's Lives.* Milton Keynes: Open University Press.

Harding, S. 1993: Rethinking standpoint epistemology, 'What is strong objectivity?' In Alcoff, L. and Potter, E. (eds), *Feminist Epistemologies.* New York: Routledge.

Harding, S. 1997: Comment on Hekman's 'Truth and method': feminist standpoint theory revisited: whose standpoint needs the regimes of truth and reality? *Signs: Journal of Women in Culture and Society* 22, 382–91.

Harley, B. 1989: Deconstructing the map. *Cartographica* 26, 1–20.

Harley, B. 1992: Deconstructing the map. In Barnes, T. and Duncan, J. (eds), *Writing Worlds.* London: Routledge.

Harris, C. 1987: *Redundancy and Recession in South Wales.* Oxford: Basil Blackwell.

Hart, A. 1995: Reconstructing a Spanish red-light district: prostitution, space and power. In Bell, D and Valentine, G. (eds), *Mapping Desire.* London: Routledge.

Hartmann, H. 1979: Capitalism, patriarchy and job segregation by sex. In Eisenstein, Z. (ed.) *Capitalist Patriarchy and the Case for Socialist Feminism.* New York: Monthly Review Press.

Hartmann, H. 1981: The unhappy marriage of marxism and feminism: towards a more progressive union. In Sargent, L. (ed.), *The Unhappy Marriage of Marxism and Feminism: A Debate on Class and Patriarchy.* London: Pluto.

Hartshorne, R. 1939: *The Nature of Geography: A Critical Survey of Current Thought in the Light of the Past.* Lancaster PA: Association of American Geographers.

Hartsock, N. 1983: The feminist standpoint: developing the ground for a specifically feminist historical materialism. In Harding, S. and Hintinkka, M.B. (eds), *Discovering Reality: Feminist Perspectives on Epistemology, Metaphysics, Methodology and Philosophy of Science.* Dordrecht: Reidel.

Hartsock, N. 1987: The feminist standpoint: developing the ground for a specifically feminist historical materialism. In Harding, S. (ed.), *Feminism and Methodology.* Bloomington, IN: Indiana University Press.

Hartsock, N. 1997: Comment on Hekman's 'Truth and method': feminist standpoint theory revisited truth or justice? *Signs: Journal of Women in Culture and Society* 22, 367–74.

Harvey, D. 1969: *Explanation in Geography.* London: Arnold.

Harvey, D. 1973: *Social Justice and the City.* Oxford: Basil Blackwell.

Harvey, D. 1982: *The Limits to Capital.* Oxford: Basil Blackwell.

Harvey, D. 1989: *The Condition of Postmodernity.* Oxford: Basil Blackwell.

Harvey, D. 1990: Between space and time: reflections on the geographical imagination. *Annals of the Association of American Geographers* 80, 418–34.

Harvey, D. 1992: Postmodern morality plays. *Antipode* 24, 300–26.

Harvey, D. 1996: *Justice, Nature and the Geography of Difference.* Oxford: Basil Blackwell.

Haugen, M. 1994: Rural women's status in family and property law: lessons from Norway. In Whatmore, S., Mardsen, T. and Lowe, P. (eds), *Gender and Rurality.* London: David Fulton.

Hayden, D. 1976: *Seven American Utopias: The Architecture of Communiarian Socialism (1790–1975).* Cambridge, MA: MIT Press.

Hayden, D. 1980: What would a non-sexist city be like? Speculations on housing, urban design, and human work. In Stimpson, C., Dixler, P., Nelson, M. and Yatrakis, K. (eds), *Women and the American City.* Chicago, IL: Chicago University Press.

Hayden, D. 1981: *The Grand Domestic Revolution: A History of Feminist Designs for American Homes, Neighbourhoods, and Cities.* Cambridge, MA: MIT Press.

Hayden, D. 1984: Redesigning the American Dream: The Future of Housing, Work and Family Life. New York and London:, W.W. Norton.

Hayden, D. 1995: *The Power of Place: Urban Landscapes as Public History.* Cambridge, MA: MIT Press.

Hayward, T. 1994: *Ecological Thought: An Introduction.* Cambridge: Polity.

Hearn, J. 1996: Is masculinity dead? A critique of the concept of masculinity. In Mac an Ghaill, M., *Understanding Masculinities.* Buckingham: Open University Press.

Heelas, P., Lash, S. and Morris, P. (eds) 1996: *Detraditionalization.* Oxford: Basil Blackwell.

Heenan, D. and Gray, M. 1997: Women, public housing and inequality: a Northen Irish perspective. *Housing Studies* 12, 157–71.

Hegel, F. 1821: *Philosophy of Right*, trans. Nisbet, H.B. Cambridge: Cambridge University Press.

Hegel, F. 1840: *George Wilhelm Freidrich Hegel's Vorlesungen Über die Philosophieder Geschichte*. Berlin: Duncker und Humblot.

Heidegger, M. 1962: *Being and Time*, MacQuarrie, J. and Robinson, E. trans. Oxford: Basil Blackwell.

Heilbrun, C. 1988: *Writing a Woman's Life*. New York: W.W. Norton.

Hekman, S. 1997: Truth and method: feminist standpoint theory revisited. *Signs: Journal of Women in Culture and Society* 22, 341–65.

Hemmings, C. 1995: Locating bisexual identities: discourses of bisexuality and contemporary feminist theory. In Bell, D. and Valentine, G. (eds), *Mapping Desire: Geographies of Sexualities*. London: Routledge.

Henriques, J., Hollway, W., Urwin, C. *et al.* 1984: *Changing the Subject*. London and New York: Methuen.

Henry, N. and Massey, D. 1995: Competitive time–space in high technology. *Geoforum* 26, 49–64.

Herdt, G. (ed.) 1992: *Gay Culture in America*. Boston, MA: Beacon Press.

Herek, G.M. and Berrill, K.T. (eds) 1992: Hate Crimes: Confronting Violence against Lesbians and Gay Men. London: Sage.

Herod, A. 1997: Notes on a spatialized labour politics: scale and the political geography of dual unionism in the US longshore industry. In Lee, R. and Wills, J. (eds), *Geographies of Economies*. London: Arnold.

Heron, L. (ed.) 1993: *Streets of Desire: Women's Fictions of the Twentieth-century City*. London: Virago.

Hewison, R. 1987: *The Heritage Industry: Britain in a Climate of Decline*. Andover, Hants: Methuen.

Hiebert, D. 1997: The colour of work: labour market segmentation in Montreal, Toronto and Vancouver, 1991. Research on Immigration and Integration in the Metropolis Working Paper Series No. 97-02, Simon Fraser University, Vancouver.

Hirst, R.P.Q. and Wooley, P. 1982: *Social Relations and Human Attributes*. London: Tavistock.

HMSO 1985: *Swann Report: Education for All*. London: HMSO.

Hobsbawm, E. 1996: Identity politics and the Left. *New Left Review* 217, 38–47.

Hobsbawm, E. and Ranger, T. (eds) 1983: *The Invention of Tradition*. Cambridge: Cambridge University Press.

Hochschild, A. 1983: *The Managed Heart: Commercialization of Human Feeling*. Berkeley, CA: University of California Press.

Hochschild, A. 1990: *The Second Shift: Working Parents and the Revolution at Home*. London: Piatkus.

Holdsworth, D. 1997: Landscapes and archives as texts. In Groth, P. and Bressi, T. (eds), *Understanding Ordinary Landscapes*. New Haven, CT: Yale University Press.

hooks, b. 1982: *Ain't I a Woman? Black Women and Feminism.* London: Pluto Press.

hooks, b. 1984: *Feminist Theory: From Margin to Center.* Boston, MA: South End Press.

hooks, b. 1986: Sisterhood: political solidarity between women. *Feminist Review* 23, 125–38.

hooks, b. 1989: *Talking Back: Thinking Feminist, Thinking Black.* Boston, MA: South End Press.

hooks, b. 1990: Marginality as a site of resistance. In Ferguson, R. *et al.* (eds), *Out there: Marginalization and Contemporary Cultures.* Cambridge, MA: MIT Press.

hooks, b. 1991a: *Yearning: Race, Gender and Cultural Politics.* Boston, MA: South End Press.

hooks, b. 1991b: Choosing the margin as a space of radical openness. In hooks, b., *Yearning: Race, Gender and Cultural Politics.* London: Turnaround Books.

hooks, b. 1992a: Representing whiteness in the Black imagination. In Grossberg, L. *et al.* (eds), *Cultural Studies.* London: Routledge.

hooks, b. 1992b: *Black Looks: Race and Representation.* Boston, MA: South End Press.

hooks, b. 1994: *Outlaw Culture: Resisting Representations.* London: Routledge.

hooks, b. and McKinnon, T. 1996: Sisterhood: beyond public and private. *Signs* 21, 814–29.

Hoskins, W.G. 1955: *The Making of the English Landscape.* London: Hodder and Stoughton.

Houston, R. and Snell, K. 1984: Proto-industrialization? Cottage industry, social change and industrial revolution. *The Historical Journal* 27, 473–92.

Hufton, O. 1995: *The Prospect before Her: The History of Women in Western Europe, Volume 1, 1500–1800.* London: HarperCollins.

Hungry Wolf, B. 1980: *The Ways of My Grandmothers.* New York: William Morrow.

Hurstfield, J. 1987: *Part-timers: Under Pressure.* London: Low Pay Unit.

Hutcheon, L. 1989: *The Politics of Postmodernism.* London: Routledge.

Hutcheon, L. 1994: *Irony's Edge: The Theory and Politics of Irony.* London: Routledge.

Hutton, W. 1996: *The State We're In.* London: Vintage.

Huws, U., Hurstfield, J. and Holtmaal, R. 1989: *Whose Flexibility? The Casualisation of Women's Employment.* London: Low Pay Unit.

Huyssen, A. 1986: *After the Great Divide: Modernism, Mass Culture, and Postmodernism.* Bloomington: Indiana University Press.

IICCG 1997: Inaugaral International Critical Conference in Geography, http://www.geog.ubc.ca/iiccg/toc.html

Imrie, R. 1996: Ableist geographies, disableist spaces: towards a reconstruction of Golledge's 'Geography and the disabled'. *Transactions of the Institute of British Geographers: New Series* 21, 397–403.

Ingham, G. 1996: Review essay: the new economic sociology. *Work, Employment and Society* 10, 549–66.

Innes, C. 1994: Virgin territories and motherlands: colonial and nationalist representations of Africa and Ireland. *Feminist Review* 47, 1–4.

Irigaray, L. 1977: *Ce sexe qui n'en est pas un*, trans. C. Porter (1985) Ithaca, NY: Cornell University Press.

Irigaray, L. 1985: *Speculum of the Other Woman*. Ithaca, NY: Cornell University Press.

Itzin, C. (ed.) 1992: *Pornography: Women, Violence and Civil Liberties*. Oxford: Oxford University Press.

Jackson, J. 1984: *Discovering the Vernacular Landscape*. New Haven, CT: Yale University Press.

Jackson, P. 1985: Urban ethnography. *Progress in Human Geography* 9, 157–76.

Jackson, P. 1989: *Maps of Meaning: An introduction to Cultural Geography*. London: Unwin Hyman.

Jackson, P. 1990: The cultural politics of masculinity: towards a social geography. Transactions of the Institute of British Geographers: New Series 16, 199–213.

Jackson, P. 1991: Mapping meanings: a cultural critique of locality studies. *Environment and Planning A* 23, 215–28.

Jackson, P. 1993: Towards a cultural politics of consumption. In Bird, J., Curtis, B., Putnam, T., Robertson, G. and Tickner, L. (eds), *Mapping the Futures: Local Cultures, Global Change*. London: Routledge.

Jackson, P. 1998: Constructions of 'whiteness' in the geographical imagination. *Area* 30, 99–106.

Jackson, P. and Thrift, N. 1995: Geographies of consumption. In Miller, D. (ed.), *Acknowledging Consumption*. London: Routledge.

Jacobs, J. 1993: The city unbound: qualitative approaches to the city. *Urban Studies* 30, 827–48.

Jacobs, J. 1996a: Edge of Empire: Postcolonialism and the City. London: Routledge.

Jacobs, J. 1996b: Earth honouring: feminism, environmentalism and indigenous knowledges. In Rose, G. and Blunt, A. (eds), *Sexual / Textual Colonisations*. New York: Guildford Press.

Jacobs, M. 1992: *Sigmund Freud*. London: Sage.

Jagger, A.M. 1983: *Feminist Politics and Human Nature*. Totowa, NJ: Rowman and Allanheld/Brighton, E. Sussex: Harvester Press.

Jameson, F. 1983: Pleasure: a political issue. In Jameson, F. (ed.), *Formations of Pleasure*. London: Routledge and Kegan Paul.

Jameson, F. 1984: Postmodernism, or the Cultural Logic of Late Capitalism. *New Left Review* 146, 53–92.

Jameson, F. 1988, Of islands and trenches: neutralization and the production of utopian discourse. In Jameson, F., *The Ideologies of Theory: Essays 1971–1986*. London, Routledge.

Jardine, A.1985: *Gynesis: Configurations of Women and Modernity*. Ithaca, NY: Cornell University Press.

Jayawardena, K. 1986: *Feminism and Nationalism in the Third World*. London: Zed Books.

Jensen, J. 1989: The talents of women and the skills of men. In Wood, S. (ed), *The Transformation of Work?* London: Routledge.

Jenson, J., Hagen, E. and Reddy, C. (eds) 1988: *Feminisation of the Labour Force: Paradoxes and Promises*. Cambridge: Polity Press.

Jessop, B. 1990: *State Theory: Putting Capitalist States in their Place*. Cambridge: Polity Press.

Johnson, L. 1987: (Un)realist perspectives: patriarchy and feminist challenges in geography. *Antipode* 19, 210–15.

Johnson, N. 1995: Cast in stone: monuments, geography, and nationalism. *Environment and Planning D: Society and Space* 13, 51–65.

Johnston, R. 1993: The rise and decline of the corporate-welfare state: a comparative analysis in global context. In Taylor, P. (ed.), *Political Geography of the Twentieth Century: A Global Analysis*. London: Bellhaven Press.

Johnston, R., Gregory, D. and Smith, D. (eds) 1994: *The Dictionary of Human Geography*. Oxford: Basil Blackwell.

Jones, B. 1996: The social constitution of labour markets: why skills cannot be commodities. In Crompton, R., Gallie, D. and Purcell, K. (eds), *Changing Forms of Employment: Organisations, Skills and Gender*. London: Routledge.

Jones, G. 1993: *Youth, Gender and Citizenship*. Buckingham: Open University Press.

Jones, J. 1996: The self as other: creating the role of Joni the ethnographer for Broken Circles. *Text Performance Quarterly*, 131–45.

Jones III, J.P., Nast, H. and Roberts, S. (eds) 1997: *Thresholds in Feminist Geography: Difference, Methodology, Representation*. Lanham, MD: Rowman and Littlefield.

Jones, K. and Jonasdottir, A. 1988: *The Political Interests of Gender*. London: Sage.

Jordan, B. 1994: *Poverty and Social Exclusion*. Cambridge: Polity Press.

Kabbani, R. 1986: *Europe's Myths of Orient; Devise and Rule*. London: Macmillan.

Kabeer, N. 1994: *Reversed Realities: Gender Hierarchies in Development Thought*. London: Verso.

Kahn, J. 1995: *Culture, Multiculture, Postculture*. London: Sage.

Kandiyoti, D. 1988: Bargaining with patriarchy. *Gender and Society* 2, 274–90.

Kanter, R.B. 1977: *Men and Women of the Organisation*. New York: Basic Books.

Kaplan, C. 1987: Deterritorializations: the rewriting of home and exile in western feminist discourse. *Cultural Critique* 6, 187–98.

Kaplan, C. 1996: *Questions of Travel: Postmodern Discourses of Displacement*. Durham, NC: Duke University Press.

Katz, C. 1992: All the world is staged: intellectuals and the projects of ethnography. *Environment and Planning D: Society and Space* 10, 495–510.

Katz, C. 1993: Growing girls/closing circles: limits on the spaces of knowing in rural Sudan and US cities. In Katz, C. and Monk, J. (eds), *Full Circles: Geographies of Women over the Life Course*. London: Routledge.

Katz, C. 1994: Playing the field: questions of fieldwork in geography. *Professional Geographer* 46, 67–72.

Katz, C. and Monk, J. (eds) 1993: *Full Circles: Geographies of Women Over the Life Course*. London: Routledge.

Katz, J.N. 1995: *The Invention of Heterosexuality*. New York: Dutton.

Katz, M. (ed.) 1993: *The 'Underclass' Debate: Views from History*. Princeton, NJ: Princeton University Press.

Kay, J. 1991: Landscapes of women and men: rethinking the regional historical geography of the United States and Canada. *Journal of Historical Geography* 17, 435–52.

Kearns, R. 1991: The place of health in the health of place: the case of the Hokianga special medical area. *Social Science and Medicine* 33, 519–30.

Kearns, R.A. 1993: Place and health: towards a reformed medical geography. *The Professional Geographer* 45, 136–47.

Kearns, R.A. 1995: Medical geography: making space for difference. *Progress in Human Geography* 19, 249–57.

Keegan, T. 1988: *Facing the Storm: Portraits of Black Lives in Rural South Africa*. London: Zed Books.

Kehily, M. and Nayak, A. 1997: 'Lads and laughter': humour and the production of heterosexual hierarchies. *Gender and Education* 9, 69–87.

Keith, M. and Pile, S. (eds) 1993: *Place and the Politics of Identity*. London: Routledge.

Keller, E.F. 1985: *Reflections on Gender and Science*. New Haven, CT: Yale University Press.

Keller, W. 1971: *Diaspora: the Post-Biblical History of the Jews*. London: Pitman.

Kelly, L. 1987: The continuum of sexual violence. In Hanmer, J. and Maynard, M. (eds), *Women, Violence and Social Control*. London: Macmillan.

Kelly, L. 1991: Unspeakable acts: women who abuse. *Trouble and Strife* 21, 13–20.

Kempson, E. 1996: *Life on a Low Income*. York: Joseph Rowntree Foundation.

Kimmel, M. (ed.) 1995: *The Politics of Manhood: Profeminist Men Respond to the Mythopoetic Men's Movement (And the Mythopoetic Leaders Answer)*. Philadelphia, PA: Temple University Press.

Kincaid, K. 1988: *A Small Place*. London: Virago.

King, A. 1991: *Culture, Globalisation and the World System*. London: Macmillan.

King, A. 1997: The politics of vision. In Groth, P. and Bressi, T. (eds), *Understanding Ordinary Landscapes*. New Haven, CT: Yale University Press.

King, E. 1994: *Safety in Numbers*. London: Cassells.

Kinnaird, V. and Hall, D. (eds) 1994: *Tourism: A Gender Analysis*. Toronto: John Wiley.

Kinsman, P. 1995: Landscape, race and national identity: the photography of Ingrid Pollard. *Area* 27, 300–10.

Kipnis, A. 1992: In quest of archetypal masculinity. In Harding, C. (ed.), *Wingspan*. New York: St Martin's Press.

Kishwar, M. 1990: Why I do not call myself a feminist. *Manushi* 61, 2–6.

Klug, F. 1989: Oh to be in England; the British case study. In Yuval-Davis, N. and Anthias, F. (eds), *Woman–Nation–State*. London: Macmillan.

Kniffen, F. 1962 [1936]: Louisiana house types. In Wagner, P. and Mikesell, M. (eds), *Readings in Cultural Geography*. Chicago, IL: University of Chicago Press.

Knopp, L. 1990: Some theoretical implications of gay involvement in an urban land market. *Political Geography Quarterly* 9, 337–52.

Knopp. L. 1992: Sexuality and the spatial dynamics of capitalism. *Environment and Planning D: Society and Space* 10, 651–69.

Kobayashi, A. 1989: A critique of dialectical landscape. In Kobayashi, A. and Mackenzie, S. (eds), *Remaking Human Geography*. London: Unwin Hyman.

Kobayashi, A. 1993: Multiculturalism: representing a Canadian institution. In Duncan, J. and Ley, D. (eds), *Place/Culture/Representation*. London: Routledge.

Kobayashi, A. 1994: Coloring the field: gender, race, and the politics of field-work. *Professional Geographer* 46, 73–80.

Kobayashi, A. and Peake, L. 1994: Unnatural discourse: 'race' and gender in geography. *Gender, Place and Culture* 1, 225–43.

Kochan, T., Katz, H. and McKersie, R. 1986: *The Transformation of American Industrial Relations*. New York: Basic Books.

Kofman, E. and Peake, L. 1990: Introduction: a gendered agenda for political geography in the 1990s. *Political Geography Quarterly* 9, 313–36.

Kofman, E. and Youngs, G. (eds) 1996: *Globalization: Theory and Practice*. London: Frances Pinter.

Kollontai, A. 1977: *Selected Writings of Alexandra Kollontai*, translation and introduction by A. Holt. London: Allison and Busby.

Kollontai, A. 1984: *Sexual Relations and the Class Struggle*. London: Socialist Workers Party.

Kolodny, A. 1975: *The Lay of the Land: Metaphor as Experience and History in American Life and Letters*. Chapel Hill, NC: University of North Carolina Press.

Kolodny, A. 1984: *The Land Before Her: Fantasy and Experience of the American Frontiers, 1630–1860*. Chapel Hill, NC: University of North Carolina Press.

Kondo, D. 1995: Bad girls: theater, women of color, and the politics of repre-sentation. In Behar, R. and Gordon, D. (eds), *Women Writing Culture*. Berkeley, CA: University of California Press.

Koptiuch, K. 1996: 'Cultural defense' and crimonological displacements: gender, race and (trans)nation in the legal surveillance of U.S. diaspora Asians. In Lavie, S. and Swedenburg, T. (eds), *Displacement, Diaspora and Geogra-phies of Identity*. Durham, NC, and London: Duke University Press.

Koskela, H. 1997: 'Bold walk and breakings': women's spatial confidence versus fear of violence. *Gender, Place and Culture* 4, 301–19.

Kramer, L. quoted in Shilts, R. 1995: And the band played on: politics, people, and the AIDS epidemic. New York: Penguin.

Kreutzner, G. 1989: On doing cultural studies in West Germany. *Cultural Studies* 3, 240–9.

Kristeva, J. 1980: *Language in Desire*, Gora, T., Jardine, A. and Roudiez, L.S. trans.; Roudiez, L.S. (ed.). New York: Columbia University Press.

Kristeva, J. 1982: *Powers of Horror: An Essay in Abjection*. New York: Columbia University Press.

Kristeva, J. 1984: *Revolution in Poetic Language*. New York: Columbia University Press.

Kristeva, J. 1987 *The Kristeva Reader*, Moi, T. (ed.). New York: Columbia University Press.

Kristeva, J. 1991: *Strangers to Ourselves*, trans. Roudiez, L.S. New York: Columbia University Press.

Kroeber, A. 1917: The superorganic. *American Anthropologist* 19, 163–213.

Kroeber, A. 1952: *The Nature of Culture*. Chicago, IL: University of Chicago Press.

Lacan, J. 1977a: *Écrits: A Selection*. trans. Sheridan, A. New York: W.W. Norton.

Lacan, J. 1977b: The Four Fundamental Concepts of Psycho-analysis. London: The Hogarth Press.

Lacan, J. 1982: *Feminine Sexuality*, trans. Rose, J.; Rose, J. and Mitchell, J. (eds). London: Macmillan.

Lacey, N. 1998: *Unspeakable Subjects: Feminist Essays in Legal and Social Theory*. Oxford: Hart Publishing.

Laclau, E. 1990: *New Reflections on the Revolution of our Time*. London: Verso.

Laclau, E. and Mouffe, C. 1985: *Hegemony and Socialist Strategy*. London: Verso.

Lamphere, L. 1996: Feminist anthropology. In Levinson. D. and Ember, M. (eds), *Encyclopedia of Cultural Anthropology*. New York: Henry Holt, 488–93.

Lapovsky Kennedy, E. and Davis, M. 1994: *Boots of Leather, Slippers of Gold*. New York: Penguin.

Laqueur, T. 1990: *Making Sex: Body and Gender from the Greeks to Freud*. London: Harvard University Press.

Lash, S. and Urry, J. 1994: *Economies of Signs and Space*. London: Sage.

Latour, B. 1993: *We Have Never Been Modern*. Cambridge, MA: Harvard University Press.

Laurie, N., Smith, F., Bowlby, S. *et al.* 1997: In and out of bounds and resisting boundaries: feminist geographies of space and place. In Woman and Geography Study Group, *Feminist Geographies: Explorations in Diversity and Difference*. Harlow, Essex: Longman.

Lavalette, M. 1994: *Child Employment in the Capitalist Labour Market*. Aldershot, Hants: Avebury.

Lavalette, M. and Kennedy, J. 1996: *Solidarity on the Waterfront: The Liverpool Lock Out of 1995/6*. Birkenhead: Liver Press.

Law, L. 1997: Dancing on the bar: sex, money and the uneasy politics of third space. In S. Pile and Keith, M. (eds), *Geographies of Resistance*. London: Routledge.

Laws, G. 1994: Aging, contested meanings, and the built environment. *Environment and Planning A* 26, 1787–1802.

Laws, G. 1997: Women's life courses, spatial mobility, and state policies. In Jones III, J. P., Nast, H. and Roberts, S. (eds), *Thresholds in Feminist Geography: Difference, Methodology, and Representation*. New York: Rowman and Littlefield.

Lazreg, M. 1994: Women's Experience and Feminist Epistemology. In Lennon, K. and Whitford, M. (eds), *Knowing the Difference: Feminist Perspectives in Epistemology*. London: Routledge.

Leacock, E. 1980: Social behaviour, biology and the double standard. In Barlow, G.W. and Silverberg, J. (eds), *Sociobiology: Beyond Nature/Nurture? Reports, Definitions and Debate*. Boulder, CO: Westview Press.

Leborgne, D. and Lipietz, A. 1990: Fallacies and open issues about post-fordism. CEPREMAP paper 9009.

Lefebvre, H. 1976 [1973]: *The Survival of Capitalism: Reproduction of the Relations of Production,* trans. F. Bryant. London: Allison and Busby.

Lefebvre, H. 1991 [1974]: *The Production of Space*, trans. D. Nicholson-Smith. Oxford: Basil Blackwell.

Lefebvre, H. 1995 [1962]: *Introduction to Modernity: Twelve Preludes*, trans. J. Moore. London: Verso.

Leighly, J. (ed.) 1967: *Land and Life: A Selection from the Writings of Carl Ortwin Sauer*. Berkeley, CA: University of California Press.

Leiss, W. 1974: *The Domination of Nature*. Boston: Beacon Press.

Lennon, M., McAdam, M. and O'Brien, J. 1988: *Across the Water: Irish Women's Lives in Britain*. London: Virago.

Leslie, D.A. 1993: Femininity, post-Fordism, and the 'new traditionalism'. *Environment and Planning D: Society and Space* 11, 689–708.

Levi-Strauss, C. 1966: *The Savage Mind*. Chicago, IL: University of Chicago Press.

Levi-Strauss, C. 1977 [1963]: *Structural Anthropology*, vols I and II. New York: Basic Books.

Levitas, R. (ed.) 1985: *The Ideology of the New Right*. Cambridge: Polity.

Lewin, E. 1993: *Lesbian Mothers: Accounts of Gender in American Culture*. Ithaca, NY: Cornell University Press.

Lewis, J. 1984: *Women in England 1870–1950: Sexual Divisions and Social Change*. Brighton, E. Sussex: Wheatsheaf.

Lewis, J. 1992: Gender and the development of welfare regimes. *Journal of European Social Policy* 2, 159–73.

Lewis, R. 1996: *Gendering Orientalism: Race, Femininity and Representation*. London and New York: Routledge.

Ley, D. 1980: Liberal ideology and the postindustrial city. *Annals of the Association of American Geographers* 70, 238–58.

Ley, D. 1993: Co-operative housing as a moral landscape: re-examining 'the postmodern city'. In Duncan, J. and Ley, D. (eds), *Place/Culture/Representation*. London: Routledge.

Light, J. 1995: The digital landscape: new space for women? *Gender, Place and Culture* 2, 133–46.

Lim, L. 1990: Women's work in export factories: the politics of a cause. In Tinker,

I. (ed.), *Persistent Inequalities: Women and World Development.* New York: Oxford University Press.

Lipietz, A. 1986: New tendencies in the international division of labour: regimes of accumulation and modes of regulation. In Scott, A.J. and Storper, M. (eds), *Production, Work, Territory.* Boston, MA: Allen and Unwin.

Lipietz, A. 1987: *Mirages and Miracles: The Crisis of Global Fordism.* London, Verso.

Lipietz, A. 1993: From Althusserianism to 'Regulation Theory'. In Kaplan, A. and Sprinker, M. (eds), *The Althusserian Legacy.* London: Verso.

Little, J. 1994: *Gender, Planning and the Policy Process.* Oxford: Pergamon Press.

Little, J., Peake, L. and Richardson, P. (eds) 1988: *Women in Cities: Gender and the Urban Environment.* Basingstoke, Hants: Macmillan Education.

Livingstone, D.N. 1992: *The Geographical Tradition. Episodes in the History of a Contested Enterprise.* Oxford: Basil Blackwell.

Lloyd, G. 1984: *The Man of Reason: 'Male' and 'Female' in Western Philosophy.* London: Methuen.

Longhurst, R. 1994: The geography closest in – the body – the politics of pregnability. *Australian Geographical Studies* 32, 214–23.

Longhurst, R. 1995: The body and geography. *Gender, Place and Culture* 2, 97–105.

Longhurst, R. 1996: Refocusing groups: pregnant women's geographical experiences of Hamilton, New Zealand/Aotearoa. *Area* 28, 143–9.

Longhurst, R. 1997: (Dis)embodied geographies. *Progress in Human Geography* 21, 486–501.

Longwe, S. 1991: Gender awareness: the missing element in the Third World development project. In Wallace, T. with March, C. (eds), *Changing Perceptions: Writings on Gender and Development.* Oxford: Oxfam.

Loomba, A. 1998: *Colonialism/Postcolonialism.* London: Routledge.

Lopez Estrada, S. 1998: Women, urban life, and city images in Tijuana, Mexico. *Historical Geography* 26, forthcoming.

Lorde, A. 1984 *Sister Outside.* New York: Crossing Press.

Lovejoy, A.O. 1964: *The Great Chain of Being.* Cambridge, MA: Harvard University Press.

Lovelock, J. 1979: *Gaia: A New Look at Life on Earth.* Oxford: Oxford University Press.

Lovelock, J. 1994: Gaia. In C. Merchant (ed.) *Ecology.* New Jersey: Humanities Press International.

Lovering, J. 1990: A perfunctory sort of post-fordism: economic restructuring and labour market segmentation in Britain in the 1980s. *Work, Employment and Society* (special issue) May, 9–28.

Low, G.C.-L. 1986: *White Skins Black Masks.* London: Routledge.

Lowe, G.S. 1987: *Women in the Administrative Revolution: The Feminisation of Clerical Work.* Toronto: University of Toronto Press.

Lowe, L. 1992: *Critical Terrains: French and British Orientalisms.* Ithaca, NY: Cornell University Press.

Lowe, P., Marsden, T. and Whatmore, S. (eds) 1994: *Regulating Agriculture*. London: David Fulton.

Lowe, S. 1986: *Urban Social Movements: The City after Castells*. Basingstoke, Hants: Macmillan Education.

Lowenthal, D. 1985: *The Past is a Foreign Country*. Cambridge: Cambridge University Press.

Lukacs, G. 1968: *History and Class Consciousness*. London: Merlin.

Lukes, S. 1974: *Power: A Radical View*. London: Macmillan.

Lunn, E. 1985: *Marxism and Modernism: An Historical Study of Lukács, Brecht, Benjamin and Adorno*. London: Verso.

Lutz, C. and Collins, J. 1991: The Photograph as an intersection of gazes: the example of *National Geographic*. *Visual Anthropology Review* 7, 134–49.

Lykke, N. and Braidotti, R. (eds) 1996: *Between Monsters, Goddesses and Cyborgs: Feminist Confrontations with Science, Medicine and Cyberspace*. London, NJ: Zed Books.

Lyons, M. 1996: Employment, feminisation, and gentrification in London, 1981–93. *Environment and Planning A* 28, 341–56.

Lyotard, J.F. 1984: *The Postmodern Condition: A Report on Knowledge*, trans. Bennington, G. and Massumi, B. Minneapolis, MN: University of Minnesota Press.

Mac an Ghaill, M. 1994: *The Making of Men*. Buckingham: Open University Press.

MacCannell, D. 1976: *The Tourist: A New Theory of the Leisure Class*. New York: Schocken.

McCarthy, P. and Simpson, R. 1991: *Issues in Post Divorce Housing*. Aldershot, Hants: Gower.

McClintock, A. 1993: Family feuds: gender, nationalism and the family. *Feminist Review* 44, 61–80.

McClintock, A. 1995: *Imperial Leather: Race, Gender, and Sexuality in the Colonial Context*. New York: Routledge.

McDermott, P. and Briskin, L. (eds) 1993: *Women Challenging Unions: Feminism, Democracy and Militancy*. Toronto: University of Toronto Press.

Macdonald, M. 1991: Post-Fordism and the flexibility debate. *Studies in Political Economy* 36, 177–201.

MacDonald, S., Holden, P. and Andener, S. 1993: *Images of Women in Peace and War*. Madison, WI: University of Wisconsin Press.

McDowell, L. 1983: Towards an understanding of the gender division of urban space. *Environment and Planning D: Society and Space* 1, 59–72.

McDowell, L. 1986: Beyond patriarchy: a class-based explanation of women's subordination. *Antipode* 18, 311–21.

McDowell, L. 1989: Gender divisions. In Hamnett, C., McDowell, L. and Sarre, P. (eds), *The Changing Social Structure*. London: Sage.

McDowell, L. 1990: Sex and power in academia. *Area* 22, 323–32.

McDowell, L. 1991a: The baby and the bathwater: diversity, deconstruction and feminist theory in geography. *Geoforum* 22, 123–33.

McDowell, L. 1991b: Life without father and Ford: The new gender order of post-Fordism. *Transactions of the Institute of British Geographers: New Series* 16, 400–19.

McDowell, L. 1992: Doing gender: feminism, feminists and research methods in human geography. *Transactions of the Institute of British Geographers: New Series* 17, 399–415.

McDowell, L. 1993a: Space, place and gender relations. Part I: feminist empiricism and the geography of social relations. *Progress in Human Geography* 17, 159–79.

McDowell, L. 1993b: Space, place and gender relations: Part II: identity, difference, feminist geometries and geographies. *Progress in Human Geography* 17, 305-18.

McDowell, L. 1995: Body work: heterosexual gender performances in city workplaces. In Bell, D. and Valentine, G. (eds), *Mapping Desire*. London: Routledge.

McDowell, L. 1996: Spatializing feminism. In Duncan, N. (ed.), *Bodyspace*. London: Routledge.

McDowell, L. 1997a: The new service class: housing, consumption, and lifestyle among London bankers in the 1990s. *Environment and Planning A* 29, 2061–78.

McDowell, L. 1997b: *Capital Culture: Gender at Work in the City*. Oxford: Basil Blackwell.

McDowell, L. 1998a: Some academic and political implications of *Justice, Nature and the Geography of Difference*. *Antipode* 30, 3–5.

McDowell, L. 1998b: *Gender, Identity and Place: Understanding Feminist Geographies*. Cambridge: Polity Press.

McDowell, L. and Massey, D. 1984: A woman's place. In Massey, D. and Allen, J. (eds), *Geography Matters!* Cambridge: Cambridge University Press.

McDowell, L. and Pringle, R. (eds) 1992: *Defining Women: Social Institutions and Gender Divisions*. Cambridge: Polity Press.

McDowell, L. and Sharp, J.P. 1997: *Space, Gender, Knowledge: Feminist Readings*. London: Arnold.

McHoul, A. and Grace, W. 1993: *A Foucault Primer: Discourse, Power and the Subject*. Melbourne: Melbourne University Press.

Mackay, H. (ed.) 1997: *Consumption and Everyday Life*. London: Sage.

McKee, L. and Bell, C. 1985: Marital and family relations in times of male unemployment. In Roberts, R., Finnegan, R. and Gallie, D. (eds), *New Approaches to Economic Life*. Manchester: Manchester University Press.

Mackenzie, C. 1986: Simone de Beauvoir: philosophy and/or the female body. In Pateman, C. and Gross, E. (eds), *Feminist Challenges: Social and Political Theory*. Sydney: Allen and Unwin.

MacKenzie, J.M. 1995: *Orientalism. History, Theory and the Arts*. Manchester: Manchester University Press.

MacKenzie, S. 1989: Women in the city. In Peet, R. and Thrift, N. (eds), *New Models in Geography*, vol. 2. London: Unwin Hyman.

MacKenzie, S. and Rose, D. 1983: Industrial changes, the domestic economy and home life. In Anderson, J., Duncan, S. and Hudson, R. (eds), *Redundant Spaces: Industrial Decline in Cities and Regions*. London: Macmillan.

MacKian, S. 1995: 'That great dust-heap called history': recovering the multiple spaces of citizenship. *Political Geography* 14, 209–16.

MacKinnon, C. 1979: *The Sexual Harassment of Working Women*. New Haven, CT: Yale University Press.

MacKinnon, C. 1982: Feminism, marxism, method and the state: an agenda for theory. *Signs* 7, 515–44.

MacKinnon, C. 1987: *Feminism Unmodified: Disclosures on Life and Law*. Cambridge, MA: Harvard University Press.

MacKinnon, C. 1989: *Towards a Feminist Theory of the State*. Cambridge, MA: Harvard University Press.

McNeil, M. 1991: Putting the Alton Bill in context. In Franklin, S., Lury, C. and Stacey, J. (eds), *Off-centre: Feminism and Cultural Studies*. London: HarperCollins Academic.

McRobbie, A. 1981: Settling accounts with subcultures: a feminist critique. in Bennett, T., Martin, G., Mercer, C. and Woollacott, J. (eds), *Culture, Ideology and Social Process*. London: Batsford.

McRobbie, A. 1991: *Feminism and Youth Culture: From* Jackie *to* Just Seventeen. London, Macmillan.

Madrell, A. 1998: Discourses of race and gender and the comparative method in geography school texts 1830–1918. *Environment and Planning D: Society and Space* 16, 81–103.

Mahon, R. 1991: From 'bringing' to 'putting': the state in late twentieth-century social theory. *Canadian Journal of Sociology* 16, 119–44.

Malinowski, B. 1922: *Argonauts of the Western Pacific*. London: Routledge.

Malos, E. (ed.) 1995: *The Politics of Housework*. Cheltenham, Glos.: New Clarion Press.

Malthus, T.R. 1817: *An Essay on the Principle of Population*. London: John Murray.

Marchand, M. and Parpart, J. (eds) 1995: *Feminism Postmodernism and Development*. London: Routledge.

Marks, E. and de Courtivron, I. (eds) 1981: *New French Feminisms*. Brighton, E. Sussex: Harvester.

Marsden, J. 1996: Virtual sexes and feminist futures: the philosophy of 'cyberfeminism'. *Radical Philosophy* 78, 6–16.

Marsh, R. 1996: *Women in Russia and the Ukraine*. Cambridge: Cambridge University Press.

Marshall, B. 1994: *Engendering Modernity: Feminism, Social Theory and Social Change*. Cambridge: Polity Press.

Marshall, T. 1963: *Class, Citizenship and Social Development*. Chicago, IL: University of Chicago Press.

Marston, S.A. 1990: Who are 'the people'?: gender, citizenship, and the making of the American nation. *Environment and Planning D: Society and Space* 8, 449–58.

Martin, B. 1988: Lesbian identity and autobiographical difference(s). In Brodzki, B. and Schenck, C. (eds), *Life/Lines: Theorizing Women's Autobiography*. Ithaca, NY: Cornell University Press.

Martin, B. 1994: Sexualities without genders and other queer utopias. *Diacritics* 24, 104–21.

Martin, E. 1987: *The Woman in the Body:A Cultural Analysis of Reproduction*. Boston, MA: Beacon Press.

Martin, M.K. and Voorhies, B. 1975: *Female of the Species*. New York: Columbia University Press.

Martin, R. 1994: Stateless monies, global financial integration and national economic autonomy: the end of geography? In Corbridge, S. Martin, R. and Thrift, N. (eds), *Money, Power and Space*. Oxford: Basil Blackwell.

Martin, R. and Rowthorn, B. 1986: *The Geography of Deindustrialization*. London: Macmillan.

Martin, R., Sunley, P. and Wills, J. 1996: *Union Retreat and the Regions: The Shrinking Landscape of Organized Labour*. London: Jessica Kingsley.

Marx, K. 1954: *Capital. Volume One*. London: Lawrence and Wishart.

Marx, K. 1968: The Eighteenth Brumaire of Louis Bonaparte. In Marx, K. and Engels, F., *Marx–Engels: Selected Works*. Moscow: Progress Publishers.

Marx, K. 1976: Theses on Feuerbach. in Marx, K. and Engels, F., *The German Ideology*. Moscow: Progress Publishers.

Marx, K. and Engels, F. 1968: *Marx–Engels Selected Works in One Volume*. Moscow: Progress Publishers.

Mascia-Lees, F., Sharp, P. and Cohen, C.B. 1989: The postmodern turn in anthropology: cautions from a feminist perspective. *Signs* 15, 7–33.

Maslow, A. 1973: *The Farther Reaches of Human Nature*. Harmondsworth: Penguin.

Massey, D. 1979: In what sense a regional problem? *Regional Studies* 13, 233–43.

Massey, D. 1983: Industrial restructuring as class restructuring: production decentralization and local uniqueness. *Regional Studies* 17, 73–89.

Massey, D. 1984 (and 1995): *Spatial Divisions of Labour: Social Structures and the Geography of Production*. London: Methuen.

Massey, D. 1991a: The political place of locality studies. *Environment and Planning A* 23, 267–281.

Massey, D. 1991b: A global sense of place. *Marxism Today* (June), 24–9.

Massey, D. 1991c: Flexible sexism. *Environment and Planning D: Society and Space* 9, 31–57.

Massey, D. 1991d: A global sense of place. In Open University, D103 Block 6, *The Making of the Regions*. Milton Keynes: Open University Press.

Massey, D. 1992: A place called home? *New Formations* 17, 3–15.

Massey, D. 1993: Politics and time–space'. In Keith, M.and Pile, S. (eds), *Place and the Politics of Identity*. London: Routledge.

Massey, D. 1994a: *Space, Place and Gender*. Cambridge: Polity Press.

Massey, D. 1994b: A global sense of place. In Massey, D., *Space, Place and Gender*. Cambridge: Polity Press, 146–56.

Massey, D. 1994c: A place called home? in *Space, Place and Gender*. Cambridge: Polity Press.

Massey, D. 1995a: The concepualization of place. In Massey, D. and Jess, P. (eds), *A Place in the World*. Milton Keynes: Open University Press.

Massey, D. 1995b: Masculinity, dualisms and high technology. *Transactions of the Institute of British Geographers* 20, 487–99.

Massey, D. and Allen, J. 1995: High-tech places: poverty in the midst of growth. In Philo, C. (ed.), *Off the Map: The Social Geography of Poverty in the UK*. London: Child Poverty Action Group.

Massey, D. and Jess, P. 1995: *A Place in the World?* Oxford: Oxford University Press.

Mathias, P. 1983: *The First Industrial Nation: An Economic History of Britain 1700–1914*. London: Routledge.

Matless, D. 1997: The geographical self, the nature of the social and geo-aesthetics: work in social and cultural geography, 1996. *Progress in Human Geography* 21, 393–405.

Mayer, T. 1989: Consensus and invisibility: the representation of women in human geography textbooks. *Professional Geographer* 41, 397–409.

Mayer, T. 1994: Heightened Palestinian nationalism: military occupation, repression, difference and gender. In Mayer, T. (ed.), *Women and the Israeli Occupation: The Politics of Change*. New York, Routledge.

Mbembe, A. 1995: Provisional notes on the postcolony. *Africa* 62, 3–37.

Mead, M. 1928: *Coming of Age in Samoa*. New York: Morrow.

Mead, M. 1935: *Sex and Temperament in Three Primitive Societies*. Pittsburgh, PA: University of Pittsburgh Press.

Meinig, D. 1976: The beholding eye: ten versions of the same scene. *Landscape Architecture* 66, 47–54.

Meinig, D.W. (ed.) 1979: *The Interpretation of Ordinary Landscapes*. New York: Oxford University Press.

Meinig, D.W. 1983: Geography as an art. *Transactions of the Institute of British Geographers* NS8, 314–28.

Meinzen, R.S., Brown, L.R., Feldstein, H.S. and Quisumbing, A.R. 1997: Gender, property rights, and natural resources. *World Development* 25, 1303–15.

Mellor, M. 1997: *Feminism and Ecology*. Cambridge: Polity Press.

Melman, B. 1992: *Women's Orients. English Women and the Middle East, 1718–1918. Sexuality, Religion and Work*. London: Macmillan.

Meono-Picado, P. 1997: Redefining the barricades: Latina lesbian politics and the creation of an oppositional public space. In Jones III, J.P., Nast, H. and Roberts, S. (eds), *Thresholds in Feminist Geography: Difference, Methodology, Representation*. Boston, MA: Rowman and Littlefield.

Merchant, C. 1980: *The Death of Nature: Women, Ecology, and the Scientific Revolution*. New York: Harper and Row.

Merchant, C. 1992: *Radical Ecology: The Search for a Liveable World*. London: Routledge.

Merchant, C. 1995: *Earthcare: Women and the Environment*. New York: Routledge.

Merrifield, A. and Swyngedouw, E. (ed.) 1995: *The Urbanization of Injustice.* London: Lawrence and Wishart.

Merriman, J.M. 1991: *The Margins of City Life: Explorations of the French Urban Frontier 1815–1851.* New York and Oxford: Oxford University Press.

Meskimmon, M. 1997: *Engendering the City: Women Artists and Urban Space.* London: Scarlet Press.

Messner, M. 1997: *The Politics of Masculinities: Men in Movements.* London: Sage.

Meulders, D., Plasman, O. and Plasman, R. 1994: *Atypical Employment in the EC.* Aldershot, Hants: Dartmouth Publishing.

Michell, B. and Draper, D. 1982: *Relevance and Ethics in Geography.* London: Longman.

Middleton, D. 1982: *Victorian Lady Travellers* (reprint). Chicago, IL: Academy Chicago Publishers.

Midgley, M. 1979: *Beast and Man: The Roots of Human Nature.* Brighton: Harvester.

Mies, M. and Shiva, V. 1993: *Ecofeminism.* London: Zed Books.

Mikesell, M. 1968: Landscape. In Sills, D. (ed.), *International Encyclopedia of the Social Sciences*, vol. 8. New York: Oxford University Press.

Miles, R. 1993: *Racism after 'Race relations'.* London: Routledge.

Miles, M. and Crush, J. 1993: Personal narratives and interactive texts: collecting and interpreting migrant life-histories. *The Professional Geographer* 45, 84–96.

Miliband, R. 1969: *The State in Capitalist Society.* London: Weidenfeld and Nicholson.

Miller, C. 1990: *Theories of Africans: Francophone literature and Anthropology in Africa.* Chicago, IL: University of Chicago Press.

Miller, D. 1987: *Material Culture and Mass Consumption.* Oxford: Basil Blackwell.

Miller, D. 1995: Citizenship and pluralism. *Political Studies* XLIII, 432–50.

Miller, D., Jackson, P., Thrift, N. *et al.* 1998: *Shopping, Place and Identity.* London: Routledge.

Millett, K. 1971: *Sexual Politics.* London: Rupert Hart-Davis.

Mills, C. 1993: Myths and meanings of gentrification. In Duncan, J. and Ley, D. (eds), *Place/Culture/Representation.* London: Routledge.

Mills, S. 1991: *Discourses of Difference: An Analysis of Women's Travel Writing and Colonialism.* London and New York: Routledge.

Mincer, J. and Polachek, S. 1974: Family investment in human capital: earnings of women. *Journal of Political Economy* 82, 2.

Mirandé, A. and Enríquez, E. 1979: *La Chicana: The Mexican American Woman.* Chicago, IL: University of Chicago Press.

Mitchell, D. 1993: Book review of *Writing Worlds*, T. Barnes and J. Duncan (eds), *Professional Geographer* 45, 474–75.

Mitchell, D. 1995: There's no such thing as culture. *Transactions of the Institute of British Geographers* 20, 102–16.

Mitchell, D. 1996: *The Lie of the Land: Migrant Workers and the California Landscape.* Minneapolis, MN: University of Minnesota Press.

Mitchell, J. 1975: *Psychoanalysis and Feminism*. New York: Vintage.

Mitchell, J. 1984: *Women: The Longest Revolution. Essays in Feminism, Literature and Psychoanalysis*. London: Virago.

Mitchell, J. (ed.) 1991: *The Selected Melanie Klein*. Harmondsworth, Middx: Penguin.

Mitchell, J. and Rose, J. 1982: *Feminine Sexuality. Jacques Lacan and the École freudienne*. London: Macmillan.

Mitter, S. 1986: *Common Fate, Common Bond: Women in the Global Economy*. London: Pluto Press.

Mohanty, C.T. 1987: Feminist Encounters: locating the politics of experience. *Copyright* 1, Fall, 30–44.

Mohanty, C.T. 1991a: Cartographies of struggle: Third World women and the politics of feminism. In C.T. Mohanty, A. Russo and L. Torres (eds), *Third World Women and the Politics of Feminism*. Bloomington, IN: Indiana University Press.

Mohanty, C. T. 1991b: Under Western eyes: feminist scholarship and colonialist discourses. In Mohanty, C.T., Russo, A. and Torres, L. (eds), *Third World Women and the Politics of Feminism*. Bloomington, IN: Indiana University Press.

Mohanty, C.T. 1992: Feminist encounters: locating the politics of experience. In Barrett, M., Phillips, A. (eds), *Destabilising Theory. Contemporary Feminist Debates*. Stanford, CA: Stanford University Press.

Mohanty, C.T., Russo, A. and Torres, L. (eds) 1991: *Third World Women and the Politics of Feminism*. Bloomington, IN: Indiana University Press.

Moi, T. 1985: *Sexual/Textual Politics: Feminist Literary Theory*. London: Methuen.

Moi, T. (ed.) 1987: *French Feminist Thought: A Reader*. Oxford: Basil Blackwell.

Momsen, J. 1991: *Women and Development in the Third World*. London: Routledge.

Momsen, J.H. and Kinnaird, V. (eds) 1993: *Different Places, Different Voices*. London: Routledge.

Momsen, J.H. and Townsend, J. 1987: *Geography of Gender in the Third World*. London: Hutchinson.

Monk, J. 1983: Integrating women into the geography curriculum. *Journal of Geography* 34, 11–23.

Monk, J. 1992: Gender in the landscape: expressions of power and meaning. In Anderson, K. and Gale, F. *Inventing Places: Studies in Cultural Geography*. Chichester, Sussex: John Wiley.

Monk, J. 1996: Braided streams: personal and professional in women geographer's lives. Paper presented at the Annual Conference of the American Association of Geographers, Charlotte, NC. Available from the author, South West Institute for Research on Women, University of Arizona, Tuscon.

Monk, J. and Hanson, S. 1982: On not excluding half of the human in human geography. *Professional Geographer* 34, 11–23.

Moon, M. 1998: *A Small Boy and Others: Imitation and Initiation in American Culture from Henry James to Andy Warhol.* Durham NC: Duke University Press.

Moore, G. and Moore, R. 1986: *Margaret Sanger and the Birth Control Movement: A Bibliography, 1911–1984.* London: Scarecrow.

Moore, H.L. 1988: *Feminism and Anthropology.* Cambridge: Polity.

Moore, H.L. 1994: *A Passion for Difference.* Cambridge: Polity Press.

Moore, R. and Gillette, D. 1990: *King, Warrior, Magician, Lover: Rediscovering the Archetypes of the Mature Masculine.* New York: HarperCollins.

Moraga, C. 1986: From a long line of vendidas: chicanas and feminism. In de Lauretis, T. (ed.), *Feminist Studies/Critical Studies.* Bloomington IN: Indiana University Press.

Moraga, C. and Anzaldúa, G. (eds) 1981: *This Bridge Called My Back: Writings by Radical Women of Color.* Watertown, MA: Persephone.

Morgan, D. 1992: *Discovering Men.* London: Routledge.

Morgan, L.H. 1907: *Ancient Society.* New York: Henry Holt.

Morgan, R. 1970: *Sisterhood is Powerful.* Harmondsworth, Middx: Penguin.

Morgen, S. 1989: *Gender and Anthropology: Critical Reviews for Research and Teaching.* Washington, DC: American Anthropological Association.

Morin, K. 1998: British women travellers and constructions of racial difference across the 19th-century American West. *Transactions of the Institute of British Geographers*, forthcoming.

Morley, D. 1992: *Television, Audiences and Cultural Studies.* London: Routledge.

Morrell, H. 1996: Women's safety. In Booth, C., Darke, J. and Yeandle, S. (eds), *Changing Places: Women's Lives in the City.* London: Paul Chapman.

Morris, J. and Winn, M. 1990: *Housing and Social Inequality.* London: Hilary Shipman.

Morris, L. 1993: Domestic labour and the employment status of married couples. *Capital and Class* 49, 37–52.

Morris, L. 1995: *Social Divisions: Economic Decline and Social Structural Change.* London: UCL Press.

Morris, M. 1988: Things to do with shopping centres. In Sheridan, S. (ed.), *Grafts: Feminist Cultural Criticism.* London: Verso.

Morris, P. 1993: *Literature and Feminism.* Oxford: Basil Blackwell.

Morris, R. 1995: All made up: performance theory and the new anthropology of sex and gender. *Annual Review in Anthropology* 24, 461–95.

Morrison, T. 1981: *Tar Baby.* London: Chatto and Windus.

Morrison, T. 1992: *Playing in the Dark: Whiteness and the Literary Imagination.* Cambridge, MA, and London: Harvard University Press.

Morrow, V. 1996: Rethinking childhood dependency: children's contributions to the domestic economy. *The Sociological Review* 44, 58–77.

Mort, F. 1987: *Dangerous Sexualities: Medico-Moral Politics in England Since 1830.* London: Routledge and Kegan Paul.

Mort, F. 1989: The politics of consumption. In Hall, S. and Jacques, M. (eds), *New Times.* London: Lawrence and Wishart.

Mosedale, S. 1978: Science corrupted: Victorian biologists consider 'the women question'. *Journal of the History of Biology* 11, 1–55.

Moser, C. 1989: Gender planning in the Third World: meeting practical and strategic gender needs. *World Development* 17, 1799–825.

Moser, C. 1993a: *Gender Planning and Development: Theory, Practice and Training*. London: Routledge.

Moser, C. 1993b: Adjustment from below: low-income women, time and the triple role in Guayaquil, Ecuador. In Radcliffe, S. and Westwood, S. (eds), *'Viva': Women and Popular Protest in Latin America*. London: Routledge.

Moss, P. 1995: Embeddedness in practice, numbers in context: the politics of knowing and doing. *Professional Geographer* 47, 442–49.

Moss, P. and Dyck, I. 1996: Inquiry into environment and body: women, work, and chronic illness. *Environment and Planning D: Society and Space* 14, 737–53.

Mouffe, C. 1991: Democratic ctizenship and the political community. In Miami Theory Collective (ed.), *Community at Loose Ends*. Minneapolis, MN: University of Minnesota Press.

Mouffe, C. 1992: Feminism, citizenship and radical democratic politics. In Butler, J. and Scott, J. (eds), *Feminists Theorize the Political*. New York: Routledge.

Mulvey, L. 1975: Visual pleasure and narrative cinema. *Screen* 16.

Mulvey, L. 1985 (reprint): Visual pleasure and narrative cinema. In Mast, G. and Cohen, M. (eds), *Film Theory and Criticism: Introductory Readings*, 3rd edn. New York: Oxford University Press.

Mulvey, L. 1989a (reprint): Visual pleasure and narrative cinema. In Mulvey, L., *Visual and Other Pleasures*. London: Macmillan, 14–26.

Mulvey, L. 1989b: *Visual and Other Pleasures*. London: Macmillan.

Munro, M. and Smith, S.J. 1989: Women and housing: broadening the debate. *Housing Studies* 4, 81–93.

Munt, S. 1993: The lesbian flâneur. In Bell, D. and Valentine, G. (eds), *Mapping Desire: Geographies of Sexualities*. London: Routledge.

Murgatroyd, L. *et al.* 1985: *Localities, Class and Gender*. London: Pion.

Murray, C. 1994: *Underclass: The Crisis Deepens*. London: IEA Health and Welfare Unit, in association with *The Sunday Times*.

Naess, A. 1995: Articles. In G. Sessions (ed.) *Deep Ecology in the Twentieth Century*. Boston: Shambala.

Nagar, R. 1997: Exploring methodological borderlands through oral narratives. In Jones III, J.P., Nast, H. and Roberts, S. (eds), *Thresholds in Feminist Geography. Difference, Methodology, Representation*. New York: Rowman and Littlefield.

Nagle, T. 1986: *The View from Nowhere*. New York: Oxford University Press.

Namaste, K. 1996: Genderbashing: sexuality, gender, and the regulation of public space. *Environment and Planning D: Society and Space* 14, 221–40.

Nandy, A. 1983: *The Intimate Enemy: Loss and Recovery of Self Under Colonialism*. Oxford: Oxford University Press.

Nash, C. 1993: Remapping and renaming: new cartographies of identity, gender and landscape in Ireland. *Feminist Review* 44, 39–57.

Nash, C. 1994: Remapping the Body/Land: New Cartographies of Identity, Gender, and Landscape in Ireland. In Blunt, A. and Rose, G. (eds), *Writing Women and Space: Colonial and Postcolonial Geographies.* New York and London: Guildford Press.

Nash, C. 1996a: Men again: Irish masculinity, nature, and nationhood in the early twentieth century. *Ecumene* 3, 427–52.

Nash, C. 1996b: Reclaiming vision: looking at landscape and the body. *Gender, Place and Culture* 3, 149–69.

Nast, H. 1994: Women in the field: critical feminist methodologies and theoretical perspectives. *Professional Geographer* 46, 54–66.

Nast, H. 1988a: The body as 'place': reflexivity and fieldwork in Kano, Nigeria. In Nast, H. and Pile, S. (eds), *Places Through the Body.* London: Routledge.

Nast, H.J. 1998b: Unsexy geographies. *Gender, Place and Culture* 5, 191–206.

Nayak, A. and Kehily, M. 1996: Playing it straight: masculinities, homophobias and schooling. *Journal of Gender Studies* 5, 211–30.

Nelson, K. 1986: Female labour supply characteristics and the subordination of low-wage office work. In Scott, A. and Storper, A. (eds), *Production, Work, Territory: The Geographical Anatomy of Industrial Capitalism.* London: Allen and Unwin.

New, C. 1991: Women's Oppression in the World and in Ourselves: a Fresh look at feminism and psychoanalysis. In Abbott, P. and Wallace, C. (eds) *Gender, Power and Sexuality.* London: Macmillan.

New, C. 1996: *Agency, Health and Social Survival: the Ecopolitics of Rival Psychologies.* London: Taylor and Francis.

Newburn, T. and Stanko, E. (eds) 1994: *Just Boys Doing Business? Men, Masculinities and Crime.* London: Routledge.

Nicholson, L.J. 1986: *Gender and History: The Limits of Social Theory in the Age of the Family.* New York: Columbia University Press.

Nicholson, L. (ed.) 1990: *Feminism/Postmodernism.* London: Routledge.

Nicholson, L. 1995: Interpreting gender. In Nicholson, L. and Seidman, S. (eds), *Social Postmodernism: Beyond Identity Politics.* Cambridge: Cambridge University Press.

Nietzsche, F. 1956: *The Genealogy of Morals.* New York: Doubleday.

Norwood, V. and Monk, J. (eds) 1987: *The Desert is No Lady: Southwestern Landscapes in Women's Writing and Art.* New Haven, CT: Yale University Press.

Nystuen, J.D. 1965: Identification of some fundamental spatial concepts. In Berry, B. and Marble, D.F. (eds), *Spatial Analysis: A Reader in Statistical Geography.* Englewood Cliffs, NJ: Prentice-Hall.

Oakley, A. 1974: *The Sociology of Housework.* Harmondsworth, Middx: Penguin.

Oakley, A. and Mitchell, J. 1997: *Who's Afraid of Feminism?* Harmondsworth, Middx: Penguin.

Oberhauser, A. 1995: Gender and household economic strategies in rural Appalachia. Gender, *Place and Culture* 2, 51–70.

O'Brien, M. 1981: *The Politics of Reproduction*. London: Routledge and Kegan Paul.

O'Brien, M. 1995: Allocation of resources within the household: children's perspectives. *The Sociological Review* 43, 501–17.

O'Brien, R. 1991: *Global Financial Integration: The End of Geography*. London: Frances Pinter.

Ohmae, K. 1990: *A Borderless World*. New York: Harper.

Okin, S. 1989: *Justice, Gender and the Family*. New York: Basic Books.

Oliver, M. 1990: *The Politics of Disablement*. London: Macmillan.

Olsen, T. 1978: *Silences*. London: Virago.

Olwig, K. 1996: Recovering the substantive nature of landscape. *Annals of the American Association of Geographers* 86, 630–53.

Ong, A. 1987: *Spirits of Resistance and Capitalist Discipline: Factory Women in Malaysia*. Albany, NY: State University of New York Press.

Ong, A. 1988: Colonialism and modernity: feminist re-presentations of women in non-western societies. *Inscriptions* 3/4, 79–93.

Ong, A. 1991: The gender and labor politics of postmodernity. *Annual Review of Anthropology* 20, 279–310.

Oppenheim, C. 1993: *Poverty: The Facts*. London: CPAG.

Ortner, S. 1974 Is female to male as nature to culture? In Rosaldo, M. and Lamphere, L. (eds), *Women, Culture and Society*. Stanford, CA: Stanford University Press.

Osborne, P. 1995: *The Politics of Time: Modernity and Avant-Garde*. London: Verso.

Osman, S. 1983: A to Z of feminism. *Spare Rib* (November), 27–30.

Ostergaard, L. (ed.) 1991: *Gender and Development: A Practical Guide*. London: Routledge.

Ó Tuathail, G. 1996: An anti-geopolitical eye: Maggie O'Kane in Bosnia, 1992–1993. *Gender, Place and Culture* 3, 171–85.

Ó Tuathail, G. and Agnew, J. 1992: Geopolitics and discourse: practical geopolitical reasoning in American foreign policy. *Political Geography* 11, 190–204.

Oudshoorn, N. 1996: A natural order of things? Reproductive sciences and the politics of othering. In Robertson, G., Mash, M., Tickner, L. *et al.* (eds), *FutureNatural*. London: Routledge.

Pacione, M. 1995: The geography of multiple deprivation in the Clydeside conurbation. *Tijdschrift Voor Economische en Sociale Geografie* 86, 407–25.

Pahl, R. (ed.) 1988: *On Work: Historical, Comparative and Theoretical Approaches*. Oxford: Basil Blackwell.

Pain, R. 1991: Space, sexual violence and social control: integrating geographical and feminist analyses of women's fear of crime. *Progress in Human Geography* 15, 415–31.

Pain, R. 1994: Kid gloves: children's geographies and the impact of violent crime. Departmental occasional paper, new series no 1, Division of Geography and Environmental Management, University of Northumbria.

Pain, R. 1997a: Whither women's fear? Perceptions of sexual violence in public and private space. *International Review of Victimology* 4, 297–312.

Pain, R.H. 1997b: Social geographies of women's fear of crime. *Transactions of the Institute of British Geographers* 22, 2.

Pain, R.H. 1997c: 'Old age' and ageism in urban research: the case of fear of crime. *International Journal of Urban and Regional Research* 21, 1.

Painter, J. 1994: Locality. In Johnston, R., Gregory, D. and Smith, D. (eds), *The Dictionary of Human Geography*. Oxford: Basil Blackwell.

Painter, J. 1995: *Politics, Geography and Political Geography. A Critical Perspective*. London: Arnold.

Painter, J. and Philo, C. 1995: Spaces of Citizenship: an introduction. *Political Geography* 14, 107–20.

Painter, K. 1992: Different worlds: the spatial, temporal and social dimensions of female victimisation. In Evans, D.J., Fyfe, N.R. and Herbert, D.T. (eds), *Crime, Policing and Place*. London: Routledge.

Pajaczkowska, C. and Young, L. 1992: Racism, representation, psychoanalysis. In Donald, J. and Rattansi, A. (eds), *'Race', Culture and Difference*. London: Sage.

Parezo, N. (ed.) 1993: *Hidden Scholars: Women Anthropologists and the Native American Southwest*. Albuquerque, NM: University of New Mexico Press.

Parpart, J.L. 1993: Who is the 'Other'?: A postmodern feminist critique of women and development theory and practice. *Development and Change* 24, 439–64.

Parr, H. 1997: Naming Names: brief thoughts on disability and geography. *Area* 29, 173–6.

Parr, H. and C. Philo 1995: Mapping 'mad' identities. In Pile, S. and Thrift, N. (eds), *Mapping the Subject*. London: Routledge.

Parr, H. and Philo, C. 1996: A forbidding fortress of locks and bars?: the locational history of Nottingham's mental health care, Historical Geography Research Series 32.

Parr, J. 1990: *The Gender of Breadwinners: Women, Men and Change in Two Industrial Towns, 1880–1950*. Toronto: University of Toronto Press.

Parry, B. 1987: Problems in current theories of colonial discourse. *Oxford Literary Review* 9, 27–58.

Parsons, E.C. 1916: *Social Rule: A Study of the Will to Power*. New York: Putnam.

Parton, N. 1985: *The Politics of Child Abuse*. London: Macmillan.

Pateman, C. 1988a: *The Sexual Contract*. Cambridge: Polity Press.

Pateman, C. 1988b: The patriarchal welfare state. In Pateman, C., *The Disorder of Women*. Stanford, CA: Stanford University Press.

Pateman, C. 1989: *The Disorder of Women: Democracy, Feminism and Political Theory*. Stanford, CA: Stanford University Press.

Pateman, C. and Grosz, E. 1987: *Feminist Challenges*. Boston, MA: North Eastern University Press.

Patton, C. 1990: *Inventing AIDS*. New York: Routledge.

Pawson, E. and Banks, G. 1993: Rape and fear in a New Zealand city. *Area* 25, 55–63.

Peake, L. 1993: Race and sexuality: challenging the patriarchal structuring of urban social space. *Environment and Planning D: Society and Space* 11, 415–32.

Peake, L. 1997: Toward a social geography of the city: race and dimensions of urban poverty in women's lives. *Journal of Urban Affairs* 19, 335–61.

Peake, L. and Trotz, A. 1998: *Guyana-ways: Placing Gender and Ethnicity.* London: Routledge.

Pearson, R. 1986: Female workers in the First and Third Worlds: the greening of women's labour. In Purcell, K., Wood, S.,Watson, A. and Allen, S. (eds), *The Changing Experience of Employment: Restructuring and Recession.* Basingstoke, Hants: Macmillan Education. Reprinted (1988) in Pahl, R. (ed.), *On Work: Historical, Comparative and Theoretical Approaches.* Oxford: Basil Blackwell.

Pearson, R. 1992: Gender issues in industrialisation In Hewitt, T., Johnson, H. and Wield, D. (eds), *Industrialisation and Development.* Oxford: Oxford University Press in association with the Open University.

Pearson, R., Whitehead, A. and Young, K. 1984: Introduction. The continuing subordination of women in the development process. In Young, K., Wolkowitz, C. and McCullah, R. (eds), *Of Marriage and the Market: Women's Subordination Internationally and its Lessons.* London: Routledge and Kegan Paul.

Peck, J. 1996: *Work–Place.* London: Guildford Press.

Peck, J. and Tickell, A. 1994: Searching for a new institutional fix: the after-Fordist crisis and global–local disorder. In Amin, A. (ed.), *Post-Fordism: A Reader.* Oxford: Basil Blackwell.

Peet, J.R. 1969: A new left geography. *Antipode* 1, 3–5.

Peet, R. 1993: Book review of *The City as Text*, J. Duncan. *Annals of the Association of American Geographers* 83, 184–87.

Penhale, B. 1993: The abuse of elderly people: considerations for practice. *British Journal of Social Work* 23, 95–112.

Penrose, J. 1992: Introduction. *Antipode* 24, 218–20.

Perrons, D. 1995: Gender inequalities in regional development. *Regional Studies* 29, 465–76.

Pepper, D. 1993: *Eco-Socialism: From Deep Ecology to Social Justice.* London: Verso.

Peters, J. and Wolper, A. (eds) 1995: *Women's Rights, Human Rights: International Feminist Perspectives.* New York and London: Routledge.

Peterson, V.S. 1992: Introduction. In Peterson, V.S. *Gendered States: Feminist (Re)visions of International Relations Theory.* London: Lynne Rienner.

Phillips, A. 1983: *Hidden Hands: Women and Economic Policies.* London: Pluto Press.

Phillips, A. (ed.) 1987: *Feminism and Equality.* Oxford: Basil Blackwell.

Phillips, A. and Rakusen, J. (eds) 1978: Boston Women's Health Collective. *Our Bodies, Ourselves: A Health Book by and for Women.* Harmondsworth: Penguin: Allen and Lane.

Phillips, A. and Taylor, B. 1980: Sex and skill: notes towards a feminist economics. *Feminist Review* 6, 79–88.

Phillips, R.S. 1995: Spaces of adventure and cultural politics of masculinity: R.M. Ballantyne and *The Young Fur Traders*. *Environment and Planning D: Society and Space* 13, 591–608.

Phillips, R.S. 1997: *Mapping Men and Empire: A Geography of Adventure*. London: Routledge.

Philo, C. 1987: Fit localities for an asylum: the historical geography of the nineteenth century 'mad business' in England as viewed through the pages of the *Asylum Journal. Journal of Historical Geography* 13, 398–415.

Philo, C. 1995a: Animals, geography, and the city: notes on inclusions and exlusions. *Environment and Planning D: Society and Space* 13, 655–81.

Philo, C. (ed.) 1995b: *Off The Map: The Social Geography of Poverty in the UK*. London: CPAG.

Philo, C. and Kearns, G. 1993: *Selling Places: The City as Cultural Capital Past and Present*. Oxford: Pergamon.

Phizacklea, A. 1990: *Unpacking the Fashion Industry: Gender, Racism and Class in Production*. London: Routledge.

Phizacklea, A. and Wolkowitz, C. 1995: *Homeworking Women: Gender, Racism and Class at Work*. London: Sage.

Phoenix, A. 1990: Theories of gender and black families. In Lovell, T. (ed.), *British Feminist Thought: A Reader*. Oxford: Basil Blackwell.

Phoenix, A., Woollett, A. and Lloyd, E. (eds) 1991: *Motherhood: Meaning. Practices and Ideologies*. London: Sage.

Pickup, L. 1985: Women's travel needs in a period of rising female employment. In Jansen, G.R.M. *et al.* (eds), *Transportation and Mobility in an Era of Transition*. Amsterdam: Elsevier.

Piercy, M. 1976: *Woman on the Edge of Time*. New York: Alfred A. Knopf.

Pieterse, J. 1996: Globalization as hybridization. In Featherstone, M., Lash, S. and Robertson, R. (eds), *Global Modernities*. London: Sage.

Pietila, H. and Vicker, J. 1994: *Making Women Matter: The Role of The United Nations*. London: Zed Books.

Pile, S. 1994: Masculinism, the use of dualistic epistemologies and third spaces. *Antipode* 26, 255–77.

Pile, S. 1996: *The Body and the City. Psychoanalysis, Space and Subjectivity*. London: Routledge.

Pile, S. and Thrift, N. (eds) 1995: *Mapping the Subject: Geographies of Cultural Transformation*. London: Routledge.

Pinch, S. 1985: *Cities and Services: The Geography of Collective Consumption*. London: Routledge.

Piven, F.F. and Cloward, R. 1971: *Regulating the Poor*. New York: Pantheon.

Plant, J. 1989: *Healing the Wounds: The Promise of Ecofeminism*. Santa Cruz, CA: New Society Publishers.

Plant, S. 1996: On the Matrix: Cyberfeminist Simulations. In Sheilds, R. (ed.) *Cultures of the Internet: Virtual Spaces, Real Histories, Living Bodies*. London: Sage.

Plant, S. 1997: *Zeros and Ones: Digital Women and the New Technoculture.* London: Fourth Estate.

Pleck, J. 1981: *The Myth of Masculinity.* Cambridge, MA: MIT Press.

Plummer, K. 1975: *Sexual Stigma: An Interactionist Account.* London: Routledge and Kegan Paul.

Plummer, K. (ed.) 1981: *The Making of the Modern Homosexual.* London: Hutchinson.

Plummer, K. 1983: *Documents of Life: An Introduction to the Problems and Literature of a Humanistic Method.* London: Allen and Unwin.

Plumwood, V. 1993: *Feminism and the Mastery of Nature.* London: Routledge.

Polachek, S. 1975: Discontinuities in labour market participation and its effects on women's market earnings. In Lloyd, C.B. (ed.), *Sex, Discrimination and the Division of Labour.* New York: Columbia University Press.

Polan, D. 1982: Towards a theory of law and Patriarchy. In Kairys, D. *The Politics of Law: A Progressive Critique.* New York: Pantheon.

Polanyi, K. 1944: *The Great Transformation.* Boston, MA: Beacon Press.

Pollock, G. 1987: Feminism and modernism. In Parker, R. and Pollock, G. (eds), *Framing Feminism: Art and the Women's Movement 1970–85.* London and New York: Pandora.

Pollock, G. 1988: *Vision and Difference: Femininity, Feminism and the Histories of Art.* London: Routledge.

Pollock, G. (ed.) 1996: *Generations and Geographies in the Visual Arts: Feminist Readings.* London: Routledge.

Porritt, J. and Winner, D. 1988: *The Coming of the Greens.* London: Fontana.

Portes, A., Castells, M. and Benton, L. (eds) 1989: *The Informal Economy: Studies in Advanced and Less Developed Countries.* Baltimore, MA: Johns Hopkins University Press.

Pratt, A. 1982: *Archetypal Patterns in Women's Fiction.* Brighton, E. Sussex: Harvester.

Pratt, A. 1995: Putting critical realism to work: the practical implications for geographical research. *Progress in Human Geography* 19, 61–74.

Pratt, G. 1986: Housing tenure and social cleavages in urban Canada. *Annals of the Association of American Geographers* 76, 366–80.

Pratt, G. 1989: Incorporation theory and the reproduction of community fabric. In Wolch, J. and Dear, M. (eds), *The Power of Geography: How Territory Shapes Social Life.* London: Unwin Hyman.

Pratt, G. 1992: Spatial metaphors and speaking positions. *Environment and Planning D: Society and Space*, 241–4.

Pratt, G. 1993: Reflections on poststructuralism and feminist empirics, theory and practice. *Antipode* 25, 51–63.

Pratt, G. 1997: Stereotypes and ambivalence: the construction of domestic workers in Vancouver, British Columbia. *Gender, Place and Culture* 4, 159–77.

Pratt, G. and Hanson, S. 1994: Geography and the construction of difference. *Gender, Place and Culture* 1, 5–30.

Pratt, G. and Hanson, S. 1995: *Gender, Work, and Space*. London and New York: Routledge.

Pratt, M.B. 1984: Identity: skin blood heart. In Burkin, E., Pratt, M.B. and Smith, B. (eds), *Yours in Struggle: Three Feminist Perspectives on Anti-Semitism and Racism*. Ithaca, NY: Firebrand Books.

Pratt, M.L. 1992: *Imperial Eyes: Travel Writing and Transculturation*. London and New York: Routledge.

Pred, A. 1984: Place as a historically contingent process: structuration and the time-geography of becoming places. *Annals of the Association of American Geographers* 74, 279–97.

Price, M. and Lewis, M. 1993: The reinvention of cultural geography. *Annals of the Association of American Geographers* 83, 1–17.

Price-Chalita, P. 1994: Spatial metaphor and the politics of empowerment: mapping a place for feminism and postmodernism in geography. *Antipode: A Radical Journal of Geography* 26, 254–63.

Pringle, R. 1989: *Secretaries Talk: Sexuality, Power and Work*. London: Verso.

Pringle, R. and Watson, S. 1992: Women's interests and the post-structuralist state. In Barrett, M. and Philips, A. (eds), *Destabilizing Theory. Contemporary Feminist Debates*. Stanford, CA: Stanford University Press

Probyn, E. 1993: *Sexing the Self: Gendered Positions in Cultural Studies*. London: Routledge.

Proctor, J. 1998: Ethics in geography: giving moral form to the geographical imagination. *Area* 30, 8–18.

Pulido, L. 1997: Community, place and identity. In Jones III, J.P., Nast, H. and Roberts, S. (eds), *Thresholds in Feminist Geography: Difference, Methodology, Representation*. New York: Rowman and Littlefield.

Pulsipher, L.M. 1993: 'He won't let she stretch she foot': gender relations in traditional West Indian houseyards. In Katz, C. and Monk, J. (eds), *Full Circles: Geographies of Women over the Life Course*. London: Routledge.

Quack, S. and Maier, F. 1994: From state socialism to market economy – women's employment in East Germany. *Environment and Planning A* 26, 1257–76.

Radcliffe, S. 1990: Ethnicity, patriarchy, and incorporation into the nation: female migrants as domestic servants in Peru. *Environment and Planning D: Society and Space* 8, 379–93.

Radcliffe, S. 1991: The role of gender in peasant migration: conceptual issues from the Peruvian Andes. *Review of Radical Political Economy* 23, 148–73.

Radcliffe, S. 1994: (Representing) post-colonial women: authority, difference and feminisms. *Area* 26, 25–32.

Radcliffe, S. 1996: Gendered nations: nostalgia, development and territory in Ecuador. *Gender, Place and Culture: A Journal of Feminist Geography* 3, 5–21.

Radcliffe, S. and Westwood, S. (eds) 1993: *'Viva': Women and Popular Protest in Latin America*. London: Routledge.

Radcliffe, S., Westwood, S. 1996: *Remaking the Nation: Place, Identity and Politics in Latin America*. London: Routledge.

Radway, J. 1985: *Reading the Romance: Women, Patriarchy and Popular Literature*. London: Verso.

Rai, S. 1996: Women and the state in the Third World; some issues for debate. In Rai, S. and Lievesy, G. (eds), *Women and the State: International Perspectives*. London: Taylor and Francis.

Ralston, M. 1996: *'Nobody Wants to Hear our Truth': Homeless Women and Theories of the Welfare State*. London: Greenwood Press.

Randall, V. 1987: *Women and Politics. An International Perspective*, 2nd edn. Basingstoke, Hants: Macmillan Education.

Rathgeber, E. 1990: WID, WAD, GAD: trends in research and practice. *The Journal of Developing Areas* 24, 489–502.

Rattansi, A. 1992: Changing the subject? Racism, culture and education. In Donald, J. and Rattansim, A. (eds), *'Race', Culture and Difference*. London: Sage.

Reichert, D. 1994: Woman as utopia. *Gender, Place and Culture* 1, 91–101.

Reimer, S. 1998: Working in a risk society. *Transactions of the Institute of British Geographers* 23, 116–27.

Reinharz, S. 1992: *Feminist Methods in Social Research*. Oxford: Oxford University Press.

Reiter, R. (ed.) 1975: *Toward an Anthropology of Women*. New York: Monthly Review Press.

Relph, E. 1976: *Place and Placelessness*. London: Pion.

Renne, T. (ed.) 1997: *Ana's Land: Sisterhood in Eastern Europe*. Boulder, CO: Westview Press.

Rich, A. 1977: *Of Woman Born*. London: Virago.

Rich, A. 1980: Compulsory heterosexuality and Lesbian existence. *Signs: Journal of Women in Culture and Society* 5, 631–90.

Rich, A. 1986: Towards a politics of location. In Rich, A. (ed.), *Blood, Bread and Poetry: Selected Prose, 1979–1985*. London: Virago.

Richardson, D. (ed.) 1996: *Theorising Heterosexuality: Telling it Straight*. Buckingham: Open University Press.

Ricoeur, P. 1971: The model of the text: meaningful action considered as text. *Social Research* 38, 529–62.

Ridd, R. 1987: Powers of the powerless. In Ridd, R. and Callaway, H. (eds), *Women and Political Conflict: Portraits of Struggle in Times of Crisis*. New York: New York University Press.

Rigby, A. 1974: *Alternative Realities: A Study of Communes and their Members*. London: Routledge and Kegan Paul.

Riley, D. 1988: *'Am I That Name?': Feminism and the Category 'Women' in History*. London: Macmillan.

Riviere, J. 1929: Womanliness as a masquerade. *The International Journal of Psychoanalysis* 10; reprinted in Burgin, V., Donald, J. and Kaplan, C. (eds) 1986: *Formations of Fantasy*. London: Methuen.

Roberts, M. 1991: *Living in a Man-made World: Gender Assumptions in Modern Housing Design*. London, Routledge.

Robertson, G. *et al.* (eds) 1994: *Travellers' Tales: Narratives of Home and Displacement.* London: Routledge.

Robertson, R. 1990: After nostalgia? Wilful nostalgia and the phases of globalization. In Turner, B. (ed.), *Theories of Modernity and Postmodernity.* London: Sage.

Robertson, R. 1996: Glocalization: time–space and homogeneity–heterogeneity. In Featherstone, M. *et al.* (eds), *Global Modernities.* London: Sage.

Roediger, D. 1992: *The Wages of Whiteness: Race and the Making of the American Working Class.* London: Verso.

Roiphe, K. 1994: *The Morning After: Sex, Fear and Feminism.* London: Hamish Hamilton.

Romero, M. and Higgenbotham, E. 1995: *Women and Work: Race, Ethnicity and Class.* Newbury Park, CA: Sage.

Rorty, R. 1979: *Philosophy and the Mirror of Nature.* Princeton, NJ: Princeton University Press.

Rorty, R. 1989: *Contingency, Irony, and Solidarity.* New York: Cambridge University Press.

Rosaldo, M. 1980: The use and abuse of anthropology: reflections on feminism and cross-cultural understanding. *Signs* 5, 389–417.

Rosaldo, M.Z. and Lamphere, L. 1974: *Women, Culture, and Society.* Stanford, CA: Stanford University Press.

Rose, D. 1989: A feminist perspective of employment restructuring and gentrification: the case of Montreal. In Wolch, J. and Dear, M. (eds), *The Power of Geography.* Boston, MA: Unwin Hyman.

Rose, G. 1991: On being ambivalent: women and feminisms in geography. In Philo, C. (ed.), *New Words, New Worlds.* Social Geography Study Group, Institute of British Geographers, London.

Rose, G. 1993a: *Feminism and Geography: The Limits to Geographical Knowledge.* Cambridge: Polity Press.

Rose, G. 1993b: Looking at landscape: the uneasy pleasures of power. In Rose, G. (ed.), *Feminism and Geography.* Cambridge: Polity Press.

Rose, G. 1995a: Tradition and paternity: same difference? *Transactions of the Institute of British Geographers* 20, 414–16.

Rose, G. 1995b: Distance, surface, elsewhere: a feminist critique of the space of phallocentric self/knowledge. *Environment and Planning D: Society and Space* 13, 761–81.

Rose, G. 1995c: Making space for the female subject of feminism: the spatial subversions of Holzer, Kruger and Sherman. In Pile, S. and Thrift, N. (eds), *Mapping the Subject.* London: Routledge.

Rose, G. 1995d: The interstitial perspective: a review essay of Homi Bhabha's *The Location of Culture. Environment and Planning D: Society and Space* 13, 365–73.

Rose, G. 1996a: As if the mirrors bled: masculine dwelling, masculinist theory and feminist masquerade. In Duncan, N. (ed.), *BodySpace.* London: Routledge.

Rose, G. 1996b: The strike at Bryant and May's match factory, East London, July 1988. In Charlesworth, A., Gilbert, D., Randall, A. *et al.* (eds), *An Atlas of Industrial Protest in Britain 1750–1990*. London: Macmillan.

Rose, G. 1997a: Situating knowledges: positionality, reflexivities and other tactics. *Progress in Human Geography* 21, 305–20.

Rose, G. 1997b: Performing inoperative community: the space and the resistance of some community arts projects. In Pile, S. and Thrift, N. (eds), *Mapping the Subject*. London: Routledge.

Rose, G., Kinnaird, V., Morris, M. and Nash, C. 1997: Feminist Geographies of environment, nature and landscape. In WGSG, *Feminist Geographies: Explorations in Diversity and Difference*. Harlow, Essex: Addison Wesley Longman.

Rose, J. 1987: *Sexuality in the Field of Vision*. London: Verso.

Rose, S. Kamin, L.J. and Lewontin, R.C. 1984: *Not in Our Genes*. Harmondsworth, Middx: Penguin.

Rosenau, J. and Czempiel, E. (eds) 1992: *Governance without Government: Order and Change in World Politics*. Cambridge: Cambridge University Press.

Rosenau, P.M. 1992: *Post-Modernism and the Social Sciences*. Princeton, NJ: Princeton University Press.

Rosenbloom, S. 1993: Women's travel patterns. In Katz, C. and Monk, J. (eds), *Full Circles: Geographies of Women over the Life Course*. London: Routledge.

Rosenbloom, S. and Raux, C. 1985: Employment, child-care and travel behaviour: France, The Netherlands and The United States. In Conference Proceedings: *Behavioural Research for Transport Policy*. Utrecht: VNU Science Press.

Roseneil, S. 1995: *Disarming Patriarchy: Feminism and Political Action at Greenham*. Buckingham: Open University Press.

Rothenberg, T. 1995: 'And she told two friends': lesbians creating urban social space. In Bell, D and Valentine, G. (eds) *Mapping Desires*. London: Routledge.

Rotundo, A. 1993: *American Manhood*. New York: Basic Books.

Routledge, P. 1997: A spatiality of resistances: theory and practice in Nepal's revolution of 1990. In Pile, S. and Keith, M. (eds), *Geographies of Resistance*. London: Routledge.

Routledge, P. and Simons, J. 1995: Embodying spirits of resistance. *Environment and Planning D: Society and Space* 13, 471–98.

Rowbotham, S. 1989: To be or not to be: the dilemmas of mothering. *Feminist Review* 31, 82–93.

Rowbotham, S. 1997: *A Century of Women: The History of Women in Britain and the United States*. London: Viking.

Rowbotham, S., Segal, L. and Wainwright, H. 1979: *Beyond the Fragments: Feminism and the Making of Socialism*. London: Merlin Press.

Rubin, G. 1975: The traffic in women: notes on the 'political economy' of sex. In Rapp, R. (Reiter, R.R.) (ed.), *Toward an Anthropology of Women*. New York: Monthly Review Press.

Rubin, G. 1984: *Thinking Sex: Notes for a Radical Theory of the Politics of Sexuality Pleasure and Danger*, C. Vance (ed.). London: Routledge and Kegan Paul.

Rushdie, S. 1991: *Imaginary Homelands: essays and criticism 1981–1991*. London: Granta Books.

Russell, D. 1982: *Rape in Marriage*. New York: Macmillan.

Russell, D. 1989: *Lives of Courage: Women for a New South Africa*. New York: Basic Books.

Russell, D. 1995: *Women, Madness and Medicine*. Cambridge: Polity Press.

Ryan, J. 1994: Visualizing imperial geography: Halford MacKinder and the Colonial Office Visual Instruction Committee, 1902–11. *Ecumene* 1, 157–76.

Ryan, J. 1997: *Picturing Empire*. London: Reaktion Press.

Sahgal, G. and Yuval-Davis, Y. 1992: *Refusing Holy Orders: Women and Fundamentalism in Britain*. London: Virago.

Sahlins, M. 1977: *The Use and Abuse of Biology*. London: Tavistock Publications.

Said, E. 1978: *Orientalism: Western Conceptions of the Orient*. London: Routledge and Kegan Paul/Harmondsworth, Middx: Penguin.

Said, E. 1983. *The World, The Text and the Critic*. Cambridge MA: Harvard University Press.

Said, E. 1993. *Culture and Imperialism*. London: Verso.

Sainsbury, D. (ed.) 1994: *Gendering Welfare States*. London: Sage.

Salleh, A. 1984: Deeper than deep ecology: the eco-feminist connection. *Environmental Ethics*, 6.

Salleh, A. 1997: *Ecofeminism as Politics: Nature, Marx and the Postmodern*. London: Zed Books.

Samuel, R. 1994: *Theatres of Memory. Volume 1: Past and Present in Contemporary Culture*. London: Verso.

Sandercock, L. and Forsyth, A. 1992: A gender agenda: new directions for planning theory. *Journal of the American Planning Association* 58, 49–58.

Sargent, L. 1981: *The Unhappy Marriage of Marxism and Feminism: a Debate of Class and Patriarchy*. London: Pluto Press.

Sargisson, L. 1996: *Contemporary Feminist Utopianism*. London: Routledge.

Sarup, M. 1993: *An Introductory Guide to Post-structuralism and Postmodernism*, 2nd edn. New York: Harvester.

Sauer, C. 1963: The Morphology of Landscape, first published in 1925. In Leighly, J. (ed.), *Land and Life: A Selection from the Writings of Carl Ortwin Sauer*. Berkeley, CA: University of California Press.

Saunders, P. 1986: *Social Theory and the Urban Question*. London: Unwin Hyman.

Saunders, P. 1984: Beyond housing classes: the sociological significance of private property rights in means of consumption. *Journal of Urban and Regional Research* 8, 202–27.

Saunderson, W. 1997: Women, cities and identity: the production and consumption of gendered urban space in Belfast. In Byrne, A. and Leonard, M. (eds), *Women and Irish Society: A Sociological Reader*. Belfast: Beyond The Pale.

Savage, M., Barlow, J., Duncan, S. and Saunders, P. 1987: Locality Research: The Sussex Programme on Economic Restructuring, Social Change and the Locality. *Quarterly Journal of Social Affairs* 4, 27–51.

Sawicki, J. 1991: *Disciplining Foucault: Feminism, Power and the Body.* New York: Routledge.

Sayer, A. 1984: *Method in Social Science: A Realist Approach.* London: Hutchinson.

Sayer, A. 1989: Postfordism in question. *International Journal of Urban and Regional Research* 13, 666–93.

Sayer, A. 1992: *Method in Social Science: A Realist Approach,* 2nd edn. London: Hutchinson.

Sayer, A. and Walker, R. 1992: *The New Social Economy: Reworking the Division of Labour.* Oxford: Basil Blackwell.

Sayers, J. 1982. *Biological Politics: Feminist and Anti-Feminist Perspectives.* London: Tavistock.

Sayers, J. 1986: *Sexual Contradictions.* London: Tavistock.

SBS 1989: *Against the Grain: A Celebration of Survival and Struggle.* London: Southall Black Sisters.

Schaffer, K. 1990: *Women and the Bush: Forces of Desire in the Australian Cultural Tradition.* Cambridge: Cambridge University Press.

Schoenberger, E. 1997: *The Cultural Crisis of the Firm.* Oxford: Basil Blackwell.

Scott, A. 1990: *Ideology and the New Social Movements.* London: Unwin Hyman.

Scott, A.M. 1991: Informal sector or female sector? Gender bias in the urban labour market models. In Elson, D. (ed.), *Male Bias in the Development Process.* Manchester: Manchester University Press.

Scott, B.K. (ed.) 1990: *The Gender of Modernism: A Critical Anthology.* Bloomington: Indiana University Press.

Scott, J. 1986: *Weapons of the Weak: Everyday Forms of Peasant Resistance.* New Haven and London: Yale University Press.

Scott, J.W. 1988a: *Gender and the Politics of History.* New York: Columbia University Press.

Scott, J.W. 1988b: Deconstructing equality-versus-difference: or, The uses of poststructuralist theory for feminism. *Feminist Studies* 14, 33–50.

Scott, J. 1992: Experience. In Butler, J. and Scott, J. (eds), *Feminists Theorise the Political.* New York and London: Routledge.

Scull, A.T. 1977: *Decarceration: Community Treatment and the Deviant – a Radical View.* Englewodd Cliffs, NJ: Prentice Hall.

Seabrook, J. 1996: *Travels in the Skin Trade: Tourism and the Sex Industry.* London: Pluto Press.

Seager, J. 1993: *Earth Follies: Feminism, Politics and the Environment.* London: Earthscan.

Seager, J. 1997: *The State of Women in the World Atlas.* New York: Penguin.

Sears, J. and Williams, W. 1997: *Overcoming Heterosexism and Homophobia: Strategies That Work.* New York: Columbia University Press.

Sedgwick, E.K. 1990: *Epistemology of the Closet.* New York: Harvester Wheatsheaf.

Segal, L. 1987 *Is the Future Female? Troubled Thoughts on Contemporary Feminism.* London: Virago.

Segal, L. 1990: *Slow Motion: Changing Masculinities, Changing Men*. London: Virago.

Segal, L. and McIntosh, M. 1992: *Sex Exposed: Sexuality and the Pornography Debate*. London: Virago.

Seidman, S. 1996: *Queer Theory/Sociology*. Malden, MA: Basil Blackwell.

Seigel, J.E. 1986: *Bohemian Paris: Culture, Politics and the Boundaries of Bourgeois Life 1830–1930*. New York: Viking.

Selden, R. 1989: *A Reader's Guide to Contemporary Literary Theory*. Lexington, KY: University of Kentucky Press.

Sen, G. and Grown, C. 1987: *Development Crises and Alternative Visions. Third World Women's Perspectives*. London: Earthscan.

Sessions, G. (ed.) 1995: *Deep Ecology in the Twenty First Century*. Boston: Shambala.

Shanley, M.and Pateman, C. (eds) 1991: *Feminist Interpretations and Political Theory*. Cambridge: Polity Press.

Sharoni, S. 1994: Homefront as battlefield: gender, military occupation and violence against women. In Mayer, T. (ed.), *Women and the Israeli Occupation, The Politics of Change*. New York: Routledge.

Sharp, J. 1996: Gendering nationhood: a feminist engagement with national identity. In Duncan, N.G. (ed.), *BodySpace: Destabilizing Geographies of Gender and Sexuality*. New York: Routledge.

Shields, R. 1991: *Places on the Margin: Alternative Geographies of Modernity*. London: Routledge.

Shields, R. (ed.) 1992: *Lifestyle Shopping: The Subject of Consumption*. London: Routledge.

Shields, R. (ed.) 1996: *Cultures of the Internet: Virtual Spaces, Real Histories, Living Bodies*. London: Sage.

Shiva, V. 1989: *Staying Alive: Women, Ecology and Development*. London: Zed Books.

Shiva, V. 1993: *Monocultures of the Mind: Perceptions on Biodiversity and Biotechnology*. London: Zed Books.

Shiva, V. and Mies, M. 1993: *Ecofeminism*. London: Zed Books.

Short, J. 1993: *An Introduction to Political Geography*, 2nd edn. London: Routledge

Showalter, E. 1977: *A Literature of Their Own: British Women Novelists from Brontë to Lessing*. Princeton, NJ: Princeton University Press.

Showalter, E. (ed.) 1985: *The New Feminist Criticism: Essays on Women, Literature, and Theory*. New York: Pantheon.

Showstack Sassoon, A. (ed.) 1987: *Women and the State. The Shifting Boundaries of Public and Private*. London: Hutchinson.

Shurmer-Smith, P. and Hannam, K. 1994: *Worlds of Desire and Realms of Power; A Cultural Geography*. London: Arnold.

Shurmer-Smith, P. 1994: Cixous' spaces: sensuous spaces in women's writing. *Ecumene* 1, 349–62.

Sibley, D. 1995a: *Geographies of Exclusion*. London: Routledge.

Sibley, D. 1995b: Gender, science, politics and geographies of the city. *Gender, Place and Culture* 2, 37–49.

Silk, J. 1995: Time rolling forward, politics rolling back? *WGSG Newsletter* 3, 19–20.

Silverblatt, I. 1988: Women in states. *Annual Review of Anthropology* 17, 427–60.

Simpson, M. 1994: *Male Impersonators*. London: Cassell.

Singer, L. 1993: *Erotic Welfare*. London: Routledge.

Sjoo, M. and Moore, B. 1987: *The Great Cosmic Mother: Rediscovering the Religion of the Earth*. San Francisco: Harper and Row.

Skocpol, T. 1992: *Protecting Soldiers and Mothers: The Political Origins of Social Policy in the United States*. Cambridge, MA: Harvard University Press.

Slater, D. 1997: Geopolitical imaginations across the North–South divide: issues of difference, development and power. *Political Geography* 16, 631–53.

Smailes, J. 1995: 'The struggle has never been simply about bricks and mortar': Lesbian's experience of housing. In Gilroy, R. and Woods, R. (eds), *Housing Women*. London: Routledge.

Smelser, N. and Swedberg, R. 1995: *The Handbook of Economic Sociology*. Princeton, NJ: Princeton University Press.

Smith, A. and Pickles, J. (eds) 1998: *Theorizing Transition: The Political Economy of Post-communist Transformations*. London: Routledge.

Smith, B., Hull, G. and Scott, P.B. (eds) 1982: *All the Women are White, All the Blacks are Men, But Some of Us are Brave: Black Women's Studies*. Old Westbury, NY: Feminist.

Smith, D. 1987: *The Everyday World as Problematic*. Milton Keynes: Open University Press.

Smith, D. 1994: *Geography and Social Justice*. Oxford: Basil Blackwell.

Smith, F.M. 1998: Between East and West: sites of resistance in East German youth cultures. In Skelton, T. and Valentine, G. (eds), *Cool Places: Geographies of Youth Cultures*. London: Routledge.

Smith, G.E. 1994: Socialism. In Johnston, R.J., Gregory, D. and Smith, D.M. (eds), *The Dictionary of Human Geography*. Oxford: Basil Blackwell.

Smith, M.P. 1980: *The City and Social Theory*. Oxford: Basil Blackwell.

Smith, N. 1986a: Gentrification, the frontier and the restructuring of urban space. In Smith, N. and Williams, P. (eds), *Gentrification of the City*. Boston, MA: Allen and Unwin.

Smith, N. 1986b: *Uneven Development*. Oxford: Basil Blackwell.

Smith, N. 1987: Dangers of the empirical turn: some comments on the CURS initiative. *Antipode* 19, 56–66.

Smith, N. 1990: *Uneven Development: Nature, Capital and the Production of Space*. Oxford: Basil Blackwell.

Smith, N. 1993: Homeless/global: scaling places. In Bird, J. *et al.* (eds), *Mapping the Futures: Local Cultures, Global Change*. London: Routledge.

Smith, N. 1994: Geography, empire and social theory. *Progress in Human Geography* 18, 491–500.

Smith, N. and Katz, C. 1993: Grounding metaphor: towards a spatialized politics. In Keith, M. and Pile, S. (eds), *Place and the Politics of Identity*. London: Routledge.

Smith, S.J. 1986: *Crime, Space and Society*. Cambridge: Cambridge University Press.

Smith, S. 1995: Citizenship: all or nothing? *Political Geography* 14, 190–93.

Smith, Y. 1997: The household, women's employment and social exclusion. *Urban Studies* 34: 1159–77.

Smyth, A. 1991: The floozie in the jacuzzi. *Feminist Review* 17, 7–28.

Smyth, C. 1992: *Lesbians Talk: Queer Notions*. London: Scarlett Press.

Snitow, A. 1990: A gender diary. In Hirsch, M. and Fox Keller, E. (eds), *Conflicts in Feminism*. London and New York: Routledge.

Snitow, A., Stansell, C. and Thompson, S. (eds) 1984: *Desire. The Politics of Sexuality*. London: The Women's Press.

Soja, E. 1980: The socio-spatial dialectic. *Annals of the Association of American Geographers* 70, 207–25.

Soja, E. 1985: The spatiality of social life: towards a transformative retheorisation. In Gregory, D. and Urry, J. (eds), *Social Relations and Spatial Structures*. London: Macmillan.

Soja, E. 1989: *Postmodern Geographies: The Reassertion of Space in Critical Social Theory*. London: Verso.

Soja, E.W. 1996. *Thirdspace: Journeys to Los Angeles and Other Real-and-Imagined Places*. Oxford: Basil Blackwell.

Somers, M.R. 1994: The narrative constitution of identity: a relational and network approach. *Theory and Society* 23, 605–49.

Sontag, S. 1967: Notes on camp. In Sontag, S. *Against Interpretation*. London: Eyre and Spottiswoode.

Soper, K. 1995: *What is Nature? Culture, Poliics and the Non-Human*. Oxford: Basil Blackwell.

Soper, K. 1996: Nature/'nature'. In Robertson, G., Mash, M., Tickner, L. *et al.* (eds), *Future Natural*. London: Routledge.

Spain, D. 1992: *Gendered Spaces*. London and Chapel Hill, NC: University of North Carolina Press.

Sparke, M. 1994: Writing on patriarchal missiles: the chauvinism of the 'Gulf War' and the limits of critique. *Environment and Planning A* 26, 1061–89.

Sparke, M. 1995: Between Deconstructing and Demythologizing the Map: Shawnadithit's Newfoundland and the Alienation of Canada. *Cartographica* 31, 1–21.

Sparke, M. 1996a: Displacing the field in fieldwork: masculinity, metaphor and space. In Duncan, N. (ed.), *BodySpace: Destabilizing Geographies of Gender and Sexuality*. London: Routledge.

Sparke, M. 1996b: Negotiating national action: free trade, constitutional debate and the gendered geopolitics of Canada. *Political Geography* 15, 615–39.

Spender, D. 1980: *Man Made Language*. London: Routledge and Kegan Paul.

Spivak, G. 1985: Three women's texts and a critique of imperialism. *Critical Inquiry* 12, 243–61.

Spivak, G.C. 1988a: Subaltern studies: deconstructing historiography. In Spivak, G.C., *In Other Worlds: Essays in Cultural Politics*. New York: Methuen.

Spivak, G. 1988b: Can the subaltern speak? In Nelson, C. and Grossberg, L. (eds), *Marxism and the Interpretation of Culture*. Basingstoke, Hants: Macmillan.

Spivak, G. 1988c: *In Other Worlds: Essays in Cultural Politics*. New York: Routledge.

Spivak, G.C. 1990a: Poststructuralism, marginality, postcoloniality and value. In Collier, P. and Geyer-Ryan, H. (eds), *Literary Theory Today*. Ithaca, NY: Cornell University Press.

Spivak, G.C. 1990b: *The Post-colonial Critic: Interviews, Strategies, Dialogues*, ed. S. Harasym. New York: Routledge.

Spivak, G.C. 1990c: Negotiating the structures of violence. In Harasym, S. (ed.), *The Post-colonial Critic: Interviews, Strategies, Dialogues*. New York: Routledge.

Spooner, R. 1996: Contested representations: black women and the St Paul's Carnival. *Gender, Place and Culture* 3, 187–204.

Spretnak, C. and Capra, F. 1985: *Green Politics*. Glasgow: Paladin.

Squires, J. (ed.) 1993: *Principled Positions: Postmodernism and the Rediscovery of Value*. London: Lawrence and Wishart.

Stack, C. 1974: *All Our Kin: Strategies for Survival in a Black Community*. London: HarperCollins.

Staeheli, L. 1993: Publicity, privacy, and women's political action. *Environment and Planning D: Society and Space* 14, 601–19.

Staeheli, L. 1994: Empowering political struggle: spaces and scales of resistance. *Political Geography* 13, 387–91.

Staeheli, L. and Cope, M. 1994: Empowering women's citizenship. *Political Geography Quarterly* 13, 443–60.

Staeheli, L. and Lawson, V. 1994: A discussion of 'women in the field': the politics of feminist fieldwork. *Professional Geographer* 46, 96–102.

Stanko, E.A. 1987: Typical violence, normal precaution: men, women and interpersonal violence in England, Wales, Scotland and the USA. In Hanmer, J. and Maynard, M. (eds), *Women, Violence and Social Control*. London: Macmillan.

Stanko, E. 1990: *Everyday Violence: Women's and Men's Experience of Personal Danger*. London: Pandora

Stanley, L. 1990: *Feminist Praxis, Research, Theory and Epistemology in Feminist Sociology*. London: Routledge.

Stanley, L. and Wise, S. 1993: *Breaking Out Again: Feminist Ontology and Epistemology*. London: Routledge.

Stanton, D. (ed.) 1987: *The Female Autograph: Theory and Practice of Autobiography from the Tenth to the Twentieth Century*. Chicago, IL: University of Chicago Press.

Starhawk 1982: *Dreaming the Dark: Magic, Sex and Politics*. Boston, MA: Beacon Press.

Stea, D. 1969: Positions, purposes, pragmatics: a journal of radical geography. *Antipode* 1, 1–2.

Stocking, Jr, G. 1987: *Victorian Anthropology*. New York: The Free Press.

Stoddart, D. 1991: Do we need a feminist historiography of geography – and if we do, what should it be? *Transactions of the Institute of British Geographers* 16, 484–7.

Stoler, A. 1995: *Race and the Education of Desire*. Durham, NC: Duke University Press.

Stoller, R.J. 1968: *Sex and gender: on the development of masculinity and femininity*. London: Hogarth Press.

Stone, A.R. 1991: Will the real body please stand up? Boundary stories about virtual culture. In Benedikt, M. (ed.) *Cyberspace: First Steps*. Cambridge, MA: MIT Press.

Strange, S. 1994: From Bretton Woods to casino economy. In Corbridge, S. Martin, R. and Thrift, N. (eds), *Money, Power and Space*. Oxford: Basil Blackwell.

Strathern, M. 1992: *After Nature: English Kinship in the Late Twentieth Century*. Cambridge: Cambridge University Press.

Strohmayer, U. and Hannah, M. 1992: Domesticating postmodernism. *Antipode* 24, 29–55.

Stubbs, J. 1984: Some thoughts on the life story method in labour history and research on rural women. *Institute of Development Studies Bulletin* 15, 34–7.

Swyngedouw, E. 1997: Excluding the Other: the production of scale and scaled politics. In Lee, R. and Wills, J. (eds), *Geographies of Economies*. London: Arnold.

Tagg, J. 1988: *The Burden of Representation*. Amherst, MA: University of Massachusetts Press.

Talalay, M., Farrands, C. and Tooze, R. 1997: *Technology, Culture and Competitiveness*. London: Routledge.

Tannen, D. 1994: *Talking Nine to Five*. London: Virago.

Taylor, A. 1991: *Prostitution*. London: Optima.

Taylor, B. 1983: *Eve and the New Jerusalem: Socialism and Feminism in the Nineteenth Century*. London: Virago.

Taylor, P. 1993: *Political Geography. World Economy, Nation-state and Locality*, 3rd edn. Harlow, Essex: Longman.

Taylor, P. 1995: *Political Geography: World-economy, Nation-state, and Locality*, 4th edn. Harlow, Essex: Longman.

Tester, K. (ed.) 1994: *The Flâneur*. London: Routledge.

The Lancaster Regionalism Group 1985: *Localities, Class, and Gender*. London: Pion.

Thirsk, J. 1961: Industries in the countryside. In Fisher, F. (ed.), *Essays in the Economic and Social History of Tudor and Stuart England*. Cambridge: Cambridge University Press.

Thomas, K. 1983: *Man and the Natural World*. London: Allen Lane.

Thomas, N. 1994: *Colonialism's Culture: Anthropology, Travel and Government*. Cambridge: Polity Press.

Thomas, W. and Znaniecki, F. 1984: *The Polish Peasant in Europe and America*. Urbana, IL: University of Illinois Press.

Thompson, E.P. 1980: *The Making of the English Working Class*. Harmondsworth, Middx: Penguin Books.

Thompson, K. (ed.) 1991: *To be a Man: In Search of the Deep Masculine*. New York: Jeremy P. Tarcher/Perigree Books.

Thrift, N. 1983: On the determination of social action in space and time. *Environment and Planning D: Society and Space* 1, 23–57.

Thrift, N. 1994: On the social and cultural determinants of international financial centres: the case of the City of London. In Corbridge, S. Martin, R. and Thrift, N. (eds), *Money, Power and Space*. Oxford: Basil Blackwell.

Thrift, N. 1996: *Spatial Formations*. London: Sage.

Thrift, N. and Leyshon, A. 1994: A phantom state? The de-traditionalization of money, the international financial system and international financial centres. *Political Geography* 13, 299–327.

Tickell, A. and Peck, J. 1995: Social regulation after Fordism: regulation theory, neo-liberalism and the global local nexus. *Economy and Society* 24, 357–86.

Tickner, J.A. 1992: *Gender and International Relations*. New York: Columbia University Press.

Tilly, L. and Scott, J. 1978: *Women, Work, and Family*. London: Holt, Rinehart and Winston.

Tilly, L. and Scott, J. 1987: *Women, Work and Family*. London: Methuen.

Tivers, J. 1985: *Women Attached: The Daily Activity Patterns of Women with Young Children*. London: Croom Helm.

Todaro, M.P. 1997: *Economic Development*, 6th edn. Harlow, Essex: Longman.

Tong, R. 1989: *Feminist Thought: A Comprehensive Introduction*. Boulder, CO: Westview Press.

Tóth, O. 1993: No envy, no pity. In Funk, N. and Mueller, M. (eds), *Gender Politics and Post-communism*. London: Routledge.

Touraine, A. 1971: *The Post-industrial Society: Tomorrow's Social History: Classes, Conflicts and Culture in the Programmed Society* (translation). New York: Random House.

Townsend, P. 1987: Deprivation. *Journal of Social Policy* 16, 125–46.

Treichler, P. 1994: Aids, identity and the politics of gender. In Bender, G. and Druckrey, T. (eds), *Culture on the Brink: Ideologies of Technology*. Seattle, WA: Bay Press.

Trigg, R. 1988: *Ideas of Human Nature*. Oxford: Blackwell.

Trilling, L. 1972: *Sincerity and Authenticity*. Cambridge, MA: Harvard University Press.

Tronto, J. 1993: *Moral Boundaries. A Political Argument for an Ethic of Care*. London: Routledge.

Troyna, B. 1990: *Education, Racism and Reform*. London: Routledge.

Truong, T.-D. 1990: *Sex, Money and Morality: Prostitution and Tourism in Southeast Asia*. London: Zed Books.

Turner, V. 1957: *Schism and Continuity in an African Society: A Study of Ndembu Village Life*. Manchester: Manchester University Press.

Twine, F.W. 1996: Brown skinned white girls: class, culture and the construction of white identity in suburban communities. *Gender, Place and Culture* 3, 205–24.

Tylor, E.B. 1871: *Primitive Culture*. London: John Murray.

Tylor, E.B. 1958: *Primitive Culture, Part I: The Origins of Culture*. New York: Harper and Brothers.

Udayagiri, M. 1996: Challenging modernization: gender and development, postmodern feminism and activism. In Marchand M. and Parpart, J. (eds), *Feminism/Postmodernism/Development*. London: Routledge.

Unger, R 1983: The critical legal studies movement. *The Harvard Law Review* 96, 320–432.

UNICEF 1997: *The State of the World's Children*. Oxford: Oxford University Press.

United Nations Development Programe (UNDP) 1995: *Human Development Report*. New York: Oxford University Press.

Urry, J. 1981a: Localities, regions and social class. *International Journal of Urban and Regional Research* 5, 455–74.

Urry, J. 1981b: *The Anatomy of Capitalist Societies: The Economy, Civil Society and the State*. London: Macmillan.

Urry, J. 1990: *The Tourist Gaze: Leisure and Travel in Contemporary Societies*. London: Sage.

Urry, J. 1995: *Consuming Places*. London: Routledge.

Valentine, G. 1989: The geography of women's fear. *Area* 21, 385–90.

Valentine, G. 1992: Images of danger: women's sources of information about the spatial distribution of male violence. *Area* 24, 22–9.

Valentine, G. 1993a: Negotiating and managing multiple sexual identities: lesbian time–space strategies. *Transactions of the Institute of British Geographers* 18, 237–48.

Valentine, G. 1993b: (Hetero)sexing space: lesbian perceptions and experiences of everyday spaces. *Environment and Planning D: Society and Space* 11, 395–413.

Valentine, G. 1996: Children should be seen and not heard: the production and transgression of adults' public space. *Urban Geography* 17, 205–20.

Valentine, G. 1997a: 'My son's a bit dizzy'. 'My wife's a bit soft': gender, children and cultures of parenting. *Gender, Place and Culture* 4, 37–62.

Valentine, G. 1997b: Making space: separatism and difference. In Jones, J.P., Nast, H. and Roberts, J. (eds), *Thresholds in Feminist Geography*. Lanham, MD: Rowman and Littlefield.

Van den Abeele, G. 1992: *Travel as Metaphor: From Montaigne to Rousseau*. Minneapolis, MN: University of Minnesota Press.

van den Berghe, P.L. 1978: Bridging the paradigms: biology and the social sciences. In Gregory, M.S, Silvers, A. and Sutch, D. (eds), *Sociobiology and Human Nature*. San Francisco, CA: Jossey-Bass.

Vanneste, D. 1997: Rural economy and indigence in mid-nineteenth century Belgium. *Journal of Historical Geography* 23, 3–15.

Varley A. 1996: Woman-headed households: some more equal than others? *World Development* 24, 505–20.

Visvanathan, N., Duggan, L., Nissonoff, L. and Wiergersma, N. (eds) 1997: *The Women, Gender and Development Reader*. London: Zed Books.

Visweswaran, K. 1994: *Fictions of Feminist Ethnography*. Minneapolis, MN: University of Minnesota Press.

Visweswaran, K. 1997: Histories of feminist ethnography. *Annual Review of Anthropology* 26, 591–621.

Von Ankum, K. (ed.) 1997: *Women in the Metropolis: Gender and Modernity in Weimar Culture*. Berkeley: University of California Press.

WAF (undated) *Women Against Fundamentalism Education Pack*, available from Women Against Fundamentalism (WAF), 129 Seven Sisters Road, London N7 7QG, England.

WAF (Women Against Fundamentalism) 1990: *Newsletter* no. 1 (November).

WAF 1992: *Statement of Principle*. London: Women Against Fundamentalism.

Walby, S. 1986: *Patriarchy at Work: Patriarchal and Capitalist Relations in Employment*. Minneapolis, MN: University of Minnesota Press.

Walby, S. 1990: *Theorising Patriarchy*. Oxford: Basil Blackwell.

Walby, S. 1997: *Gender Transformations*. London: Routledge.

Walker, A. 1984a: *In Search of our Mothers' Gardens*. London: The Women's Press.

Walker, A. 1984b: *The Colour Purple*. London: The Women's Press.

Walker, B. 1994: Disaster response: special issue. *Focus on Gender* 2, 2–6 (editorial).

Walker, D. and McDowell, L. 1993: Editorial. *Antipode* 25, 1–3.

Walker, J. 1988: Women, the state and the family in Britain: Thatcher economics and the experience of women. In Rubery, J. (ed.), *Women and Recession*. London: Routledge.

Walklate, S. 1995: *Gender and Crime*. Hemel Hempstead, Herts: Prentice Hall.

Wallace, T. 1993: Refugee women: their perspectives and our responses. *Focus on Gender* 1, 17–23.

Wallace, T. with March, C. 1991 *Changing Perceptions: Writings on Gender and Development*. Oxford: Oxfam.

Walter, B. 1995: Irishness, gender, and place. *Environment and Planning D: Society and Space* 13, 35–50.

Walter, B. 1996: Gender, 'race', and diaspora: racialized identities of emigrant Irish women. In Jones, J P., Nast, H. and Roberts, S. (eds), *Thresholds in Feminist Geography: Difference, Methodology, Representation*. Lanham, MD: Rowman and Littlefield.

Walter, N. 1998: *The New Feminism*. London: Little, Brown.

Ward, M. 1989: *Unmanageable Revolutionaries*. London: Pluto Press.

Warde, A. 1991: Gentrification as consumption: issues of class and gender. *Environment and Planning D: Society and Space* 9, 223–32.

Ware, V. 1992: *Beyond the Pale: White Women, Racism and History*. London: Verso.

Warner, M. 1985: *Monuments and Maidens: The Allegory of the Female Form*. London: Picador.

Warren, M. 1980: *The Nature of Women: An Encyclopaedia and Guide to the Literature*. Inverness, CA: Edgepress.

Watson, S. 1990: *Playing the State*. London: Verso.

Watson, W. 1955: Geography: a discipline in distance. *Scottish Geographical Magazine* 71, 1–13.

Watts, M. 1983: *Silent Violence*. Berkeley, CA: University of California Press.

Weaver, J. (ed.) 1996: *Defending Mother Earth: Native American Perspectives on Environmental Justice*. New York: Orbis Books.

Weeks, J. 1977: *Coming Out: Homosexual Politics in Britain from the Nineteenth Century to the Present*. London: Quartet.

Weeks, J. 1985: *Sexuality and its Discontents: Meaning, Myths and Modern Sexualities*. London: Routledge.

Weeks, J. 1986: *Sexuality*. London: Horwood and Tavistock.

Weeks, J. 1991: *Against Nature: Essays on Sexuality, History and Identity*. London: Rivers Oram Press.

Wekerle, G. 1988: *Women's Housing Projects in Eight Canadian Cities*. Ottawa: Canada Housing and Mortgage Corporation.

Wekerle, G. and Peake, L. 1996: New social movements and women's urban activism. In Caulfield, J. and Peake, L. (eds), *City Lives and City Forms: Critical Research and Canadian Urbanism*. Toronto: University of Toronto Press.

West, C. and Zimmerman, D. 1987: Doing gender. *Gender and Society* 1, 125–51.

Western, J. 1993: Ambivalent attachments to place in London: twelve Barbadian families, *Environment and Planning D: Society and Space* 11, 147–70.

Weston, J. 1986: *Red and Green*. London: Pluto Press.

WGSG 1984: *Geography and Gender: An Introduction to Feminist Geography*, Women and Geography Study Group of the Institute of British Geographers. London: Hutchinson.

WGSG 1997: *Feminist Geographies: Explorations in Diversity and Difference*, Women and Geography Study Group of the Royal Geographical Society; with the Institute of British Geographers. Harlow, Essex: Longman.

Whatmore, S. 1991: *Farming Women: Gender, Work and Family Enterprise*. Basingstoke, Hants: Macmillan.

Whatmore, S., Lowe, P. and Marsden, T. (eds) 1994: *Gender and Rurality*. London: David Fulton.

Wheatley, P. 1971: *The Pivot of the Four Quarters: A Preliminary Enquiry into the Origins and Character of the Ancient Chinese City*. Edinburgh: Edinburgh University Press.

Wheelock, J. 1990: *Husbands at Home: The Domestic Economy in a Post-industrial Society*. London: Routledge.

Whitford, M. 1991: *Luce Irigaray: Philosophy in the Feminine*. London: Routledge.

Williams, A. 1987: Untitled. *Ten*. 8, 6–11.

Williams, R. (ed.) 1968: *May Day Manifesto 1968*. Harmondsworth, Middx: Penguin.

Williams, R. 1976: *Keywords: A Vocabulary of Culture and Society*. London: Fontana.

Williams, R. 1977: *Marxism and Literature*. Oxford: Oxford University Press.

Williams, R. 1983: *Keywords: A Vocabulary of Culture and Society*. London: Fontana.

Williamson, J. 1986: *Consuming Passions*. London: Marion Boyars.

Willis, P. 1977: *Learning to Labour: How Working Class Kids Get Working Class Jobs*. Farnborough, Hants: Saxon House.

Wills, J. 1995: *Geographies of Trade Union Tradition*. Unpublished PhD thesis, Open University, Milton Keynes.

Wills, J. 1996: Geographies of trade unionism: translating traditions across space and time. *Antipode* 28, 352–78.

Wills, J. 1998: The discreet charm of a new Labour Government. *Environment and Planning A* 30.

Wilson, E. 1977: *Women and the Welfare State*. London: Tavistock Publications.

Wilson, E. 1981: Psychoanalysis: psychic law and order. *Feminist Review* 8, 63–78.

Wilson, E. 1987: Thatcherism and women: after seven years. In Miliband, R., Panitch L. and Saville J. (eds), *The Socialist Register*. London: The Merlin Press.

Wilson, E. 1991: *The Sphinx in the City: Urban Life, the Control of Disorder, and Women*. London: Virago.

Wilson, E. 1992: The invisible flâneur. *New Left Review* 191, 90–110.

Wilson, E. 1997: Looking backward: nostalgia and the city. In Westwood, S. and Williams, J. (eds), *Imagining Cities: Scripts, Signs, Memory*. London: Routledge.

Wilson, F. 1993: Workshops as domestic domains: reflections on small-scale industry in Mexico. *World Development* 21, 67–80.

Wilton, T. 1997: *Engendering AIDS: Deconstructing Sex, Text and Epidemic*. London: Sage.

Winkler, G. (ed.) 1990: *Frauenreport '90*. Berlin: Verlag Die Wirtschaft.

Winnicott, D.W. 1950: *The Family and Individual Development*. London: Tavistock Publications.

Winnicott, D.W. 1956: *Collected Papers: Through Paediatrics to Psycho-analysis*. London: Hogarth.

Winnicott, D.W. 1964: *The Child the Family and the Outside World*. Harmondsworth, Middx: Penguin.

Winnicott, D.W. 1965: *The Maturational Processes and the Facilitating Environment*. London: Hogarth.

Winnicott, D.W. 1971: *Playing and Reality*. London: Tavistock Publications.

Wise, S. and Stanley, L. 1987: *Georgie Porgie: Sexual Harassment in Everyday Life*. London: Pandora.

Wittig, M. 1971 [1969]: *Les Guérillères*, trans. D. Le Vay. London: David Owen.

Wittig, M. 1992: *The Straight Mind and Other Essays*. Hemel Hempstead, Herts: Harvester Wheatsheaf.

Wolch, J. 1990: *The Shadow State: Government and Voluntary Sector in Transition*. New York: The Foundation Center.

Wolch, J., Dear, M. and Atika, A. 1988: Explaining homelessness. *Journal of the American Planning Association* 54, 443–53.

Wolf, D. 1996: *Feminist Dilemmas in Fieldwork*. Boulder, CO: Westview Press.

Wolf, M. 1992: *A Thrice Told Tale: Feminism, Postmodernism and Ethnographic Responsibility*. Stanford, CA: Stanford University Press.

Wolf, N. 1991: *The Beauty Myth: How Images of Beauty are Used Against Women*. New York: Doubleday.

Wolf, N. 1994: *Fire with Fire: The New Female Power and How It Will Change the 21st Century*. London: Vintage.

Wolff, J. 1985: The invisible flâneuse: woman and the literature of modernity. *Theory, Culture and Society* 2, 37–46.

Wolff, J. 1990: Feminism and modernism. In Wolff, J., *Feminine Sentences: Essays on Women and Culture*. Cambridge: Polity Press.

Wolff, J. 1993: On the road again: metaphors of travel in cultural criticism. *Cultural Studies* 7, 224–39.

Wollstonecraft, M. 1792: *Vindication of the Rights of Women*, reprinted 1969. London: Everyman.

Wollstonecraft, M. 1982: *Vindication of the Rights of Women*. Harmondsworth, Middx: Penguin.

Wolpert, J. 1967: Distance and directional bias in inter-urban migratory streams. *Annals of the Association of American Geographers* 57, 605–16.

Woolf, V. 1938: *Three Guineas*. London: Hogarth.

Wright, E (ed.) 1992: *Feminism and Psychoanalysis*. Oxford: Basil Blackwell.

Wright, P. 1985: *On Living in an Old Country: The National Past in Contemporary Britain*. London: Verso.

Wrigley, N. and Lowe, M. (eds) 1996: *Retailing, Consumption and Capital: Towards the New Retail Geography*. London: Longman.

Yearly, S. 1992: *The Green Case: A sociology of Environmental Issues, Arguments and Politics*. London: Routledge.

Yeung, H.W. 1997: Critical realism and realist research in human geography: a method or a philosophy in search of method? *Progress in Human Geography* 21, 55–74.

Young E. 1992: Hunter–gatherer concepts of land and its ownership in remote Australia and North America. In Anderson, K. and Gale, F. (eds), *Inventing Places: Studies in Cultural Geography*. Melbourne: Longman Cheshire.

Young, I. 1987: Impartiality and the civic public. In Benhabib, S. and Cornell, D. (eds), *Feminism as Critique*. Minneapolis, MN: University of Minnesota Press.

Young, I.M. 1989a: Throwing like a girl. In Allen, J. and Young, I.M. (eds), *The Thinking Muse: Feminism and Modern French Philosophy*. Bloomington, IN: University of Indiana Press.

Young, I.M. 1989b: Polity and group difference: a critique of the ideal of universal citizenship. *Ethics* 99, 250–74.

Young, I.M. 1990a: *Throwing Like a Girl and Other Essays in Feminist Philosophy and Social Theory*. Bloomington, IN: University of Indiana Press.

Young, I.M. 1990b: The scaling of bodies and the politics of identity. In Young, I.M., *Justice and the Politics of Difference*. Princeton, MJ: Princeton University Press.

Young, I.M. 1990c: *Justice and the Politics of Difference*. Princeton, NJ: Princeton University Press.

Young, I.M. 1990d: The ideal of community and the politics of difference. In Nicholson, L. (ed.), *Feminism/Postmodernism*. London: Routledge.

Young, I.M. 1998: Harvey's complaint with race and gender struggles: a critical response. *Antipode* 30, 36–42

Young, R. 1990: *White Mythologies: Writing History and the West*. London: Routledge.

Young, R. 1995: *Colonial Desire. Hybridity in Theory, Culture and Race*. London: Routledge.

Yuval-Davis, N. 1992: Fundamentalism, multiculturalism and women in Britain. In Donald, J. and Rattansi, A. (eds), *'Race', Culture and Difference*. London: Sage.

Yuval-Davis, N. 1997: *Gender and Nation*. London: Sage.

Yuval-Davis, N. and Anthias, F. 1989: *Woman–Nation–State*. London: Macmillan.

Zaretsky, E. 1997: Bisexuality, capitalism and the ambivalent legacy of psychoanalysis. *New Left Review* 223, 69–89.

Zelinksy, W. 1973: *The Cultural Geography of the United States*. Englewood Cliffs, NJ: Prentice-Hall.

Zukin, S. and DiMaggio, P. (eds) 1990: *Structures of Capital: The Social Organisation of the Economy*. Cambridge: Cambridge University Press.

Zwerman, G. 1994: Mothering on the Lam: politics, gender fantasies and maternal thinking in women associated with armed, clandestine organization in the United States. *Feminist Review* 47, 35–56.